Environmental Protection in the European Union

Volume 4

Series Editors

Michael Schmidt, Brandenburg University of Technology Cottbus-Senftenberg, Cottbus, Germany

Lothar Knopp, Brandenburg University of Technology Cottbus-Senftenberg, Germany

For further volumes:
http://www.springer.com/series/5487

Environmental Protection in the European Union

Volume 1
M. Schmidt, L. Knopp
Reform in CEE-Countries with Regard to European Enlargement
2004, XII, 205 pages
ISBN 978-3-540-40259-6

Volume 2
M. Schmidt, E. João, E. Albrecht
Implementing Strategic Environmental Assessment
2005, XXXII, 742 pages
ISBN 978-3-540-20562-3

Volume 3
M. Schmidt et al.
Standards and Thresholds for Impact Assessment
2008, XXIX, 493 pages
ISBN 978-3-540-31140-9

Eike Albrecht • Michael Schmidt •
Magdalena Mißler-Behr • Simon P.N. Spyra
Editors

Implementing Adaptation Strategies by Legal, Economic and Planning Instruments on Climate Change

Springer

Editors
Eike Albrecht
Civil and Public Law with References
to the Law of Europe and the Environment
Brandenburg University of
Technology Cottbus-Senftenberg
Cottbus, Germany

Michael Schmidt
Environmental Planning
Brandenburg University of
Technology Cottbus-Senftenberg
Cottbus, Germany

Magdalena Mißler-Behr
Planning and Innovation Management
Brandenburg University of
Technology Cottbus-Senftenberg
Cottbus, Germany

Simon P.N. Spyra
Civil and Publis Law
Brandenburg University of
Technology Cottbus
Cottbus, Germany

ISBN 978-3-662-51859-5 ISBN 978-3-540-77614-7 (eBook)
DOI 10.1007/978-3-540-77614-7
Springer Heidelberg New York Dordrecht London

Softcover reprint of the hardcover 1st edition 2014

Printed on acid-free paper

Springer is part of Springer Science+Business Media (www.springer.com)

Foreword

The 5th Assessment Report of the Intergovernmental Panel on Climate Change (IPCC) published in 2014 consolidated the certainty that mankind will have to contend with significant challenges of climate change. In this regard, the reduction of greenhouse gas emissions is of fundamental significance. The Federal Republic of Germany has, through consistent promotion of electricity generation from renewable energies, contributed enormously to the transition to a non-fossil fuel energy generation. Particularly, through the creation of a large entry market for industrial mass production of alternative energy systems in Germany. Hence, the drastic drop in prices for photovoltaic systems of more than 70 % within a few years is a result of the massive amount of assistance from the German Renewable Energy Sources Act (EEG). Consumers in Germany have, through the EEG surcharge, exclusively financed the technological development and launching costs of these innovative, clean and unlimited sources of energy. Unfortunately, the number of countries that consistently try to reduce their greenhouse gas emissions is rather small. International commitments in the 2nd Commitment Period of the Kyoto Protocol have been received only from Australia, the EU and some other European countries. Meanwhile the time for effective mitigation measures is running out. Climate change is in full swing and will continue to worsen. In this respect, adaptation measures are essential. This book deals with the adaptation to climate change in different areas, regions and countries of the world and, thus, represents a significant contribution to possible adaptation measures, but, however, does not relieve the states of the world and its citizens from responsibility to reduce their emission of greenhouse gases. Sun and wind don't send us a bill. With the 100 % change of energy we can generate not only ecological advantages, but also economical profits. Germany is renewable, Europe is renewable, the world is renewable.

Franz Alt is a German journalist and author

Foreword

Climate change is transforming the world we live in and coping with the changes is unavoidable. While preventing climate change from worsening must be a key priority, making sure we're ready to deal with the consequences is equally as pressing.

The latest reports by the International Panel for Climate Change (IPCC) are unequivocal. Action to reduce greenhouse gas emissions must be taken now to prevent more climate-related damages in the future. The EU has already taken bold measures: it is bound to overachieve its commitments under the Kyoto protocol and now observes the lowest emissions on record. But with only 10 % of the global emissions, the EU cannot act alone. All countries—large, small, developed, and developing—have a role to play and agree on a global agreement on climate change in 2015.

This book is testimony of what an all-encompassing and complex task is before us. Adaptation strategies are needed at all levels of governance and administration. Most measures will take place at the local, regional, and national levels to address the immediate impacts on our communities and ecosystem. On the other hand, trans-boundary effects of climate change—like when entire river basins are affected—make it clear that efforts also need to be stepped up at the European and international levels. Each of these levels has a specific role to play, but we can only tackle climate change efficiently if this is done in a coordinated manner. Countries and regions need to learn from one another and set up examples of good practices to guide new adaptation measures around the world.

Here, the EU's comprehensive adaptation strategy is a good example of how working together with all EU countries is helping us to strengthen our climate change defenses. The EU adaptation strategy adopted in April 2013 makes it easier for Member States to work on joint initiatives. It also makes sure that adaptation is considered when taking decisions in other policy areas like agricultural planning or construction of energy systems.

But of course, the EU is not alone. The contributions in this book give a good overview of adaptation initiatives from around the world and document best practices in different economic sectors. With input from a range of international scholars and environmental experts, the book sheds light on current discussions on adaptation and outlines challenges to further global action.

Forewarned is forearmed, an old expression says. This holds especially true for climate change. The earlier we take action to prevent the risks of climate change, and the sooner we prepare to face its impacts now, the more we can avoid the damages to our environment and our society.

Connie Hedegaard

Foreword

Humanity stands in an unthinkable moment in time, facing unequivocal climate change and an imminent threat of uncontrollable planetary heating. Policy responses pale in comparison to the magnitude of danger confronting global society. Though most leaders have set a goal of limiting Earth's temperature increase to 2 °C, actions taken thus far are grossly inadequate to meet that goal. Moreover, recent science indicates that even 2 °C heating may lead to disastrous consequences for human civilization, compromising the basic habitability of the planet.[1]

Alarmingly, carbon emissions from human activity have caused the global atmospheric carbon dioxide (CO_2) concentration to rise over 400 parts per million (ppm) in 2014—a concentration that has not been exceeded in millions of years.[2] Without a swift and ambitious global transition away from fossil fuels, atmospheric CO_2 concentrations could pass a cataclysmic threshold, triggering dangerous feedback loops that create "self-amplifying" climate change to which there is no

[1] James Hansen *et al. Assessing "Dangerous Climate Change": Required Reduction of Carbon Emissions to Protect Young People, Future Generations and Nature*, 8 PLOS ONE e81648 (2013) (presenting science indicating need to restore the atmosphere to a CO_2 concentration of 350 ppm to maintain a planet that is conducive to human habitation). *See also* Brief for Scientists Amicus Group as Amici Curiae Supporting Plaintiffs-Appellants at 16, Alec L. v. McCarthy, No. 13-5192 (D.C. Cir. Nov. 12, 2013) ("Effective action remains possible, but delay in undertaking sharp reductions in emissions will undermine any realistic chance of preserving a habitable climate system, which is needed by future generations no less than by prior generations."), *available at* http://ourchildrenstrust.org/sites/default/files/FiledScienceAmicus.pdf; *see also* Joel Smith *et al.*, *Assessing Dangerous Climate Change Through an Update of the International Panel on Climate Change (IPCC) "Reasons for Concern,"* 106 Proceedings of the National Academy of Sciences 4133 (2008).

[2] For interpretation, *see* "*NASA Scientists React to 400 ppm Carbon Milestone*," NASA, GLOBAL CLIMATE CHANGE, http://climate.nasa.gov/400ppmquotes/; Mark Fischetti, "*2-Degree Global Warming Limit is Called a 'Prescription for Disaster*,'" SCIENTIFIC AMERICAN BLOG (Dec. 6, 2011). http://blogs.scientificamerican.com/observations/2011/12/06/two-degree-global-warming-limit-is-called-a-prescription-for-disaster/.

practical prospect of adaptation.[3] The future well-being of the world's youth and their descendants hinges on present society's willingness to redefine and reconstruct the current disaster-track socioeconomic structure.

While slashing carbon emissions stays crucial, the concomitant challenge of adaptation becomes imperative in response to the climate chaos already under way. Due to the carbon pollution remaining in the atmosphere from prior emissions, there exists heating "in the pipeline" that cannot be called back. As economist Thomas Friedman states, a climate approach must both "manage what is unavoidable and avoid what is unmanageable."[4] Managing the unavoidable (the climate change already occurring) is known as *adaptation*, while avoiding what is unmanageable (lowering carbon emissions to stave off global catastrophe) is known as *mitigation*. This volume focuses on the adaptation challenge by providing a pragmatic and analytical approach to managing impacts from global warming.

The degree of change faced by humanity may be nearly unfathomable to decision makers operating in a context of industrialized society that has enjoyed relatively stable climate conditions. The average global temperature in 2012 was only 0.85 °C warmer than what it was in 1880.[5] To a layperson, this may sound negligible. However, five times as many natural disasters—including floods, hurricanes, droughts, and wildfires—occurred during 2001–2010 as compared with 1971–1980.[6] Rising sea levels, extreme water scarcity, ferocious storms, and plummeting crop yields are already destroying human communities and causing deaths across the world. While different regions face different forms of upheaval, one thing is certain: there is no safe haven on the planet safe from global climate change. Projections indicate that by 2050, between 50 million and 350 million people could be forced to migrate due to the impact of global warming.[7] Climate change will literally redraw the geopolitical boundaries that presently exist.

The adaptation challenge calls forth the basic duty of government to provide for the safety and welfare of the people. Officials in countries throughout the world must rise to these unprecedented circumstances. Leaders today occupy a crucial moment in history, holding unparalleled responsibility for the welfare of both

---

[3] Durwood Zaelke, *As Climate Impacts Accelerate, Speed of Mitigation Becomes Key*, HUFFINGTON POST (July 15, 2014), http://www.huffingtonpost.com/durwood-zaelke/as-climate-impacts-accele_1_b_5588113.html.

[4] Thomas L. Friedman, *The Scary Hidden Stressor*, THE NEW YORK TIMES (Mar. 2, 2013), http://www.nytimes.com/2013/03/03/opinion/sunday/friedman-the-scary-hidden-stressor.html?_r=0.

[5] International Panel on Climate Change, 2013: *Climate Change 2013: The Physical Science Basis. Contribution of Working Group I to the Fifth Assessment Report of the Intergovernmental Panel on Climate Change* [Stocker, T.F., D. Qin, G.-K. Plattner, M. Tignor, S. K. Allen, J. Boschung, A. Nauels, Y. Xia, V. Bex and P.M. Midgley (eds.)]. Cambridge University Press, Cambridge, United Kingdom and New York, NY, USA.

[6] World Meteorogical Organization, *Atlas of Mortality and Economic Losses from Weather, Climate and Water Extremes (1970—2012)*, WMO—No. 1123 (2014).

[7] United Nations, General Assembly, *Climate change and Its Possible Security Implications: Report of the Secretary-General*, A/64/350 (11 September 2009).

present and future generations of citizens. Today's crisis requires such leaders to act with vision and courage in face of uncertainty and turbulence. The impacts of climate change once considered distant possibilities are happening now, several generations earlier than indicated by even the most pessimistic predictions of the past. Leaders of today's world must learn to be proactive, rather than reactive. Entire coastlines sit vulnerable to rising seas, for example, a phenomenon that will destroy homes, flood entire cities, and unleash massive toxins from existing port facilities and nuclear plants situated near the ocean. A systematic relocation and preemptive cleanup will save lives, communities, and economies. But the challenge is immense, and society must act swiftly in the waning window of climate stability to accomplish such massive restructuring. The task ahead is to imagine the unimaginable and act in bold and effective ways based on the best available science and analysis—untainted by vested political and economic interests that have strong profit interests in assuring no change at all.

Ultimately, leaders must pursue adaptation and mitigation urgently as one integral goal. With smart planning, a community can fortify its ability to adapt to climate change while also accomplishing measurable carbon emission reduction. Rather than being mutually exclusive, the adaptation and mitigation goals can be—and in every case should be—jointly reinforcing and synergistic. This great societal transition should take shape around the basic aim of building local resilience and self-sufficiency by creating fossil-fuel free economies. This effort will entail unraveling many of the economic systems that perpetuate dependency on distant global markets and multinational corporations. Such an effort will require re-localizing food systems, creating local manufacturing systems, and designing transportation options independent of fossil fuels. Action on these levels should achieve significant co-benefits, by infusing local economies with investments and creating more political autonomy for communities. But the importance of lifestyle choices cannot be underestimated. If the world's youth are to inherit a habitable planet, citizens of developed nations must decide to reject high-carbon lifestyles that have become intolerably destructive, and citizens of developing nations must form a new vision that does not simply model the excesses of the industrialized world. In all cases, this re-envisioning of how we live and how we work requires willpower and imagination, but perhaps most of all the fortitude to challenge the entrenched assumptions governing our current decision-making processes.

This book provides a platform for building a concrete vision of local and global adaptation. It boldly confronts the imminent change facing humanity, yet its reasoned and practical approach allows readers to maintain the optimism and determination necessary to work towards a future with a habitable climate. It is designed for a broad range of readers, including educators, journalists, scientists, citizens, business entrepreneurs, religious leaders, and government officials. The challenge ahead calls forth people from all sectors of society to take initiative and innovate change in their own distinct realm of society. True pioneers in this pivotal age will create new systems and economies that improve the condition of humanity. But we must all act with extreme urgency. There is not a moment to lose.

Mary Christina Wood author of Nature's Trust: Environmental Law for a New Ecological Age Cambridge University Press, 2013 is the Philip H. Knight Professor of Law at the University of Oregon School of Law and serves as Faculty Director of the school's nationally ranked Environmental and Natural Resources Law (ENR) Program. She expresses gratitude for the contribution of Rance Shaw, Bowerman Fellow for the ENR Program's Conservation Trust Project.

Eugene, OR Mary Christina Wood

Preface

Climate change is considered a fact; it is no longer an illusion and not seriously disputed. Its effects are global and can be experienced in different ways and in different intensities. Human activities have led to a significant increase in the concentrations of greenhouse gases in the atmosphere which enhances the natural greenhouse effect and leads to an additional warming of the average temperature of the earth's surface and the atmosphere. It is also a fact that the majority of greenhouse gases emitted worldwide in the past and in the present come from developed countries and that the per capita emissions are still relatively low in developing countries.

But the social needs for development will have to be satisfied. As a consequence the proportion of greenhouse gases emitted from developing countries will rise worldwide. Therefore, climate change will increase and the impact on natural ecosystems and on people will intensify. So it is not surprising that the upcoming changes in the climate of the earth and its adverse effects are met with concern by all humanity.

As varied as the causes and effects of climate change, so varied are the proposed solutions and approaches to deal with the problem and to do something against it. Taking into account the global character of climate change, a comprehensive global cooperation is called for that leads to an effective and appropriate international action—in accordance with the respective responsibilities, common but at the same time differing depending on the capabilities and the social and economic situation of the respective actors.

The 19 contributions to this book are presenting some ideas, approaches, and tools about the adaptation to climate change. In addition to (existing) legal instruments, these contributions also focus on the implementation of economic instruments and planning tools as well as their (further) development. Here, Wätzold shares with us the economic perspective on climate change adaptation while Afroz and Naser give an overview of global adaptation politics. Gawor portrays the advantages of the Life Cycle Assessment in climate change mitigation. Andrew Long discusses the possibilities of REDD+; Zschiegner and Wanki introduce the German Renewable Energy Source Act (EEG), whereas Burleson depicts the opportunities of innovation in climate change law.

Apart from the strategy to counteract climate change by avoiding emissions, the general approach to adapt to climate change is pursued as well. In this context, Camacho writes on the learning infrastructure in the United States' federal system while Send, Riedel, and Hansch identify the role of crowdsourcing in climate change politics.

The responsibility of dealing with the effects of climate change is identified as a central theme by Piroch (liability for damage caused by climate change), Krause and Egute (risk management and climate change), as well as Mißler-Behr and Mehicic (strategy development and risk management in emission rights trading).

A number of articles deal with country-specific problems related to climate change and the respective possible adaptation strategies. The countries in focus are the Netherlands (Gupta, Klostermann, Bergsma, and Jong), Nigeria (Ogbonna), Syria (Ibrahim), and—treated in three different articles—Cameroon (Somah and Schmidt, Lambi and Kometa, Egute and Albrecht). On a more global level, Clouting, Douven, Ostrovskaya, Schwartz, and Pataki analyze the institutional capacity for wetland management. Spyra and Albrecht discuss concepts for the future—beside adaptation.

As the climate change conference COP19 in Warsaw lately pointed out, adaptation is becoming more important and the need for specific adaptation strategies has to become more urgent.

The key points of the climate conference in Warsaw deal with the global climate agreement: There is a specific timetable in order to develop the global climate treaty until 2015 in Paris. The conference has also created a solid framework in order to protect forests. Poorer countries should earn money when they protect their jungle and thus contribute to climate protection. Furthermore, the conference implemented the "Warsaw mechanism." Thus, developed countries should help developing countries in climate change related loss and damages.

The industrialized countries had already stated an increase in climate finance in poorer countries up to 100 billion dollars (74 billion euros) annually in 2020. Although poorer countries had asked for it, specific intermediate steps have not been made in Warsaw. A major source of this increase is the Green Climate Fund.

Based on the already existing and financial difficulties, the adaptation fund for developing countries to the consequences of climate change will get as early as 2013, a cash injection of around 100 million dollars (74 million euros) by a few industrialized countries in order to be able to stay operational. This is primarily a signal to developing countries so that they will not lose confidence in the negotiating process.

The editors would like to give their hearty thanks to the 34 authors from ten different countries around the world for their contributions and their cooperation.

Cottbus, Germany

Eike Albrecht
Michael Schmidt
Magdalena Mißler-Behr
Simon Spyra

Contents

Chapter 1
Adaptation to Climate Change in the International Climate Change Regime: Challenges and Responses

Tanzim Afroz and Mostafa Mahmud Naser

1.1 Introduction

In international law, a regime includes the entirety of rules and practices within one or several interrelated international treaties (Verheyan 2002). The climate change regime, for the purpose of this chapter, has been connoted as the collection of principles, norms, rules, and decision-making procedures in international climate change negotiations (Paavola and Adger 2006, see also Krasner 1982; Young 1994). This regime basically has emerged in the context of the United Nations Framework Convention for Climate Change (UNFCCC) of 1992 which provides a framework in international law for mitigation as well as adaption as climate response strategy. The term 'adaptation', though widely used in climate change agenda, is generally considered as an underdeveloped part of the legal regime of climate change (Linnerooth-Bayer and Meckler 2006). This is because, till date, international climate discourse is mainly focused on mitigation policy to reduce greenhouse gas emissions for addressing climate change impacts. In that respect, the developed countries agreed, under Article 3.1 of the Kyoto Protocol, for emission reductions by at least 5 % from 1990 between 2008 and 2012. However, scientific research already proved that even the most stringent mitigation efforts cannot avoid severe impacts of climate change in the next few decades (Srinivasan 2006). Most alarmingly, several impacts of climate change have already been evident in many ecosystems and economic sectors as reported by the Intergovernmental Panel on Climate Change (IPCC) in its Third Assessment Report (TAR). To cope with these consequences and moderate the impacts of climate change, adaptation as a policy has become prominent in recent climate discourses. These

T. Afroz (✉)
Department of Law, University of Dhaka, Dhaka, Bangladesh
e-mail: tanzimafroz@gmail.com

M.M. Naser
Department of Law, University of Chittagong, Chittagong, Bangladesh
e-mail: mostafa.m.naser@gmail.com

E. Albrecht et al. (eds.), *Implementing Adaptation Strategies by Legal, Economic and Planning Instruments on Climate Change*, Environmental Protection in the European Union 4, DOI 10.1007/978-3-540-77614-7_1,

concerns drag the focus on legal aspects of adaptation in the international climate change regime.

1.2 Legal Response in Adaptation to Climate Change

The UNFCCC is the basic legal document for adaptation to climate change. Beside this Convention, several pertinent provisions of the Kyoto Protocol and the decisions of the Conferences of the Parties (Melkas 2002; Verheyan 2002) govern the basic legal issues of adaptation in international climate regime. However, there is a wider legal framework for adaptation beyond this international climate regime. There are, for instance, other international laws concerned with human rights and protection of nature (e.g. the UN Convention to Combat Desertification and the Convention on Biological Diversity) which provide detailed rules as to when and to what extent adaptation measures might be necessary or mandatory to protect human beings and the environment (Verheyan 2002). International customs as well as national legislations (Paavola and Adger 2006) also have roles to play. But this broader legal framework has not been addressed in this chapter. The following discussion highlights the emergence and challenges of adaptation only in the legal regime of global climate change.

1.2.1 Adaptation Under the UNFCCC Realm

The law on adaptation is still considered in its infancy (Verheyan 2002) as mitigation has been dealt widely in climate change regime so far. As for instance, the main objective of the UNFCCC (from hereafter also: the Convention) is mitigation by stabilizing greenhouse gas concentrations in the atmosphere at a level. If that could be achieved, there was no necessity of adaptation (Verheyan 2002). Realizing the dubiousness, the Convention recognizes the limitation of resources that can be allocated for mitigating the greenhouse gas emissions. Article 2 of the Convention mentions that the mitigation strategy should not compromise food production and sustainable economic growth. That is why in this objective clause a time-frame has been stated for allowing ecosystems 'to adapt naturally to climate change'. Thus adaptation also becomes a part of the UNFCCC framework (Grasso 2010).

However, Article 4 of the Convention is the pivotal section for undertaking adaptation and enhancing adaptive capacity in climate change regime. Paragraph 1(b) of Article 4 provides that parties must formulate and implement national or regional programs containing 'measures to facilitate adequate adaptation to climate change'. Article 3(3) has complemented this clause committing the parties to 'take precautionary measures to anticipate, prevent or minimize the causes of climate change'. Thus the Convention obliges all state-parties to address adaptation in a precautionary and strategic way through programs, not simply relying on autonomous adaptation by nature (Verheyan 2002). Moreover, in paragraph 1(e) of Article 4, all parties commit to cooperate in preparing for adaptation to climate impacts. This international collaboration has been stressed in several sensitive fields like coastal zone management, water resources and agriculture, protection and rehabilitation of areas affected by droughts, desertification and floods. Beside these initiatives, Article 4, paragraph 1(f) focuses on the careful crafting of adaptation policies in economic, social and environmental sectors of state parties to prevent adverse effects of climate change (Grasso 2010). All state parties also underline a national reporting obligation in that response as per Article 12(1) of the Convention. But these initiatives obviously require 'new and additional financial resources' for developing countries. In this regard, under Article 4(3), the developed state parties and other developed parties included in Annex II (Western Organization for Economic Cooperation and Development—OECD members are considered as Annex II countries under the UNFCCC, Annex I lists all Annex II countries plus countries with economies in transition in Central and Eastern Europe as well as Russia and the Ukraine) commit to provide 'full incremental costs' and transfer of adaptive technologies as required by the developing state parties (non-Annex I party, as mentioned in the Convention). This is because common but differentiated responsibility has been adopted as one of the main principles of the Convention in Article 3(1). More specifically, the developed countries commit to assist particularly vulnerable developing countries in meeting their adaptation costs under Article 4(4). While taking decisions regarding financial assistance and adaptive technology transfer, Article 4(8) and (9) demands special attention to the specific needs of developing countries, small island states, countries with low-lying coasts and prone to natural disaster, arid countries, least developed countries (LDCs) and so on.

Nevertheless, Article 4 of the UNFCCC comprises several ambiguous and problematic issues pertaining to adaptation. For example, paragraph 1(b) of Article 4 does not make clear what constitutes an 'adequate adaptation' (Klein 2002; Grasso 2010) and what is meant by 'facilitate' adaptation. In the absence of any specific definition, donor states only prefer economic efficiency or the cost effectiveness of any adaptive measure to be adequate. But there are other non-economic issues, like the environmental sustainability, technical feasibility, administrative or legal admissibility or acceptability of the measures (Klein 2002; Grasso 2010; Verheyan 2002) which can also determine the adequacy of adaptive measures. Thus the provision legally gives state-parties discretion to judge what is adequate and, ultimately, not to take every possible measure to prevent climate change damages. Rather, by using a non-predefined term such as 'adequate', the Convention leaves a margin of discretion for each party to choose between preventing residual damages or accepting them (Verheyan 2002). Similar criticism can be drawn regarding the term 'facilitate' adaptation. Literally, the word means 'to help, aid or to assist' (Oxford Concise Dictionary 1999) and is often used in conjunction with planning. So it does not cover the full catalogue of measures that could be predicted for making a system climate-safe (Verheyan 2002). Moreover, the role of private sector in enhancing adaptive capacity is the most unexplored but important issue in the framework. In many vulnerable economies, private companies actually possess the technological, financial or human capacity to increase the adaptive capacity within a community. But Article 4.1(b) of the UNFCCC governs only public adaptation measures because, as an international treaty, the Convention cannot address private entities directly (Verheyan 2002).

There are more ambiguities in Article 4(3) and 4(4) of the Convention regarding the interpretation and application of terms like 'incremental costs' and 'resources...needed' (Werksman 1993; Bodansky 1993). The incremental cost concept is linked to the fact that the UNFCCC only applies to human-induced climate change and not on current (natural) climate change and variability (Grasso 2010). Therefore, the Convention only finances for incremental costs of impacts produced by anthropogenic climate change, whereas the general costs (and benefits) of adapting to normal climate is left aside. Moreover, the institutional infrastructure for channelling assistance to developing countries, as mentioned in the Convention, is very general in term. It does not oblige any party to carry out any particular kind or type of adaptive measure. Neither does it specify any time-frame in which the duty arises. As a consequence, the climate change regime, so far, has failed to fully materialize the assistances ensured by the Convention (Paavola and Adger 2006).

1.2.2 Adaptation Under the Kyoto Protocol and COPs Decisions

Some directives of the UNFCCC have been further specified and dealt with in detail in the Kyoto Protocol and in several decisions of the Conferences of the Parties (from hereafter, COPs) to the Convention. As for instance, Article 3(14) of the Kyoto Protocol commits the Annex 1 countries (developed states as enlisted in Annex I of the UNFCCC) to meet their emission reduction targets for minimizing 'adverse social, environmental and economic impacts on developing countries'. Article 10, paragraph 1(b) of the Protocol directs the non-Annex I parties to formulate, publish and regularly update national programs for adaptation to climate change (Paavola and Adger 2002, 2006). In this regard, National Adaptation Programs of Action (NAPAs) is a process to be used for generating these plans. Marrakech Accords (COP7 2001) specified the guidelines for the preparation of NAPAs to unify the national priorities for adaptation. These guidelines require multi-disciplinary and public consultation in the preparation of NAPAs (Decision 29/CP7). In Least Developed Countries (LDCs—a group of 50 countries with roughly 11 % of the world's population), NAPAs are prepared with the institutional support of the LDC Expert Group (LEG) and financed by a dedicated fund, the LDCF (Grasso 2010). These NAPAs are a partial implementation of Article 4.1 (b) of the UNFCCC and so far, these are the only documents within the Convention which deal solely with adaptation (Verheyan 2002).

In climate change regime, adaptation gained further attention in 2004 at COP10 in Argentina. The Buenos Aires Programs of Work on Adaptation and Response Measures was adopted during this Conference (Decision 1/CP10). This program included further scientific assessments of vulnerabilities and options for adaptation, support of the NAPAs of LCDs, new workshops and technical papers on climate change risk and adaptation, and support for mainstreaming adaptation into sustainable development planning (Srinivasan 2006). These objectives and scopes of the Buenos Aires Programs of Work were further specified by a detailed 5-year program of work at COP11 in Montreal in 2005. This 5-year program of work mainly focused on impacts, vulnerabilities and adaptation to climate change with an aim that these will assist state parties, in true sense, to make informed decisions on implementation of adaptation measures (Decision 2/CP11). The Bali Roadmap (COP13 2007) fostered alliances between the North and the South to promote adaptation in the developing world (Grasso 2010). Finally, the Copenhagen Accord (COP15 2009) concentrates on the urgent necessity of enhanced action and international cooperation for adaptation initiatives. Most significantly, the importance of adaptation for implementing the objectives of the Convention has been admitted during this Conference (Decision 3/CP15) and the Copenhagen Adaptation Framework has been established.

1.2.3 Evolving Focus on Adaptation Outside the UNFCCC Realm

Despite the COPs, there are other significant ongoing activities in the field of adaptation within as well as outside the realm of the UNFCCC. The Adaptation Policy Framework (APF) is such a program initiated by the United Nations Development Program (UNDP). The principal aim of this Framework is to incorporate adaptation into countries' national development strategies. In this regard, international donor organizations are also sharing supportive views as several reports in recent years have made them concerned that climate change is going to influence their activities in poorer countries. In 2003 the Vulnerability and Adaptation Research Group (VARG) published a report titled 'Poverty and Climate Change' (AfDB 2003). This report highlighted the importance of integrating climate change into development programming. Another report was published in the same year by the Asian Development Bank. It also focused on how to mainstream adaptation in Asian Development Bank (ADB) project operations (ADB 2003). Thus the donor agencies are supporting adaptation strategies remaining outside the Convention's realm.

Besides, several environmental and conservation institutions around the world are also turning their attention to adaptation to climate change vigorously. The Convention on Biological Diversity (CBD), for example, is working on the identification of opportunities to adapt to climate impacts in a way that protects

Table 1.1 Examples of different types of adaptation

Sector	Reactive	Anticipatory
Private	• Moving home • Changing insurance premiums • Buying air-conditioning systems	• Changing architecture of building • Buying hazard insurance • Devising new customer products
Public	• Offering compensation or subsidies • Enforcing building codes • Beach nourishment	• Installing early warning systems • Establishing new building codes • Constructing dykes

Source: UNFCCC (2006)

biodiversity (Grasso 2010). The Worldwide Fund for Nature (WWF) and the World Conservation Union (IUCN) are dealing with effective eco-system management to amplify the adaptive capacity of vulnerable communities. Even many humanitarian organizations, like the Red Cross or Red Crescent, which mainly use their expertise in disaster management, are practicing to adapt to climate impacts. A number of national initiatives, such as the United Kingdom's Climate Impact Program or Canada's Climate Change Impacts and Adaptation Program, are especially dealing with adaptation to climate change (Grasso 2010). Thus adaptation, though in a creeping manner, finally got attention in the climate change regime.

But the problem is that there is no uniform definition of adaptation, and various stakeholders define and interpret it quite differently. The IPCC, for example, considers a wide range of various adaptations according to intention, time of action and type of actors involved, such as autonomous vs. planned adaptation, anticipatory vs. reactive adaptation, and public vs. private adaptation (Srinivasan 2006); a few examples are incorporated in Table 1.1 But the UNFCCC interprets adaptation within a very limited scope and emphasizes only on human-induced long-term climate change. This difference may seem small but it is sufficient enough to lead widespread confusions regarding funding in international negotiations. This financial implication of adaptation is another area of concern which we will consider next.

1.3 Adaptation Funding: Status and Challenges

The climate change regime is significant in both legal and political terms, as it provides developing countries with a legal basis to claim funds from developed countries for the purpose defined in the UNFCCC (Verheyan 2002). But regarding adaptation finance, there is no legally binding quantitative obligation (Bouwer and Aerts 2006). All such funds as described in Table 1.2 comprise voluntary contributions from developed countries.

1.3.1 Adaptation Funding Under the UNFCCC

The UNFCCC contains two categories of financial obligations in Article 4(3) and 4(4) respectively. The first one is generally aimed at helping developing countries

Table 1.2 Adaptation funds under UNFCCC and Kyoto protocol

Fund	Created under	Funding source	Beneficiaries	Benefits	Legal basis for funding
GEF Trust Fund	UNFCCC	GEF	Developing countries	Incremental cost	UNFCCC Art. 4.3, 1/CP11, 5/CP7, GEF/C23/Inf8
Strategic Priority on Adaptation (SPA)	UNFCCC	GEF	Developing countries	Incremental cost	6/CP7, GEF/C23/Inf8
Special Climate Change Fund (SCCF)	UNFCCC	Developed country discretionary pledge	Developing countries	Additional costs of adaptive measures Uses sliding scale	5/CP7, 7/CP7, 5/CP9, GEF/C24/12, GEF/C25/4/Rev1
Least Developed Countries (LDC) Fund	UNFCCC	Developed country discretionary pledge	Least developed countries	Additional costs of adaptive measures Uses sliding scale	5/CP7, 7/CP7, 27/CP7, 28/CP7, 29/CP7, 6/CP9, 3/CP11, 4/CP11, GEF/C24/Inf7, GEF/C24/Inf8/Rev1, GEF/C25/4/Rev1
Adaptation Fund	Kyoto Protocol	2 % Share of proceeds from CDM	Particularly vulnerable developing countries	To be determined	5/CP7, 10/CP7, 17/CP7, 28/CMP1

Source: Mace (2005)

Table 1.3 Staged process under GEF

Stage	Stage I: planning (short term)	Stage II: preparation (medium term)	Stage III: initiation (long term)
Parties involved	All	Particularly vulnerable countries or regions	Particularly vulnerable countries or regions
Activities	• Studies of possible impacts of climate change • Appropriate capacity-building • Identification of particularly vulnerable countries or regions	Measures to prepare for adaptation, including further capacity-building and development of appropriate adaptation plans	Measures to facilitate adaptation, including insurance and other adaptation measures

Source: Bouwer and Aerts (2006)

to implement their duties under the Convention whereas the second one obliges adaptation costs for particularly vulnerable developing countries. Pursuant to Article 21(3) and various COP decisions, Convention funding is provided through the Global Environment Facility (GEF) Trust Fund. The GEF has been supporting developing countries on adaptation to climate change through a staged process as shown in Table 1.3 Stage I was to support studies and planning, Stage II to support detailed planning and capacity building as envisaged in Article 4.1(e) and Stage III to support actual adaptation, *inter alia*, insurance measures and other measures to implement Article 4.1(b) and 4(4). Most developing countries have already carried out the initial assessment (or Stage I) studies on adaptation. A few Stage II studies, for example in the Caribbean, Pacific and Bangladesh, have also been initiated (Adger et al. 2003). But the Stage III activities have been troubled as the guidance given at COP6*bis* is not clear enough. There are other hardships for adaptation costs under this fund because of its operational strategy. It stipulates that the projects financed or co-financed by the GEF shall result in 'global benefits' whereas adaptation often benefits only the region or country in which measures are undertaken (Bouwer and Aerts 2006; Verheyan 2002). Moreover, the Convention does not define the term 'particularly vulnerable' which keeps open the problem of interpretation. Thus the GEF does not have any clear guidance on which countries are eligible recipients for funding (Verheyan 2002).

In July 2004, the Strategic Priority on Adaptation (SPA) of the GEF Trust Fund was launched to support pilot projects that could demonstrate how climate change adaptation and planning could be integrated into country policy and sustainable development planning (Francisco 2008). However, SPA never graduated beyond its 'pilot' phase. It is available to developing countries on application, subject to a complex assessment of their capacity. Because of this tough application procedure, the expenditure under SPA has been and remains excruciatingly slow. An original allocation of US$ 50 million to the SPA had not been spent by the end of the initial pilot period and no further fund was added for the next 'replenishment' period that is from 2007 to 2010 (FCCC/CP/2007/3 para 8). Moreover, many developing countries are not aware of what is on offer or how to access these funds. According to GEF's latest report, for example, only one of 10 GEF-supported climate change projects in financial year 2006–2007 concerned adaptation through the SPA, amounting to just US$1 million of a total US$81 million spent on climate change projects (FCCC/CP/2007/3 paras 16–17). The rest was geared towards mitigation while developing countries do not have mitigation obligations.

Beside this SPA fund, there are two special funds under the Convention to support adaptive efforts. These are the Special Climate Change Fund (SCCF) and the Least Development Countries (LDC) Fund. Both of these funds were created by the Bonn Agreement in 2001 at second session of COP6 (Decision 5/CP6) and confirmed at COP7. They are based on the funding made available by the Annex I countries and managed by the GEF (Paavola and Adger 2006). The SCCF was created to support adaptation activities and capacity building but, until recently, this fund remained inactive. However, seven SCCF projects were finally approved in 2006–2007, which involved only eight countries whereas there are 121 developing country parties to the UNFCCC. The LDC Fund was entrusted to support the work program of the LDCs under the Convention, including the preparation of NAPAs (Decision 5/CP7). In this project it has provided little more than US$10 million so far (Bouwer and Aerts 2006) and 32 NAPAs have been finished to date. On the basis of these NAPAs existing at the time, the Stern Review (2006) projected that US$1.3 billion would be required for the immediate adaptation needs of the 47 Least Developed countries (LDCs). So far, nothing close to this amount is forthcoming.

1.3.2 Adaptation Funding Under the Kyoto Protocol

Article 12(8) of the Kyoto Protocol provides that a share of the proceeds of Clean Development Mechanism (CDM) projects should be used to assist particularly vulnerable developing countries to meet the costs of adaptation. In that response, COP6 created an Adaptation Fund to replenish from a 2 % levy on CDM projects (Table 1.2). Procedures for its management were eventually approved in Bali Roadmap at COP13 in 2007. One positive outcome is that this fund involves a Board with strong developing country representation. However, the World Bank estimates this fund is likely to remain small and uncertain, with funds anywhere between US$ 270 and 600 million by 2012 (World Bank 2006).

1.3.3 Future Potential Funding of Adaptation

The costs of adapting to an approximately 2 °C warmer world by 2050 were estimated and the range lies between US$ 75 and 100 billion per year from 2010 to 2050 (World Bank 2009). However, the total available annual funding for adaptation under the Convention or the Protocol up to 2012 ranges from US$ 20 million to US$ 300 million at best (Srinivasan 2006). From this viewpoint, it is unrealistic to cover all the costs related to adaptation. So the Copenhagen Accord (COP15 2009) focuses on a wide variety of funding sources, public and private, bilateral and multilateral, including alternative sources of finance (Decision 8/CP15). Foreign Direct Investment (FDI), Official Development Assistance (ODA), insurance and disaster pooling, public expenditure including public–private partnerships (PPPs), non-compliance fund, disaster relief and risk reduction can be such alternative sources. The Copenhagen Accord also established a new Copenhagen Green Climate Fund. However, it remains to be seen how this instrument relates to the already existing adaptation funding mechanisms under the Convention and the Protocol (Schalatek et al. 2010).

1.4 Conclusion

Adaptation to current and future climate change regime is a cross-cutting issue (Verheyan 2002). Defining a new approach to address adaptation in the post-2012 regime is a challenge but its necessity has been recognized by the international community in the meantime. Such a new approach may evolve into the establishment of a separate protocol for adaptation in the long run (Srinivasan 2006). A suitable mechanism for effectively monitoring the transfer of new and additional funding from developed to developing countries is also necessary. Options for establishing a mandatory global funding scheme, which is tied to both past and current greenhouse gas emissions by various countries, should be explored as a high priority. If that can be achieved, all countries, both developing and developed, can contribute to and benefit from such a scheme based on the principle of common but differentiated responsibility (Srinivasan 2006).

Biography Tanzim Afroz is Lecturer at the Department of Law of the University of Dhaka. Currently she is doing Ph.D. at Macquarie Law School, Macquarie University, Australia.
Dr. Mostafa Mahmud Naser is an Assistant Professor at the Department of Law of the University of Chittagong.

References

ADB—Asian Development Bank (2003) Guidelines to adaptation on mainstreaming for pacific development operations. Manila. Internet: http://www.adb.org/REACH/draft_guidelines_ampdo.pdf. Last accessed 28 June 2010

Adger WN, Huq S, Brown K, Conway D, Hulme M (2003) Adaptation to climate change in the developing world. Prog Dev Studies 3(3):179–195

AfDB—African Development Bank, Asian Development Bank, Department For International Development—UK, Directorate-General Development European Commission, Federal Ministry of Foreign Development Cooperation—he Netherlands, Organization for Economic Cooperation and Development, United Nations Development Program and The World Bank (2003) Reducing the vulnerability of the poor through adaptation. United Nations Development Program, New York, NY
Bodansky D (1993) The UN framework convention on climate change: a commentary. Yale Int'l Law 18:451–558
Bouwer LM, Aerts JCJH (2006) Financing climate change adaptation. Disasters 30(1):49–63
Francisco HA (2008) Adaptation to climate change: needs and opportunities in southeast Asia. ASEAN Econ Bull 25(1):7–19
Grasso M (2010) Justice in funding adaptation under the international climate change regime. Springer, New York, NY
Klein RJT (2002) Adaptation to climate variability and change: what is optimal and appropriate? In: Giupponi C, Schechter M (eds) Climate change and the Mediterranean: socio-economics of impacts, vulnerability and adaptation. Edward Elgar, Northampton, MA
Krasner SD (1982) Structural causes and regime consequences: regimes as intervening variable. Int Org 36:185–205
Linnerooth-Bayer J, Meckler R (2006) Insurance for assisting to climate change in developing countries: a proposed strategy. Climate Policy 6:1–17
Mace MJ (2005) Funding for adaptation to climate change: UNFCCC and GEF developments since COP-7. Rev Eur Commun Int Environ Law 14(3):225–246
Melkas E (2002) Sovereignty and equity within the framework of the climate regime. Rev Eur Commun Int Environ Law 11:115–128
Oxford Concise Dictionary (1999) Clarendon Press
Paavola J, Adger WN (2002) Justice and adaptation to climate change. Working Paper 23. Tyndall Centre for Climate Change Research
Paavola J, Adger WN (2006) Fair adaptation to climate change. Ecol Econ 56:594–609
Schalatek L, Bird N, Brown J (2010) Where's the money?-The status of climate finance post—Copenhagen. In the Copenhagen accord, UNFCCC negotiations and a look at the way forward. Heinrich Böll Foundation, North America
Srinivasan A (2006) Adaptation to climate change. In: Srinivasan A (ed) Asian aspirations for climate regime beyond 2012. SATO Printing, Japan, pp 77–100
Stern N (2006) The economics of climate change: the review. Cambridge University Press, Cambridge
UNFCCC (2006) Technologies for adaptation to climate change. UNON Publishing Services Section, Nairobi
Verheyan R (2002) Adaptation to the impacts of anthropogenic climate change—the international legal framework. Rev Eur Commun Int Environ Law 11(2):129–143
Werksman J (1993) Incremental costs under the climate change convention: the legal Context. Field working paper
World Bank Environmentally and Socially Sustainable Development and Infrastructure Vice Presidents (2006) Clean energy and development: towards an investment framework. Washington, DC
World Bank (2009) Costs to developing countries of adapting to climate change. New methods and estimates: the global report of the economics of adaptation to climate change study-consultation draft
Young OR (1994) International governance: protecting the environment in a stateless society. Cornell University Press, Ithaca

Chapter 2
Advancing Forest-Related Adaptation: Options for Adaptation-Oriented REDD+

Andrew Long

2.1 Introduction

The single most promising legal and political opportunity for simultaneously preserving tropical forests and combating climate change in the near future is the developing REDD+ mechanism (for summaries of REDD+ proposals see Parker et al. 2009). REDD+ has been understood primarily, if not exclusively, as a mitigation mechanism by most commentators because it has the potential to significantly reduce the nearly 20 % of global greenhouse gas emissions attributable to the forest and land use sector. However, as REDD+ becomes established, it will inevitably have a significant impact on adaptation in tropical forest regions. This impact will not necessarily be beneficial, but design choices made now—in the early stages of REDD+ development—can shape the mechanism to become a profoundly valuable instrument for advancing adaptation in tropical forest countries (for in-depth discussion see Long 2011; for a discussion focusing on biodiversity benefits see Long 2009). This chapter discusses the options for designing REDD+ to maximize adaptation benefit while retaining its mitigation value.

2.2 Climate Change and Forests

2.2.1 Adaptation in the Forest Context

The Intergovernmental Panel on Climate Change (IPCC) defines adaptation as "[i] nitiatives and measures to reduce the vulnerability of natural and human systems against actual or expected climate change effects" (IPCC 2007 p. 76). In the context

A. Long (✉)
Florida Coastal School of Law, 8787 Baypine Road, Jacksonville, FL 32256, USA
e-mail: along@fcsl.edu

E. Albrecht et al. (eds.), *Implementing Adaptation Strategies by Legal, Economic and Planning Instruments on Climate Change*, Environmental Protection in the European Union 4, DOI 10.1007/978-3-540-77614-7_2, © Springer-Verlag Berlin Heidelberg 2014

of tropical forests, adaptation requires adjustments to management, institutions, and other factors influencing the impacts that climate change will have on the forests (Glück et al. 2009 p. 187). As suggested by the IPCC definition, adaptation can be measured and discussed in essentially two ways: adaptation of natural systems and adaptation of human communities. These two types of adaptation will frequently overlap—activities beneficial to the maintenance of natural systems will often support the well-being of the human communities who interact with those systems. However, the two types of adaptation may also appear in conflict, at least in the short term. For example, human communities may seek forest clearing to expand agricultural operations as a means of counteracting reduced agricultural productivity caused by climate change. Thus, a major consideration of policy makers should be ensuring that supported adaptation activities provide benefits for sustaining both human and natural systems through climate change.

Adaptation of natural systems can be assessed in various ways, but two factors appear most significant. First, natural systems adaptation is frequently closely related to preservation of the species comprising the ecosystem in question (Secretariat of the Convention on Biological Diversity 2010 p. 58). Thus, natural systems adaptation will frequently involve a focus on preservation of biodiversity. Second, natural systems adaptation can be assessed according to its impact on the ecosystem services of a particular region. As ecosystem services are frequently underlain by biodiversity, these two measures will often yield similar assessments of the wisdom of particular activities aimed at promoting adaptation of natural systems in tropical forests. In nearly all cases, activities designed for natural systems adaptation will align with practices supported by the concept of sustainable forest management (SFM) (Adaptation of Forests and People to Climate change, Executive Summary and Key Messages, at 12; Broadhead et al. 2009 pp. 60–61).

Adaptation of human communities can be assessed according to a variety of measures, but perhaps the most comprehensive is human well-being. Well-being includes consideration of not only economic status, but also health and quality of life (for a general discussion of well-being analysis see Bronsteen et al. 2010; for an application to international climate change law see Vandenbergh et al. 2009 pp. 336–337). In tropical forest areas, as elsewhere, human well-being is at least partially linked to the maintenance of the services provided by the forest ecosystem (Secretariat of the Convention on Biological Diversity 2010 p. 9). Therefore, activities aimed at human adaptation may in fact be aimed at preserving ecosystem services, such as hydrological regulation, through climate change. Nonetheless, human community adaptation also contains institutional considerations that underlie effective natural systems adaptation, especially improving governance to support effective forest management and creating sustainable livelihoods to alleviate poverty as a means of reducing the human toll of climate change effects.

2.2.2 *Inadequacy of Existing Adaptation and SFM Frameworks*

Despite numerous references to adaptation in climate regime treaties and agreements, including general and specific international commitments to support developing country adaptation, the climate regime has thus far fallen short of delivering a well-functioning approach to adaptation. Instead, adaptation initiatives remain limited and underfunded. In large part, this appears to be attributable to the temporal and spatial disconnect of the sources of adaptation assistance and the benefits received (see generally Yamin and Depledge 2004 pp. 213–247). In short, international adaptation efforts ask developed countries to pay now for benefits that will be received by developing countries in the future.

In the forest context, international law has thus far been largely unsuccessful in achieving conservation and development goals. Despite decades of efforts, the international community has not reached agreement on a binding forests convention and the rate of tropical deforestation has remained alarmingly high (Badiozamani 2007; FAO 2009 Annex Table 2; Butler and Laurance 2008 pp. 469–472; Secretariat of the Convention on Biological Diversity 2010 p. 50). Perhaps the most significant effort to reach agreement on a global forests convention came in the lead-up to the 1992 UNCED in Rio. This effort failed due to a combination of developing country concerns that international regulation of forests would impinge upon their sovereign control of natural resources, developed country resistance to funding measures to reduce tropical deforestation, and secondary concern of developing countries that forest protection efforts could cause them to bear a disproportionate share of the costs of addressing climate change, among other reasons.

While there has been some meaningful progress on addressing forestry issues through international forums such as the UN Forum on Forests (UNFF), these institutions have not effectively fostered SFM in developing countries (see FAO 2009 p. 73). International forestry institutions in their current forms, therefore, appear profoundly unable to address the major challenge of adapting forests and forest users to a changing climate. Instead, forests are likely to suffer from the combined, and perhaps synergistic, effects of multiple existing threats and the growing impacts of climate change.

2.3 Future Climate Change and Forest Policy

While reforestation and afforestation activities may qualify as Clean Development Mechanism projects, reduced deforestation projects do not qualify as emissions reductions under the Kyoto Protocol. To address this shortcoming, the Coalition of Rainforest Nations proposed a mechanism for recognizing reduced emissions from deforestation in 2005 at COP-11 (see UNFCCC-SBSTA 2005). Creation of a reduced deforestation mechanism gained a strong endorsement as an important mitigation mechanism in the Stern Report, which suggested that reduced emissions from the forestry sector could provide a low cost means of enabling significant emissions reductions in the near term (Stern 2007 pp. 244–245, 603–620). At COP-13, the SBSTA endorsed the REDD concept. In the Bali Action Plan, the full COP recommended further action to support reduced emissions from the forestry sector, as well as enhanced conservation and reforestation activities (UNFCCC 2008). This vision of REDD has come to be known as REDD+. By COP-15 in Copenhagen, negotiations on REDD+ represented an area of relatively widespread agreement, although specific policies were not adopted. At that time official agreement on REDD+ has progressed only to a draft decision created by the AWG-LCA, (UNFCCC AWG-LCA 2009). In the 2010 Cancun Agreements, the parties adopted a decision containing several paragraphs encouraging REDD+ activities and specifying safeguards for implementation (UNFCCC 2010). The following year, the parties adopted a decision with procedures intended to ensure compliance with the safeguards agreed to in Cancun (UNFCCC 2011). At the same time, progress on demonstration projects for the voluntary market has been significant. The United Nations REDD Program, World Bank Carbon Finance Unit, and numerous international NGOs are actively involved in major REDD+ initiatives, while several developed countries have provided or pledged significant funding.

2.3.1 *The Hope of REDD+*

REDD+ offers an opportunity to avert the stumbling blocks of prior international efforts to improve tropical forest management by overcoming developing countries' sovereignty concerns and developed countries' hesitation to provide funding. If it develops into a voluntary market-based mechanism for attaching economic value to standing tropical forests, REDD+ has great potential to change the financial incentives driving a significant portion of current tropical deforestation activities. Many commentators and some country proposals have also noted that REDD+ could have significant co-benefits (e.g. Innes et al. 2009 p. 153). In particular, avoided deforestation may have particular value for biodiversity preservation, maintenance of ecosystem services, and improvements in human systems related to forest management, such as governance (Levin et al. 2008).

2.3.2 *The Risks of REDD+*

Despite the promise of REDD+ for reducing tropical deforestation on a global basis, deep concerns exist regarding the mechanism if its financial incentives remain tied exclusively to GHG emissions reductions. These concerns correlate directly with the two types of adaptation considerations identified above.

First, mounting evidence suggests that REDD+ may not yield net co-benefits for biodiversity if it is viewed primarily as a low-cost means of generating carbon offsets (Ebeling and Yasue 2008). The pressure for cost-effectiveness may disincentivize holistic projects that embrace SFM principles and encourage development of low-cost "carbon farming" projects (Sangster and Dudley 2007 p. 215). Such projects will not reduce the pressures on forests of high ecological or socio-economic value. In fact, they may involve practices that undermine maintenance of ecosystem services and fail to produce biodiversity benefit. On a global basis, a purely mitigation-focused REDD+ mechanism may direct relatively little financing toward areas of critical concern for biodiversity because these are not necessarily the most cost-effective areas for reducing forestry emissions.

Second, socioeconomic concerns regarding REDD+ are growing. In particular, indigenous peoples organizations and others express fear that REDD+ may exacerbate current inequities by further empowering elites within developing countries and blocking access of traditional forest users to areas critical to their well-being. Further, the areas most in need of governance and poverty alleviation co-benefits may, because of poor governance or lack of resources, be the areas least able to create well-planned projects that yield such benefits and attract investors (Ebeling and Yasue 2008 p. 1920).

Given the well-founded concerns over REDD's potential to realize the hoped for co-benefits, its promise for adaptation may be minimal if it remains an exclusively mitigation-driven mechanism. Accordingly, if REDD+ is to realize the promise of co-benefits and adaptation, elements to directly promote these benefits must be added.

2.3.3 Legal Approaches to Establish REDD+ as an Adaptation Mechanism

At this early stage in the development of REDD, several options exist that may steer the mechanism toward realization of greater adaptation benefits. Four broad approaches are possible. First, as UNFCCC negotiators appear to envision at present, agreements establishing REDD+ can include language promoting SFM in projects (as in the Cancun Agreements). Second, as some commentators suggest, co-benefits could be mandated in REDD+ projects, perhaps by circumscribing REDD+ financing to an international fund designed to support only the most deserving projects. Third, REDD+ could be designed to provide international financial incentives to high co-benefit projects. Finally, developed nations can use domestic legislation to require preferences for high co-benefit projects in the use of internationally-generated carbon offset credits to satisfy national requirements.

2.3.3.1 "Soft Law" Approach: Encouraging Co-benefits

The decisions adopted during COP-16 and COP-17 demonstrate awareness of co-benefit issues in REDD. The Cancun Agreements, for example, provide a list of safeguards that "should be promoted and supported," such as ensuring that actions are "consistent with the conservation of natural forests and biological diversity" measures (UNFCCC 2010, Appendix I). These safeguards were then strengthened somewhat by a COP-17 decision stating that countries undertaking REDD+ measures should provide a summary of how the safeguards are being implemented (UNFCCC 2011). On the whole, the safeguards provide guidance designed to prevent REDD+ activities from harming existing ecosystems, but they do not provide detailed requirements or incentives to promote improvements to ecosystems or support adaptation. It is, of course, possible in the further elucidation of REDD+ policy, but there is little indication that the negotiations are moving in that direction. Thus, one must be concerned that this approach to REDD+ will fall upon the same stumbling blocks that have limited prior international efforts to combat tropical deforestation.

2.3.3.2 "Hard Law" Approach: Requiring Co-benefits

To counteract the concern that the current approach does not go far enough toward promoting co-benefits, UNFCCC negotiators could require high levels of co-benefits as a prerequisite to REDD+ funding. This approach could be realized by limiting a formal UNFCCC-based REDD+ program, as some commentators and countries have suggested, to a fund-based mechanism. This would allow greater control of financing specifications for REDD+ activities by empowering an international body to make funding decisions in accord with stringent adaptation or co-benefit requirements. This approach is flawed, however.

First, it appears unlikely that an international fund could generate the level of funding required to eliminate the majority of deforestation. The Eliasch Review, for example, estimates that halving deforestation by 2030 would cost approximately $17 billion to $33 billion per year (Eliasch 2008 p. 75). The Paris-Oslo process seeking to jump-start REDD+ with public funding was unable to secure it target of $6 billion in funding, while models predict that a market in forestry-derived carbon credits could fund reductions of 22 % of emissions by 2020 and 75 % of emissions by 2030 (Eliasch 2008 pp. 182–183). Thus, employing a market-based approach vastly increases the probability of generating sufficient funds for REDD+ to significantly impact deforestation.

Second, a mandate for co-benefits may be perceived by many forested developing countries as overly intrusive into their sovereign right to control natural resources decisions within their borders. This concern could become particularly acute if the mechanism were to require particular types of governance or implementation of specific policies for poverty alleviation. Accordingly, a co-benefit mandate (whether embedded in a fund-based form of REDD+ or built into a market-based instrument) may deter tropical forested countries from participating.

In sum, REDD+ with a co-benefit mandate might ensure that a higher percentage of activities meet adaptation goals, but the number of participating countries will likely be far lower because this form of REDD+ is highly vulnerable to the stumbling blocks of prior unsuccessful efforts to combat deforestation—underfunding and the reluctance of tropical forest countries to cede control of their forests to international regulatory mandates. Therefore, limiting REDD+ projects to those complying with a strict co-benefit mandate might significantly limit the overall mitigation and adaptation benefit of the mechanism.

Having shown the inadequacy of both the extreme "soft law" approach to co-benefits (simple encouragement) and the extreme "hard law" approach (an international mandate), we are left to consider the options between these extremes. There are essentially two: international incentives to promote co-benefits and national preferences for projects that are rich in co-benefits among investor countries. Both of these options warrant closer attention in the design of REDD.

2.3.3.3 International Incentives for Adaptation-Related Co-benefits

At the international level, incentives for adaptation could be created in one of two forms: through incentives generated by mechanisms external to the climate regime or through internal incentives built into the climate regime. The external approach has been suggested by commentators and has been demonstrated in several REDD+ projects (Karousakis 2009). This approach has the advantage of targeting specific benefits—most notably biodiversity benefits—that can be built into REDD+ projects and providing a separate source of funding for them. The alignment of biodiversity funders with REDD+ projects that build in additional biodiversity benefits may, however, create significant transaction costs. In some instances, this could raise the overall costs of the project beyond the value of benefits received. This concern can be overcome in the voluntary market, at least partially, by "bundling" carbon and biodiversity credits through entities such as the CCBA and VCS that certify benefits created. It is unclear that this method of reducing costs could be scaled-up, however, to have a significant impact on the shape of REDD+ projects in a regulatory market. Therefore, it is important to explore options for incentivizing co-benefits in a regulatory market.

International incentives for REDD+ co-benefits can be created within the climate regime to provide appropriate scale and predictability and influence the shape of regulatory market-driven REDD+ projects, if the co-benefits are directly tied to adaptation. Building adaptation incentives into the REDD+ mechanism represents an opportunity to meet both mitigation and adaptation goals of the cli-mate regime through a single mechanism, which is likely to provide significant cost saving and other efficiencies.

Key to designing REDD+ with adaptation incentives is the ability to identify and reward projects that provide significant adaptation benefits. This can be achieved through certification and public financing. Certification will require development of standards for the types of benefits that can properly be understood as adaptation measures. In particular, projects that reduce the impacts of climate change on biodiversity and ecosystem services or projects that improve the ability of human systems to tolerate climate change impacts should be certified. Further, the regime could draw upon pre-existing certification systems—such as the Carbon, Community and Biodiversity Alliance (CCBA), Forest Stewardship Council, and Voluntary Carbon Standard—to create an auditing protocol for assessing projects against the standards.

Upon certification, a benefit might accrue to projects through a price premium in the carbon market paid for credits from projects enhancing adaptation capabilities. A much stronger incentive could be created, however, through the provision of direct and guaranteed benefits by the climate regime. By setting these benefits at a level that reflects a combination of the incremental costs of adding and certifying adaptation benefits to a project, as well as the reduction of the future adaptation financing likely to be required in the area, this system would counteract market preference for low cost carbon credits and reward the adaptation benefit actually provided. In general, early actions to support adaptation in REDD+ project areas hold a strong potential to provide cost-savings when compared with actions that would otherwise be required after climate change impacts become severe. The funding for the incentives to promote certification of adaptation benefits should be drawn from the planned "readiness" funding as well as funds that would otherwise be directed toward general adaption measures.

Properly designed, a REDD+ mechanism that certifies and rewards adaptation-oriented projects holds powerful potential to encourage co-benefits in REDD+ and, thereby, maximize its potential for both mitigation and adaptation (for an in-depth discussion and proposal for developing this approach see Long 2011). This approach would effectively respond to the criticisms of REDD+ that have been raised from ecological and socioeconomic perspectives. It would also prioritize public international funding for the projects that provide the most benefit, while retaining a market benefit for all verifiable reduced emissions from forestry projects.

2.3.3.4 National Preferences for High Co-benefit Projects

At the national level, investor countries can adopt rules that embrace a similar approach to the international incentives discussed above. Specifically, national legislation that permits use of REDD+ credits as an emissions offset can specify a preference for credits generated through projects that attain high co-benefits or adaptation benefits (for in-depth discussion see Long 2010). This approach could also be facilitated through reliance on existing certification regimes for REDD+—such as the CCBA standards. If this preference approach were implemented by large investors, such as the United States, these rules could significantly impact the shape of REDD+ projects. Further, this national incentive option can provide value irrespective of whether a similar incentive-based approach is embraced at the international level.

2.4 Conclusion

REDD+ provides an opportunity to lay the groundwork for successful adaptation in many tropical forest areas. The opportunity is critically important to sustaining millions of forest users, preserving the world's biodiversity, and ensuring sustainable forest governance for the long term. The course of international negotiations, however, appears likely to miss this chance by concentrating on mitigation to the exclusions of adaptation in designing a financial incentive system. REDD+ design is, however, at an early enough stage that the mechanism can still be geared toward realizing significant adaptation benefits. The strongest strategies for reaching this goal are those that incorporate both public financing and market incentives aimed at promoting adaptation into the core REDD+ mechanism. The potential to design REDD+ as both a mitigation and adaptation mechanism offers a chance to secure a wide range of environmental benefits more efficiently than separate consideration or ad hoc financial arrangements could deliver. Accordingly, these approaches—as well as appropriate implementation strategies for operationalizing them—warrant much greater attention in the literature and in negotiations than they have thus far received.

Biography Andrew Long is an Associate Professor of Law. Prior to joining the Florida Coastal School of Law faculty in 2008, Professor Long taught at the University of Louisville Louis D. Brandeis School of Law as a Visiting Assistant Professor, served as a Senior Court Attorney with the New York Court of Appeals and practiced in Oregon. Long received his LL.M. from New York University School of Law, where he served as an advisor to the NYU Environmental Law Journal, and his J.D. from Willamette University College of Law, where he served on the Law Review.

References

Badiozamani G (2007) Addressing deforestation and forest degradation through international policy. In: Freer-Smith P, Broadmeadow M, Lynch J (eds) Forestry and climate change. CABI, Wallingford, pp 197–206

Broadhead J, Durst P, Brown C (2009) Climate change: will it change how we manage forests? EFTRN News 50:57–65

Bronsteen J, Buccafusco C, Masur J (2010) Well-being analysis. Georgetown Law J 98: 1583–1642

Butler RA, Laurance W (2008) New strategies for conserving tropical forests. Trends Ecol Evol 23:469–472

Ebeling J, Yasue M (2008) Generating carbon finance through avoided deforestation and its potention to create climactic, conservation and human development benefits. Philos Trans R Soc 363:1920–1924

Eliasch J (2008) Climate change: financing global forests. UK Office of Climate Change

Food & Agricultural Organization of the United Nations (2009) State of the world's forests 2009

Glück P, Rayner J, Berghäll O, Braatz S, Robledo C, Wreford A (2009) Governance and policies for adaptation. In: Seppälä R, Buck A, Katila P (eds) Adaptation of forests and people to climate change: a global assessment report. International Union of Forest Research Organizations, Vienna, pp 187–210

Innes J, Joyce L, Kellomäki S, Louman B, Ogden A, Parrotta J, Thompson I, Ayres M, Ong C, Santoso H, Sohngen B, Wreford A (2009) Management for adaptation. In: Seppälä R, Buck A, Katila P (eds) Adaptation of forests and people to climate change: a global assessment report. International Union of Forest Research Organizations, Vienna, pp 135–186

Intergovernmental Panel on Climate Change (2007) Climate change 2007: synthesis report

Karousakis K (2009) Promoting biodiversity co-benefits in REDD. OECD Environment Working Papers, No. 11. OECD Publishing. doi:10.1787/220188577008

Levin K, McDermott C, Cashore B (2008) The climate regime as global forest governance: can reduced emissions from deforestation and forest degradation (REDD) initiatives pass a 'dual effectiveness' test? Int Forestry Rev 10(3):538–549

Long A (2011) Global climate governance to enhance biodiversity and well-being: integrating non-state networks and public international law in tropical forests. Environ Law 41:95–164

Long A (2010) Tropical forest mitigation projects and sustainable development: designing U.S. law for a supportive role. William Mitchell Law Rev 36:101–124

Long A (2009) Taking adaptation value seriously: designing REDD to protect biodiversity. Carbon Climate Law Rev 3:314–323

Parker C, Mitchell A, Trivedi M, Mardas N (2009) The little REDD+ book: an up-dated guide to governmental and non-governmental proposals for reducing emissions from deforestation and degradation. Global Canopy Program, Oxford

Sangster M, Dudley M (2007) Governance and climate change. In: Freer-Smith P, Broadmeadow M, Lynch J (eds) Forestry and climate change. CABI, Wallingford, pp 214–217

Secretariat of the Convention on Biological Diversity (2010) Global biodiversity outlook 3

Stern N (2007) The economics of climate change: the Stern review. Cambridge University Press, New York

United Nations Framework Convention on Climate Change—Subsidiary Body for Scientific and Technological Advice (2005) Reducing emissions from deforestation in developing countries: approaches to stimulate action, submissions from parties. UN Doc. FCCC/CP/2005/MISC.1

United Nations Framework Convention on Climate Change (2008) Bali, Dec. 3–15, 2007, part two: action taken by the conference of the parties at its thirteenth session. U.N. Doc. FCCC/CP/2007/6/Add.1

United Nations Framework Convention on Climate Change—Ad Hoc Working Group on Long-Term Cooperative Action Under the Convention (2009) Copenhagen, Dec. 7–15, 2009, Policy Approaches and Positive Incentives on Issues Relating to Reducing Emissions from Deforestation and Forest Degradation in Developing Countries; and the Role of Conservation, Sustainable Management of Forests and Enhancement of Forest Carbon Stocks in Developing Countries, Draft Decision (addendum). U.N. Doc. FCCC/AWGLCA/2009/L.7/Add.6

United Nations Framework Convention on Climate Change (2010) Cancun, Nov. 29–Dec. 10, 2010, The Cancun Agreements: Outcome of the Work of the Ad Hoc Working Group on Long-term Cooperative Action under the Convention. U.N. Doc. FCCC/CP/2010/7/Add.1

United Nations Framework Convention on Climate Change (2011) Durban, Nov. 28–Dec. 9, 2011, Guidance on Systems for Providing Information on how Safeguards are Addressed and Respected and Modalities Relating to Forest Reference Emission Levels and Forest Reference Levels as Referred to in Decision. U.N. Doc. FCCC/CP/2011/9/Add.2

Vandenbergh M, Ackerly B, Forster F (2009) Micro-offsets and macro-transformation: an inconvenient view of climate change justice. Harvard Environ Law Rev 33:303–348

Yamin F, Depledge J (2004) The international climate change regime: a guide to rules, institutions, and procedures. Cambridge University Press, Cambridge

Chapter 3
Innovation, Adaptation and Climate Change Law

Elizabeth Burleson

3.1 Introduction

Energy innovation is a cutting edge and crucial element of addressing climate change. Environmentally sound technology can be commercially deployed over the next decade to achieve greenhouse gas emission reductions in line with scientific recommendations. Global greenhouse gas emissions have risen approximately 50 % since 1970 and are projected to rise by another half by 2030 (Krittivas 2009). In our scramble to understand the recent financial crisis, the international community continues to respond inadequately to the larger threat that climate change poses to security. International indecision regarding who will pay for climate change mitigation and adaptation has exceeded rational behavior. Good governance enhances sustainable development, participation, human rights, rule of law, anti-corruption, transparency, accountability, and access to information (UNESCO 2005). Procedural and substantive measures are at the core of the international community's ability to agree upon a binding legal framework on climate change.

3.2 The International Climate Negotiations

The international community is engaged in a collective struggle to respond to its scientific understanding of climate change with an effective and equitable legal framework. Expansion of water as oceans absorb heat from the atmosphere contributes to sea level rise, as does melt water from ice caps and glaciers

E. Burleson (✉)
Burleson Institute Inc., P.O. Box 224, Cos Cob, CT 06807, USA
e-mail: ENBurleson@aol.com

E. Albrecht et al. (eds.), *Implementing Adaptation Strategies by Legal, Economic and Planning Instruments on Climate Change*, Environmental Protection in the European Union 4, DOI 10.1007/978-3-540-77614-7_3, © Springer-Verlag Berlin Heidelberg 2014

(Palca 2008).[1] Deep cuts in global greenhouse gas emissions are needed to avert further climate destabilization (UNFCCC 2009). 'You can expect that as you have droughts, as you have scarcity of resources...it will increase tensions and it will increase conflict,' (Rowling 2008; Dobbie 2008) according to the U.N. Deputy High Commissioner for Refugees, L. Craig Johnstone. He explains that climate change is likely to displace six million people each year, forcing up to 250 million people to become refugees by 2050 (Rowling 2008). Climate change, acting as a 'multiplier of existing health risks,' requires coordinated efforts to increase the efficiency of water use, to deal with infectious disease transmission caused by the presence of dams, and to address food security impacts of rapid development of biofuels (WHO 2009).

The 1992 UN Framework Convention on Climate Change (UNFCCC) commits member states to 'common but differentiated responsibilities' and the 1997 Kyoto Protocol sets mandatory emission reduction goals (UNFCCC 1997).[2] The Kyoto Protocol entered into force in 2005 and currently has 192 parties. Member states to both the original convention and the subsequent protocol meet annually, gathering in forums that bring together perspectives from governments, inter-governmental organizations, nongovernmental organizations, and individual members of civil society.

3.3 Sustainability Through Law: Facilitating Energy Innovation and Cooperation

Legal measures that respond effectively and equitably to climate change can facilitate sustainable development, security, and energy innovation. OUR COMMON FUTURE defined sustainable development as 'development that meets the needs of the present without compromising the ability of future generations to meet their own needs' (World Commission on Environment and Development 1987). Energy security and innovation can be enhanced through laws that address climate change.

3.3.1 Energy Security and Innovation

Several important ingredients of energy security include: diversity of energy sources, security of infrastructure, stable investment in sustainable technologies, and security of transit (Haghighi 2007: 419). Much of the global oil trade navigates

[1] The ten cities at the highest risk for flooding are: Mumbai, Guangzhou, Shanghai, Miami, Ho Chi Minh City, Calcutta, greater New York City, Osaka-Kobe, Alexandria, and New Orleans.

[2] The treaty notes that the greenhouse gases are carbon dioxide (CO_2), methane (CH_4), nitrous oxide (N_2O), hydrofluorocarbons (HFCs), perfluorocarbons (PFCs), and sulphur hexafluoride (SF_6). Since 2013, nitrogen trifluoride (NF_3) is regulated as an additional greenhouse gas. See also Burleson (2008a, b).

such strategic points as the Strait of Hormuz (Deutch and Schlesinger 2006: 23), leading to heavy naval protection of the sea-lanes that transport oil (ibid.: 30). As global energy demand rises, the need to facilitate energy cooperation among consuming, exporting and transit jurisdictions has become crucial. At the same time, there is an equally vital need to enhance energy efficiency and expand use of renewable energy. Sanam Haghighi calls for an Energy Cooperation Framework (2007), highlighting the potential of the European Energy Charter Treaty (1995). The ECT focuses on energy investment, trade, and transit.[3] The treaty seeks to balance the concerns of importing and exporting nations by preserving sovereignty over natural resources while ensuring protection of investment in exploration/production (Haghighi 2007: 423). Such legal frameworks can balance economic, environmental, and social elements of energy trade.

Reducing energy imports and diversifying global energy supply can be accomplished by reducing use of fossil fuels, making more efficient use of remaining fossil fuel products; enhancing the efficiency and security of energy infrastructure; and significantly ramping up environmentally sound energy investment (Deutch and Schlesinger 2006: 31). Countries can use energy, materials, and labor that optimize efficiency in energy generation. Hybrid-electric vehicles can reduce fossil fuel use as long as the base electricity load is not based on such fossil fuels as coal. Zillman et al. (2008) note that: 'Coal combustion wastes are created in large columns (over 100 million tones from coal and oil in the USA each year). A portion of these can be recycled in manufacturing uses (cement, wallboard), but the vast majority is land disposed as solid waste, frequently in landfills that are unlined and leak. These wastes contain toxic metals like mercury as well as radioactive material naturally found in coal. Despite industry claims of working towards 'clean coal,' the technologies for controlling most of these environmental pollutants are viewed by many as at best years away and at worse a 'myth'.'

Furthermore, transport of coal by rail, truck, barge, or pipeline results in air pollution, herbicide use on rights of way, and dredging of waterways (Zillman et al. 2008: 19).

Worldwide, approximately US$235 billion annually subsidizes coal, oil, natural gas, and nuclear energy with the result that renewable energy is seen as uncompetitive in contrast. Ottinger, Mathews and Czachor note that: 'subsidies for production, R&D, and insurance for fossil and nuclear energy go to the most affluent corporations in the world, granted primarily because of political pressures, often induced by political contributions . . . Renewable energy in the United States is

---

[3] 'This treaty guarantees protection of foreign investment, and offers energy-producing countries the essential energy technologies and upgrading of energy exploration and production mechanisms, and eventually guarantees the uninterrupted flow of energy through secure transit routes. One major obstacle in the path of the ECT's development, however, is the lack of adequate membership of the major energy-producing countries (only three important countries, namely Turkmenistan, Azerbaijan and Kazakhstan have joined).'

being inhibited today by direction of 95 % of the subsidies and R&D funding to the established energy companies' (ibid.: 186).[4]

Allowing independent power producers to sell their power onto the grid is an important means by which to diversify the energy sector and establish an even playing field for renewable resources (ibid.: 189). High transmission access rates and standby rates as well as fixed unavoidable charges constitute unreasonable interconnection requirements of utilities on new producers. Ottinger, Mathews and Czachor explain that 'intermittent-generator exit fees are often fictitious and [] commitments of availability and fees are unrealistic for intermittent energy producers' (ibid.). Similarly, government approval requirements such as for liability insurance or performance of payment guarantees are not adequately geared towards interconnection by renewable energy producers (ibid.). At the same time, intermittent resources are not given credit for increasing security via energy diversity, emissions reductions, reducing peak-load, and helping to prevent power surges (ibid.: 190).

Broader use of renewable energy can be facilitated through such market incentives as subsidies as well as cap and trade programs. Government programs can encourage deployment of renewables, while utility regulatory requirements and programs can avoid advantaging fossil fuel sources to the exclusion of intermittent energy producers. A 20 % renewable standard by 2020 can result in the development of significant renewable energy capacity as can citizen-suit enforcement measures (ibid.: 192). Net metering provisions support the expansion of small renewable energy producers since each is able to feed the grid, allowing their electricity meters to run backward when they can produce more energy than they consume.

Feed-in tariffs, such as those pioneered by Germany, require utilities to grant renewable energy producers access to the grid. They also force energy utilities to purchase energy produced at a fixed price, which can depend on the source of the renewable energy. As a result, certain technologies can receive support based upon maturity and cost of production, supporting renewable energy development (ibid.: 192).[5] While such an approach helps small local energy producers enter the market, it can be difficult to identify optimal feed-in tariff levels that are not too low to support production nor too high to result in windfall profits (ibid.: 193).

In contrast, Renewable Portfolio Standards (RPS) can mandate a set quantity of renewable energy that will be produced (ibid.: 194).[6] Renewable Portfolio

[4] 'Indeed, political reform is one of the major imperatives for elimination of unjustified subsidies. Where there are no effective limits on lobbying or campaign contributions, as in the US, the powerful can readily corrupt the political process with contributions, gaining unjustified direct subsidies and favorable financing, depreciation and tax treatment that result in policies that favor them but harm the general public.'

[5] 'Ideally the tariffs are guaranteed for a certain period of time and then, reduced, allowing time for investors to earn returns but ultimately providing incentives for cost reduction. The cost of the tariffs is usually spread among all energy consumers as an electricity surcharge.'

[6] 'Production tax credits are another common method of subsidizing renewable energy production, in which tax benefits are tied to the amount of renewable energy production. Alternatively, production payments may be given per unit of energy produced.'

Standards requiring 20 % renewable energy by 2020 can facilitate the requisite transition from coal to more efficient and environmentally sound energy sources. Programs such as the United States Energy Star system that highlights efficient consumer options can help individuals and organizations make conscientious energy decisions. Labeling programs work effectively in helping countries transition to goods and services that are equitable and environmentally sustainable. Legislatures can facilitate stable, large-scale RD&D that rises above bringing pork home to given jurisdictions (Deutch and Schlesinger 2006: 46).

Denmark has reduced its greenhouse gas emissions by 14 % since 1990, raising gross domestic product by over 40 % while keeping energy consumption constant. It has done so through cap and trade, energy/carbon taxes, environmentally sound building codes, and labeling. Energy tax revenues return to industry, subsidizing environmental innovation in addition to mass transit and energy efficiency. Energy technologies constitute 11 % of Danish exports while renewables supply 30 % of Denmark's electricity mix (Friedman 2009).

Like Denmark, Texas has recently established laws that facilitate renewable energy capacity. Texas passed a renewable portfolio standard (RPS) in 1999 that required the state's competitive electric providers to bring online 2,000 MW of new renewable energy capacity by 2009.[7] This standard was achieved in a little over 6 years, prompting the Texas Legislature to increase the RPS to 5,880-MW by 2015 and 10,000-MW by 2025.[8] A Renewable Energy Credit trading program will operate through 2019. Requiring electricity providers to include such renewables as wind, solar, biomass, hydropower and geothermal power in the energy mix that they sell to retail customers significantly facilitates the increase in renewable energy generation. Texas has also required its Public Utility Commission to designate Competitive Renewable Energy Zones as well as mandating the construction of electric transmission infrastructure with which to bring renewable energy from these zones to areas of high demand.[9]

Comprehensive energy innovation programs can be enhanced by a central figure at the national level focused on facilitating the transition to greater efficiency and environmentally sound energy innovation. This can place energy on equal footing with defense policy and international economic directorates. Deutch and Schlesinger argue that:

'Global dependence on oil is rapidly eroding U.S. power and influence because oil is a strategic commodity largely controlled by regressive governments and a cartel that raises prices and multiplies the rents that flow to oil producers. These rents have enriched and emboldened Iran, enabled President Vladimir Putin to undermine Russia's democracy, entrenched regressive autocrats in Africa, forestalled action against genocide in Sudan, and facilitated Venezuela's campaign

[7] S.B. 7, 76th Leg., Reg. Sess. (Tx. 1999).

[8] S.B. 20, 79th Leg., Reg. Sess. (Tx. 2005).

[9] S.B. 20, 79th Leg., Reg. Sess. (Tx. 2005).

against free trade in the Americas. Most gravely, oil consumers are in effect financing both sides of the war on terrorism' (Deutch and Schlesinger 2006: 60).

Strengthening energy security and mitigating climate change require a focus upon legal and technological measures to facilitate a transition to renewable energy sources. Environmentally sound energy sources are geographically broadly available while producing few greenhouse gas emissions. Yet legal barriers and lack of funding have hindered technological advances that can lower costs. Approximately 20 % of future emissions can be eliminated by switching to renewables for cooling/heat, electricity, and transport (Froggatt and Levi 2009). Beyond regenerative braking, vehicles can store electricity through onboard batteries. Yet, security issues are not resolved outright by transitioning from fossil fuel to electric transport. Materials such as the lithium used in batteries present dependency concerns since much of the world's supply of lithium has thus far been found in Bolivia (Taylor 2009). Froggatt and Levi note the need for 'rapid and widespread diffusion of renewable technologies and the introduction of new energy systems and infrastructures that are more suited to the more dispersed and intermittent generation associated with renewable energy' (Froggatt and Levi 2009: 1134). Yet, they go on to caution that, 'if managed poorly, intermittent renewable electricity sources could degrade power supplies. And some express concerns that the long-distance power transmission required for some renewable energy schemes (such as solar generation for Europe based in North Africa) would introduce infrastructure-related security vulnerabilities' (ibid.: 1135). Energy diversity sustains resilience, while its absence decreases security. Patrice Kunesh notes that 'resilience is weakened or lost when systems lose the ability to foresee and adapt to external variability. . . . The ability to withstand destabilizing forces and the development of foresight potential are critical facilities of resilient systems' (Kunesh 2009: 14). Energy innovation cooperation can increase resilience across the globe.

3.3.2 Committing to Sustainability

While sustainable development is widely respected as a goal, the international community continues to struggle with the legal complexities of maintaining international peace and security generally and addressing climate change in particular. Until recently, attention focused upon Blackwater and water boarding to the exclusion of water insecurity and climate change. While natural resources are at the core of many of the world's armed conflicts, the environmental components of security have yet to be sufficiently integrated into efforts to define and maintain security. The role of reciprocity cannot be overestimated in building consensus on climate and energy policy. Applying customs of peace to such global public goods as climate can lead to reciprocal commitments to mitigate and adapt to climate change.

Energy, environmental and economic concerns can as easily draw us together as splinter us into factions. Both for ill and for good, the last several decades have seen recognition of the growing importance of non-state actors. Although terrorism continues to rattle the international framework of decision-making and threaten the emergence of open societies, access to information has expanded, and this development has enabled civil society to speak on behalf of sustainable development generally and on ecological integrity in particular. Good governance depends upon a free press and an informed citizenry. Good governance also depends upon a culture in which news is not first and foremost entertainment. Access to information is also a key to economic development. It is equally critical to protecting human rights and the environment. Regrettably, access to information about ordinary citizens has increased while at times access to information about government has been curtailed.

Over the past decade, much ink has been spilled debating the balance between participation and surveillance. Distinguishing between potential transnational terrorists and individuals displaced by climate change will likely be difficult. Irrespective of the legal status of people who have fled, recent experiences of US Gulf coast residents and Darfur villagers indicate that we are not presently prepared for the scale of humanitarian response that is likely to be requisite over the next several decades. The term 'climate refugee' has yet to become a legal term at international law and the absence of a legal framework with which to address climate displacement will likely further threaten international peace and security (see Burleson 2010: 19).

The institutional integrity of the United Nations and nongovernmental organizations that have been working with countries can continue to lead to broad ratification of treaties that can play a role in conflict prevention. Rod Beckstrom notes that, '[w]hen a fairly centralized player gets attacked by a decentralized force, like al-Qaeda, the first reaction is to centralize further, and that's usually a strategic mistake' (Williamson interview with Beckstrom 2007). Governments have wavered in their commitment to the absolute prohibition of torture and the minimization of counter-terrorism measures encroachment upon human rights. Discussing the flow of information between government and civil society, Beckstrom notes that, '[t]he people living in any community have the best sense of what is really going on in that community. They have local intelligence' (ibid.). Local wisdom and information regarding local events remains valuable to governments during armed conflicts. If governments use violent means by which to acquire information, violence will be sustained. Facilitating access to education and employment across the Middle East could break the cycle of violence if pursued with mutual respect (ibid.). 'When cultures feel insulted, people can become radicalized', Beckstrom points out. Building consensus between cultures requires an exchange of perspectives and a genuine search for common ground.

Post-conflict peace building is vital to achieving sustainability (UN News 2008). Armed conflict has negatively impacted the international effort to achieve sustainable development. The World Bank notes that: 'While the number of civil wars has declined worldwide, 80 % of the world's poorest countries have suffered a major civil war in the past 15 years. On average, countries coming out of war face about a 40 % chance of relapsing in the first five years of peace. Even with rapid progress after peace, it can take a generation or more just to return to pre-war living standards. Civil wars have generated or intensified poverty, and contributed to the global drugs problem, the rise of terrorist organizations, and the spread of diseases such as AIDS and malaria' (World Bank Group 2006).

In contrast, June Fletcher has noted that there exists a McMansion glut in the United States that has left few people able to afford US$5,000 a year to heat and cool a 5,000-square-foot house: '[n]ationwide, electricity rates have risen 12 % over the past three years, while the price of natural gas for heating has risen 43 % in the same period' (Fletcher 2006). As Henry David Thoreau (1894) once observed: 'What's the use of a fine house if you haven't got a tolerable planet to put it on?'

But what is a 'tolerable planet'? What can be tolerated? We are a highly adaptable species but that does not mean that individuals or even nation states are highly adaptable in short timeframes. Undoubtedly, accelerating climate change will be tolerated by some life on Earth. Yet which species will tolerate it, and for how long, should not be the question. A far better question is how to avert catastrophic climate change through sustained legal cooperation that will achieve environmentally sound sustainable development. This requires the largest emitters to come together and agree to collective action. China and the United States are the two most strategic players in this process. As permanent members of the United Nations Security Council, both countries hold the unilateral power with which to address sustainability and security in a legally integrated manner.

All avenues of cooperation should be used at once. The following do not need to occur in any given order. The Security Council should agree to act to address the threat that migration, loss of sovereignty, and a wide range of climate change ramifications have on international peace and security. A resolution that recognizes that climate change threatens human security on the scale of warfare is crucial. Such a resolution can facilitate a legal response to displacement, statelessness, and a host of other climate-induced security dilemmas for individuals and nation states alike.

In addition to Security Council leadership, negotiations culminating in effective and equitable mitigation, adaptation, funding, and technology transfer should continue. Civil society participation is crucial to achieving international agreement in keeping with scientific climate consensus (UNFCCC 2007).[10] Combining mitigation, adaptation, funding, and environmentally sound technology transfer in an effective and equitable manner can jumpstart genuine sustainable development on a global scale. Doing so requires political will on the part of nation states to commit to numerical targets, as well as to follow through on sufficient funding and

[10] See also Akanle et al. (2008), noting that '[t]his roadmap includes 'tracks' under the Convention and the Kyoto Protocol. Negotiations on the follow-up to the Convention Dialogue resulted in agreement on a Bali Action Plan that established the AWG-LCA, which was mandated to launch a comprehensive process on long-term cooperative action under the Convention. The Bali Action Plan identifies four key elements: mitigation, adaptation, finance and technology. The Plan also contains a non-exhaustive list of issues to be considered under each of these areas and calls for articulating a 'shared vision for long-term cooperative action."

technology assistance. These are not new debates. While nations and non-state actors alike became preoccupied, international agreements such as the UNFCCC as well as Agenda 21 and others demonstrate that the international community has known for quite some time what should be done. Moving beyond normative pronouncements to positive commitments is legal process long overdue.

Implementation of the above program of action can occur through the coordinated expertise of non-state actors. In particular, innovation centers can play an important role. Universities already have the infrastructure and can gain the insight and inclination to become environmentally sound technology powerhouses (Shane 2009). From cooperative data collection (Eilperin 2009) to regional and international energy research centers (Economist 2009) collaboration is already underway.

Technology transfer has been a central factor in ongoing climate network coordination efforts since the drafting of the United Nations Framework Convention on Climate Change (UNFCCC). Article 4.5 of the Convention requires developed countries to 'take all practicable steps to promote, facilitate and finance, as appropriate, the transfer of, or access to environmentally sound technologies and know-how to other Parties, particularly developing country parties to enable them to implement the provisions of the Convention'.[11] The Intergovernmental Panel on Climate Change defines technology transfer as 'a broad set of processes covering the flows of know-how, experience and equipment for mitigating and adapting to climate change amongst different stakeholders such as governments, private sector entities, financial institutions, non-governmental organizations (NGOs) and research/education institutions' (IPCC 2000). The new Technology Mechanism, established in the December 2010 Cancún Agreements, can coordinate a catalytic transgovernmental network that among other activities shares best practices.[12]

A Technology Mechanism has been established under the UNFCCC Conference of the Parties. It will consist of (1) a Technology Executive Committee (TEC) and (2) a Climate Technology Centre and Network (CTCN). Implementing the technology transfer framework of Article 4, paragraph 5 of the UNFCCC will be a priority for the TEC. The CTCN will gather regional, state, sectoral and civil society technology network participants to enhance technology transfer. The CTCN will facilitate existing networks, organizations, and initiatives in order to provide assistance to developing countries on technology need identification, technology implementation, and deployment of existing technologies.[13] The UNFCCC

[11] UNFCCC Article 4.5 supra note 20; see also Article 4.7 establishes a clear link between the extent to which developing countries will implement their commitments under the UNFCCC and the effective implementation by developed countries of their commitments relating to financial resources and the transfer of technology.

[12] Outcome of the work of the Ad Hoc Working Group on Long-term Cooperative Action under the Convention (United Nations Framework Convention on Climate Change Secretariat, December 12, 2010), chapter IVB, http://unfccc.int/files/meetings/cop_16/application/pdf/cop16_lca.pdf.

[13] Decision 1/CP.16 includes the outcome of work by the Ad Hoc Working Group on Long-Term Cooperative Action under the Convention (AWG-LCA) and covers the main elements of the Bali

Conference of the Parties has also decided to strengthen capacity building support via networks for sharing communication, education, information, public awareness, training and stakeholder participation.[14]

Best practices in innovation cooperation include knowledge sharing as well as joint research, development and demonstration (Tawney and Weischer 2011). Network leadership can involve UNFCCC's Climate Technology Center and Network, Clean Energy Ministerial, IRENA and other innovative collaboration. With a core mission of renewable energy, the new International Renewable Energy Agency (IRENA)[15] may be able to catalyze a more timely pace of action than has occurred within the UNFCCC. Developing country change agents such as IRENA and UNDP supported projects in the field, can begin to address the enabling environment rational with which developed countries have based slow implementation of technology transfer.

IPRs can be balanced with innovation sharing in a manner that neither stems the flow of environmentally sound technology transfer nor divests property rights wholesale. It is clear that IPRs remain a political stumbling block for countries such as the United States where large innovative corporations have influenced state negotiating. Furthermore, technology transfer has been held back, despite oven ready language, as a bargaining chip with which to negotiate more contentious issues such as measurable, reportable, and verifiable (MRV) mitigation commitments (Burleson and Romano 2011).

The international community can build upon the new technology mechanism under the United Nations Framework Convention on Climate Change (UNFCCC)[16] to share environmentally sound technologies such as solar lamps, solar cookers, drip irrigation, and rainwater harvesting. Such innovations can lower fuel consumption per unit of power generated and expand energy access.

An international database housing green technologies and best practices can enhance implementation of technology transfer and such enabling activities as technical training, capacity-building, and R&D cooperation. The technology mechanism should facilitate sectoral technology cooperation by sharing best practices and best available technologies, both current and emerging.

Road Map. Decision 1/CMP.6 reflects the outcome of the work undertaken by the Ad Hoc Working Group on Further Commitments for Annex I Parties under the Kyoto Protocol (AWG-KP). At 121. (The TEC will consider and recommend actions to promote environmentally sound technology transfer; provide guidance on policy and program priorities; facilitate collaboration between governments, the private sector, NGOs, and academic and research communities; recommend actions to address barriers to technology transfer; and catalyze development and use of technology road maps or action plans.)

[14] Id.

[15] http://www.irena.org/home

[16] See UNFCCC Article 4.5; cf. A Universal Declaration of Human Rights, G.A. Res. 217 (III) A, art. 27(1), U.N. Doc. A/RES/217(III), (Dec. 10, 1948), available at http://www.un.org/en/documents/udhr/index.shtml#a27 ('Everyone has the right freely to participate in the cultural life of the community, to enjoy the arts and to share in scientific advancement and its benefits.')

A ramping up of collective willpower, integrity, imagination, and intelligence can relegate energy scarcity and environmental externality concerns to historical anecdotes. This is not to say that a magic bullet is guaranteed to be hiding just around the bend. However, assembling the rails and capacity to look around the bend increases the likelihood of recognizing and implementing a sound energy policy. That policy may include increased use of concentrated solar power. Water may be seen as a valuable energy storage solution without displacing millions of people. Facilitating a transition to energy-water use that minimizes climate destabilization can and should be central in international, national, local, and individual actions going forward. This can be accomplished by becoming informed decision-makers able to work together across disciplines, gender differences, age, ethnicity, race, disability, nationality and all the other ways that have so often hindered collaboration. Consensus-building comes from shared perspectives being listened to and distilled into a common vision for the future. This can be an inclusive process that maximizes human security through legal coordination to achieve sustainable development.

To date, global attention has been drawn to energy security. The United States, for instance, does not have enough domestic oil supply to satisfy its rapidly growing demand for energy. This geopolitical problem has led to a focus upon the Middle East. Wars were once fought over salt for lack of awareness of salt's abundant availability in the earth's crust. Similarly, armed conflicts over fossil fuels can be averted by timely investment in environmentally sound renewable energy sources, storage and transport. The international community possesses the requisite capacity to craft a framework convention addressing energy and a Manhattan Project Plus commitment to research, development, and deployment of environmentally sound energy technologies. UNDP notes that, '[t]he United States has successfully used tax incentives to encourage the development of a vibrant wind power industry. However, while the rapid growth of renewable energy has been encouraging, overall progress falls far short of what is possible – and of what is required for climate change mitigation'. The best way to increase the efficiency of lighting and appliances as well as the use of renewable energy is to make sure that building codes facilitate efficiency; new technologies become readily available, and tax incentives help bring efficiency and environmentally sound technology into mainstream use. A comprehensive mix of regulations and market incentives can facilitate efficiency and environmentally sound technologies. Doug Struck notes that more than one-third of all energy used in the United States goes to heat, cool, and power buildings: 'Federal, state and local governments can set standards for more efficient buildings, for example, or more efficient cars or appliances. Lawmakers can require industries to curb greenhouse gas pollution. Subsidies, taxes, incentives and fees can be structured by governments to change the economic equations. . .to end tax breaks for fossil fuel exploration in favor of subsidies for alternative energy development' (Struck 2008).

Careful but timely legal and technological action can strengthen security of energy supply, mitigate greenhouse gas emissions, and provide jobs.

3.3.3 International Innovation Cooperation

Technology collaboration offers an important means by which the international community can find middle ground in climate negotiations. Resolution in the following areas could be instrumental to addressing climate change:

(a) Promote joint R&D activities in the context of South-South, North-South and triangular cooperation;
(b) Promote the transfer of environmentally sound technologies to developing country Parties;
(c) Stimulate capacity-building, in particular for endogenous technologies;
(d) Improve access to information on existing and new technologies
(e) {Promote the sharing of IPRs}.[17]

Intellectual property rights (IPRs) remain one of the many areas in need of greater dialogue but could be resolved through funding that could establish licenses to place key technologies in the public domain. A technology action plan could 'include specific policies, actions, and funding requirements for technologies in the public domain, patented technologies and future technologies.'[18] Parties can also facilitate innovative environmentally sound technologies and international cooperation by sharing national technology road maps.[19]

Similarly, voluntary technology agreements could enhance 'cooperative R&D and large-scale demonstration projects, technology deployment projects, cooperation on specific sectors or gases, and cooperation on climate observation and warning systems for enhancing resilience.'[20] Technology information could be

[17] United Nations Framework Convention on Climate Change (2009) ch. IV, para. 197.

[18] United Nations Framework Convention on Climate Change (2009) ch. IV, para. 183.

[19] United Nations Framework Convention on Climate Change (2009) ch. IV, para. 185. Further noting that national technology roadmaps can include: '(a) Identification of technological options for specific sectors; (b) Obstacles to the development and transfer of identified technological options; (c) Policy instruments and infrastructure required for the deployment, diffusion and transfer of identified technological options; [and] (d) Capacity-building needs ...'

[20] United Nations Framework Convention on Climate Change (2009) ch. IV, para. 192. The Text further notes that: [f]or the purpose of meeting its quantified emission limitation and reduction commitments and requirements for monitoring, reporting and verification, a Party may transfer to, or acquire from, other Parties emission reduction units resulting from projects and programs that accelerate the diffusion or transfer of environmentally sound technologies, provided that:

(a) Voluntary participation is approved by each Party involved;
(b) Any such project results in measurable, reportable and verifiable reductions of GHG emissions by sources or enhancements of removals by sinks;
(c) The project contributes to the achievement of the technology targets and objectives of the host Party;
(d) The host Party has allocated assigned amount units or environmentally sound technology rewards (ESTRs) to the project or program;
(e) The project/program is registered under the Convention;
(f) Participants in the ESTR mechanism may involve private and public companies. United Nations Framework Convention on Climate Change (2009) ch. IV, para. 191.

distributed through the development of an international database housing green technologies and best practices.[21] A new body on technology transfer could also implement technology transfer mechanisms and such related enabling activities as technical training, capacity-building, and R&D cooperation.[22] Such a technology body could facilitate sectoral technology cooperation by sharing best practices and best available technologies, both current and emerging.[23] It could also help diffuse and transfer environmentally sound technologies to all relevant sectors.[24]

3.4 Conclusion

Climate resilient communities can be achieved with the support of global research, development, deployment, and diffusion of environmentally sound energy technologies and processes (Climate Action Network 2009). United Nations Secretary General Ban Ki-moon has acknowledged that post-Copenhagen, the current climate mitigation commitments offered by countries fail 'to meet the scientific bottom line' (BBC News 2009). Averting catastrophic climate change security risks can be accomplished if the international community substantially strengthens its commitment to mitigate, adapt, fund, and innovate.

Universally shared values provide a foundation for dialogue and a means by which a lasting culture of cooperation can take root (UNESCO 2001). Genuine security hinges upon our commitment to transnational, interdisciplinary, inclusive decision-making. Innovation cooperation can help emerging climate change law achieve sustainability.

Biography Professor Elizabeth Burleson has a LL.M. from the London School of Economics and Political Science and a J.D. from the University of Connecticut School of Law. She has written reports for United Nations.

References

Akanle T, Appleton A, Bushey D, Kulovesi K, Spence C, Yamineva Y (2008) Earth Negotiations Bulletin, International Institute for Sustainable Development IISD 12:395. Available at: http://www.iisd.ca/download/pdf/enb12395e.pdf. Accessed 21 July 2012

BBC News (2009) Copenhagen deal: key points, p 1. Available via BBC News. http://news.bbc.co.uk/2/hi/science/nature/8422307.stm. Accessed 21 July 2012

[21] United Nations Framework Convention on Climate Change (2009) ch. IV, para. 195.

[22] United Nations Framework Convention on Climate Change (2009) ch. IV, para. 196.

[23] United Nations Framework Convention on Climate Change (2009) ch. IV, para. 196.

[24] United Nations Framework Convention on Climate Change (2009) ch. IV, para. 196.

Burleson E (2008a) Multilateral climate change mitigation. Univ San Francisco Law Rev 41:373
Burleson E (2008b) A climate of extremes: transboundary conflict resolution. Vermont Law Rev 32:477
Burleson E (2010) Climate change displacement to refuge. J Environ Law Litigat 25:19
Burleson E, Romano C (2011) Cancun climate negotiations. Am Soc Int Law Insight 15:41
Climate Action Network (2009) Submission to AWG-LCA on finance for mitigation and low carbon and climate-resilient development. Available via Climate Action Network. http://unfccc.int/resource/docs/2008/smsn/ngo/093.pdf. Accessed 21 July 2012
Deutch J, Schlesinger J (2006) National security consequences of U.S. oil dependency, Independent Task Force report no. 58, p 23. Available at http://www.cfr.org/publication/11683/national_security_consequences_of_us_oil_dependency.html?breadcrumb=%2Fissue%2F18%2Fenergy_security. Accessed 21 July 2012
Dobbie A (2008) Climate change could force millions from homes. Reuters
Economist (2009) America, China and climate change: let's agree to agree. Available at http://www.economist.com/world/unitedstates/displaystory.cfm?story_id=14915108&fsrc=rss. Accessed 21 July 2012
Eilperin J (2009) U.S. and China reach accord on data collection, p 1. Available at http://www.washingtonpost.com/wp-dyn/content/article/2009/11/18/AR2009111803058.html. Accessed 21 July 2012
Energy Charter Treaty (1995) ILM 34:381, reprinted in Energy Charter Secretariat, The energy charter treaty and related documents: a legal framework for international energy cooperation (2004). Available at http://www.encharter.org/. Accessed 21 July 2012
Fletcher J (2006) The McMansion glut. Wall Street Journal, 16 June 2006
Friedman T (2009) The Copenhagen that matters. Available at http://www.nytimes.com/2009/12/23/opinion/23friedman.html?ref=todayspaper. Accessed 21 July 2012
Froggatt A, Levi MA (2009) Climate and energy security policies and measures: synergies and conflicts. Int Aff 85:1134, available at http://www.cfr.org/publication/20922/climate_and_energy_security_policies_and_measures.html?breadcrumb=%2Fpublication%2Fby_type%2Fjournal_article. Accessed 21 July 2012
Haghighi SS (2007) Energy security: the external legal relations of the European Union with major oil and gas supplying countries. Hart, Oxford
IPCC (2000) Methodological and technological issues in technology transfer – a special report of WG III. Available at: http://www.ipcc.ch/pdf/special-reports/spm/srtt-en.pdf
Krittivas M (2009) U.N. climate chief says industry keen on deal. Reuters
Kunesh PH (2009) Constant governments: tribal resilience and regeneration in changing times. Kan J Law Publ Pol 19(8)
Palca J (2008) In a strategic reversal, Dutch embrace floods. Available at http://www.npr.org/templates/story/story.php?storyId=18229027. Accessed 21 July 2012
Rowling M (2008) U.N. says climate change may uproot 6 million annually. Available at http://planetark.org/wen/50823. Accessed 21 July 2012
Shane S (2009) Can academics be entrepreneurial? Available at http://boss.blogs.nytimes.com/2009/11/16/can-academics-be-entrepreneurial/?partner=rss&emc=rss. Accessed 21 July 2012
Struck D (2008) A big drop in emissions is possible with today's technology. Available at http://www.washingtonpost.com/wp-dyn/content/article/2008/01/20/AR2008012001171.html?nav=rss_email/components. Accessed 21 July 2012
Tawney L, Weischer L (2011) Innovation and technology transfer: supporting low carbon development with climate finance international innovation system 1 WRI. Available at http://pdf.wri.org/working_papers/innovation_and_technology_transfer.pdf
Taylor M (2009) Bolivia holds key to lithium, the battery car metal. Reuters
Thoreau HD (1894) Familiar letters of Henry David Thoreau. Houghton Mifflin, Boston, New York, p 416
UNESCO (2001) Dialogue among civilizations. Available via UNESCO. http://www.unesco.org/dialogue2001/. Accessed 30 Sept 2005

UNESCO (2005) Media and good governance. Available at http://portal.unesco.org/ci/en/ev.phpURL_ID=18598&URL_DO=DO_TOPIC&URL_SECTION=201.html. Accessed 21 July 2012

United Nations Framework Convention on Climate Change (1992) Article 4, U.N. Doc. FCCC/INFORMAL/84. Available at http://unfccc.int/resource/docs/convkp/conveng.pdf. Accessed 21 July 2012

United Nations Framework Convention on Climate Change (1997) Kyoto protocol, art. 3, U.N. Doc. FCCC/CP/1997/L.7/add.1. Available at http://unfccc.int/resource/docs/convkp/kpeng.pdf. Accessed 21 July 2012

United Nations Framework Convention on Climate Change (2007) Bali, 3–14 Dec 2007. Available at http://unfccc.int/meetings/bali_dec_2007/meeting/6319.php. Accessed 21 July 2012

United Nations Framework Convention on Climate Change (2009) Negotiating text U.N. Doc. FCCC/AWGLCA/2009/8, ch. I, para. 3. Available at http://unfccc.int/resource/docs/2009/awglca6/eng/08.pdf. Accessed 21 July 2012

UN News (2008) UN Chairperson of the Peacebuilding Commission Yukio Takasu, Peacebuilding efforts producing results on the ground, Security Council told, p 1. Available at http://www.un.org/apps/news/story.asp?NewsID=28659&Cr=peacebuilding&Cr1=commission. Accessed 21 July 2012

Williamson E Interview with Beckstrom R A (2007) Follow the leader, or think like a starfish? Available at http://www.washingtonpost.com/wp-dyn/content/article/2006/12/31/AR2006123100832.html?nav=rss_print/asection. Accessed 21 July 2012

World Bank Group (2006) The World Bank and conflict-affected countries, p 1. Available at http://web.worldbank.org/WBSITE/EXTERNAL/NEWS/0,,contentMDK:20042303%7emenuPK:34480%7epagePK:64257043~piPK:437376~theSitePK:4607,00.html. Accessed 11 July2006

World Commission on Environment and Development (1987) Our common future 1, A/43/427. Available at http://www.un-documents.net/ocf-02.htm#I. Accessed 21 July 2012

World Health Organization (2009) Protecting health from climate change: global research priorities 6. Available at http://www.searo.who.int/LinkFiles/World_Health_Day_2008_Madrid_report_130509.pdf. Accessed 21 July 2012

Zillman DN, Redgwell C, Omorogbe YO, Barbera-Hernández LK (eds) (2008) Beyond the carbon economy: energy law in transition. Oxford University Press, Oxford

Chapter 4
Managing Adaptation: Developing a Learning Infrastructure in the United States' Federal System

Alejandro E. Camacho

4.1 Introduction

Though there is much solid evidence that anthropogenic climate change has already had and will increasingly have substantial adverse effects on biota and ecological processes (Staudinger et al. 2012), the chief barrier to effective natural resource governance over the next few decades will likely be the exceptional uncertainty that accompanies attempts to adapt to the effects of climate change on natural systems. The global scale of climate change, the array and complex interaction of variables inherent in climate, and the limited study of its effects raise massive obstacles for effective climate regulation (Siegel 2009). Moreover, efforts at climate change adaptation (dealing with the inevitable effects of climate change) as compared to mitigation (that seeks to reduce greenhouse gas emissions to prevent further climate change) face yet another order of uncertainty. The various difficulties involved in 'downscaling' ecosystem modeling—localizing global projections down to a scale necessary to inform and assist adaptation efforts—raise uncertainty to a level that environmental regulators have never attempted to manage.

Consequently, the effective adaptation of natural resource management to climate change hinges on the development of a regulatory infrastructure that provides public and private actors the capacity to assess and manage uncertainty. This chapter briefly sketches the options originally considered for natural resource governance in the United States, their insufficiency in the key function of managing uncertainty, and how to build a more effective federal system for managing natural resources in preparation for climate change. Uncertainty underscores the value of regulatory experimentation and learning in a largely decentralized and overlapping federal system, and suggests a crucial role for national governments and

A.E. Camacho (✉)
University of California, Irvine School of Law, 401 East Peltason Drive, Law 4500-A, Irvine, CA 92697-8000, USA
e-mail: acamacho@law.uci.edu

E. Albrecht et al. (eds.), *Implementing Adaptation Strategies by Legal, Economic and Planning Instruments on Climate Change*, Environmental Protection in the European Union 4, DOI 10.1007/978-3-540-77614-7_4,

international institutions of promoting agency learning and inter-jurisdictional information sharing.

4.2 Common Regulatory Models for Natural Resource Governance in the United States

4.2.1 Rely on Specialized, Fragmented Decision-Making

The paradigmatic approach to natural resource governance in the United States has been one of highly fragmented and overlapping authority, which has hindered agency action for adapting to climate change (Craig 2010). Natural resources agencies have typically been allocated limited authority to regulate or manage specific resource problems. As a result, natural systems in the United States are subject to an extensive patchwork of piecemeal management, divided based on the protected resource (e.g. water quality, water use, freshwater species, marine species, national park, national forest), the level of government, the branch of government, and the regulatory stage (Buzbee 2005). As one example, over 148 different federal programs involving ten federal agencies have been created to manage natural resources in the Great Lakes basin, not to mention the scores of local, state, and regional agencies within the United States, Canadian federal and provincial agencies, and international institutions with jurisdiction over the Great Lakes' natural resources (United States General Accounting Office 2003). Even intra-agency fragmentation is prevalent, with environmental agencies often consisting of fairly autonomous divisions and regional offices (Geltman and Skroback 1998). Thus, regulatory fragmentation is not just the allocation of authority between a national government and constituent states; it exists within each sovereign and even within individual agencies.

In addition to other potential critiques,[1] some scholars have pointed out that particularly for large-scale, broadly dispersed resource issues, regulatory fragmentation can have the effect of impeding beneficial regulatory action. Ruhl and Salzman (2010), for example, discuss the reticence of agencies to address complex problems, which, by their nature, cannot be completely addressed under the jurisdiction of a single agency. Even more so than mitigation activities, adaptation to climate change is a good example of such a collective action problem or 'regulatory commons' (Buzbee 2003). The causes and significant effects of climate change are widely recognized but cross a multitude of jurisdictional borders (Craig 2008), and federal adaptation strategies are more difficult to implement than mitigation

[1] These include claims that fragmented and overlapping governance can result in an inefficient lack of finality, poor regulator accountability (Schapiro 2005), and the potential for inefficient over-regulation when 'numerous regulators are confronted with a more particularized project or proposal with localized and discernible effects' (Buzbee 2005, p. 349).

strategies because they involve an even greater number and range of government authorities (United States Congressional Budget Office 2005). Because of a number of factors present in such a splintered regulatory setting—including the information costs of developing a regulatory response, likely diluted credit,[2] biases toward the regulatory status quo, and the risk aversion of regulators—isolated regulators have little incentive to proactively devote resources to gather information on or regulate to curtail the localized effects of global climate change (Buzbee 2003). With limited jurisdiction and information, government programs have little capacity to learn and adapt.

Indeed, reports have consistently concluded that natural resources institutions in the United States are still poorly equipped for climate change adaptation (Government Accountability Office [GAO] 2007, 2009, 2011; Intergovernmental Panel on Climate Change 2007; White House Council on Environmental Quality 2010; Pew Center on Global Climate Change [Pew Report] 2010). The United States Congress has yet to establish a comprehensive regulatory program directed at climate change adaptation, and for many years government agencies failed to engage in meaningful adaptation efforts. Increasingly, some federal and state resource agencies are finally investigating and beginning to develop adaptation plans and responses (U.S. Fish and Wildlife Service 2012; National Climate Assessment and Development Advisory Committee 2013; GAO 2013), but many remain modest, in draft form, or conceptual (GAO 2013; White House Council on Environmental Quality 2010; Pew Report 2010). Despite an initial increased emphasis on adaptation planning and funding (GAO 2011), resources have been cut (GAO 2013). Many agency activities continue to ignore adaptation, while many of those actions that address it focus on merely studying the problem and gathering information (GAO 2009, 2011). At least some state and federal officials have asserted that their narrow jurisdiction leaves them lacking the information or authority to engage in adaptation (GAO 2007, 2009, 2011).

4.2.2 Consolidate or Centralize 'Substantive' Decision-Making Authority

In response to the incentives toward inaction caused by regulatory fragmentation, a familiar response is to call for changes in natural resource governance that centralize decision-making authority (Fortney 2006; Buzbee 2003). Consolidation of decision-making might take any variety of forms. It might include a fundamental reassembly of regulatory authority, such as through federal preemption of state regulation in a particular field or the reorganization of fragmented authority. Examples of the latter include the creation of the United States Department of

[2] In fragmented regulatory contexts, a regulator has less incentive to act because credit will most likely be divided among all regulators (Buzbee 2003).

Homeland Security—in which a suite of federal agencies were combined and rearranged—or calls for the integration of United States intelligence-gathering responsibilities, which traditionally has been fragmented among many federal agencies (O'Connell 2006).

The centralization of authority might also involve a more restrained 'partial de-fragmentation'. This might involve the consolidation of a couple of resource programs, or the establishment of substantive review by a central governmental authority (such as the President) of another's functions (Buzbee 2003; Farber 2009). In any event, the intent is to consolidate substantive decision-making authority in fewer and more central institutions.

In the United States, some have recommended the consolidation of climate change adaptation planning power in the hands of a central federal authority. The most prominent example of a proposed partial integration was included in the only comprehensive climate change bill adopted to date by either chamber of the United States Congress—Waxman-Markey's American Clean Energy and Security Act of 2009 (ACES).[3] The ultimately unsuccessful bill sought to significantly increase executive oversight and control over both federal and state natural resource adaptation by consolidating authority in the President and Secretary of the Interior. ACES would have established 'an integrated Federal program' (ACES § 471) including a Natural Resources Climate Change Adaptation Panel headed by the Chair of the Council on Environmental Quality (CEQ), a division of the Executive Office of the President (ACES § 475). The Panel, which would have included the heads of federal public land and natural resources agencies, would be tasked with developing and implementing a National Resources Climate Change Adaptation Strategy (ACES § 476). Though ACES would not have adopted a more fundamental consolidation such as requiring a single federal agency to carry out all natural resource adaptation planning,[4] it would have required the adoption of an adaptation plan by each federal natural resources agency that would implement and be consistent with the Strategy, as determined by the President (ACES § 478). The institution of more centralized strategic planning, combined with mandated executive review of individual agency plans, would have marked a substantial shift toward more centralized natural resource management in the United States.

The broadest adaptation initiative actually implemented in the United States to date, however, involves a much less centralized regime for adaptation planning. Under Presidential Executive Order 13514, the CEQ, Office of Science and Technology Policy, and National Oceanic and Atmospheric Administration (NOAA) were charged with co-chairing a federal Interagency Climate Change Adaptation Task Force, which included more than 20 Federal government agencies (White House Council on Environmental Quality 2010). White House CEQ Implementing

[3] Though this bill was initially adopted by the House of Representatives, it was never adopted by the Senate.

[4] In contrast, a less integrated structure might require each federal agency to develop an adaptation plan, but not require such plans to be consistent with a centralized adaptation strategy.

Instructions (2011b) required federal agencies to submit information to CEQ demonstrating that the agency is engaging in adaptation planning by a series of deadlines, and CEQ has collected and makes available draft adaptation plans developed pursuant to this process. This initiative thus provided the CEQ an (albeit limited) oversight role in adaptation planning by other federal agencies. Presidential Executive Order 13653 (2013) replaced this Task Force with the Council on Climate Preparedness and Resilience, which is comprised of 32 federal agencies directed, in coordination with the Council, to identify and remove or reform barriers that discourage climate resilience.

Increased centralization could include substantive limitations on the discretion of states to engage in adaptation activities as well. For example, ACES would not only have shifted responsibility for setting federal natural resource goals from Congress or individual federal agencies to the Panel, but also from states to the Panel. Any state that needed federal funding to assist it to adapt its natural resources to the effects of climate change would have had to submit an adaptation plan to the Secretary of the Interior, who could have disapproved the state plan if it deemed this plan to be inconsistent with any goal, priority, or standard established by the Panel under its broad authority to adopt a federal Strategy (ACES § 479). Though states would do much of the natural resource adaptation work, the big-picture decision-making would have been primarily federal.

However, as a growing number of scholars have reasoned, a more uniform approach that consolidates or closely integrates decision-making institutions may impede a number of the benefits of more decentralized governance. In addition to the implausibility of eliminating already extensive regulatory segmentation (Buzbee 2003; Ruhl 2010), many scholars have detailed the undesirability of doing so (Ruhl and Salzman 2010; Buzbee 2003; Adelman and Engel 2007; Schapiro 1999). While there may be benefits to reducing fragmentation in certain circumstances,[5] a decentralized federal system may allow for a diversity of tailored approaches, promoting innovative management experimentation and creating the opportunity for learning about the advantages and disadvantages of particular management strategies (Adelman and Engel 2007). Initial local standards also can serve as a proving ground or can catalyze more centralized authorities to promulgate regulations (Engel 2006; Carlson 2009), ensuring regulatory progress in spite of pressures from special interest groups (Adler 2005). As stated by Adelman and Engel (2009): 'Adaptive federalism simultaneously sustains competitive legislation and administrative processes that promote the refinement of policies...and processes that produce a diverse range of policy options' (p. 290).

Some scholars also argue that the redundancy that occurs through multiple overlapping but independent jurisdictions can prevent 'group think', agency capture, and promote agency competition that may yield better outcomes than

[5] For example, such a case might exist if there is directly conflicting or redundant authority by regulators without a discernable difference in subject matter competence. In addition, 'local research into health effects, safe exposure thresholds, and potential control strategies could be duplicative. Accordingly, such research may be conducted more efficiently at the federal level' (Adler 2005, p. 148).

coordination (O'Connell 2006, p. 290; Schapiro 2005). Designed correctly, concurrent jurisdiction also has the potential to improve management decisions by allowing a range of specialized subject matter competencies to be brought to bear on a resource problem. Furthermore, each regulatory authority can serve as an accountability check on the others.

A key endeavor for any framework for adapting to the effects of climate change in the United States, then, is promoting the management innovation and opportunities for learning associated with decentralized governance while reducing and overcoming the inefficiencies that create collective action problems (Adelman and Engel 2007). As stated by Adelman and Engel (2009): 'The challenge is to maintain a process of optimization, which leads to specialization and efficiencies while cultivating a diversity of backup options in the wings' (p. 290). Other related challenges include determining when redundancy serves as a valuable fail safe or merely an impediment to necessary action, and reconciling regulatory fragmentation's potential for decreasing agency accountability to the public with its potential for increasing interagency accountability.

4.2.3 Create Regional Collaborative or Coordinating Networks

How can the United States' federal system best accommodate these competing objectives? Some assert considerable benefits from inter-governmental competition (Adelman and Engel 2007), and decry the inefficiencies that come with too much coordination (O'Connell 2006). Many, however, have emphasized the value of agency dialogue and collaboration. Various natural resources scholars (Bardach 1998; Ruhl and Salzman 2010) and regulatory officials (U.S. Climate Change Science Program 2008) have called for the development of networks that promote interagency collaboration and coordination. Proponents of these decentralized regional institutions emphasize the need to focus resource management on regulatory linkages around particular ecosystems or landscapes (Grumbine 1994; Karkkainen 2008). Rather than combining institutions or providing one government authority substantive oversight of another, the aim of these regional or ecosystem-based institutions is to serve as a venue for agencies to discuss and possibly coordinate regulatory activity. The expectation is that such opportunities for communication and synchronization will reduce the effects of fragmentation without resort to the consolidation of decision-making authority.

Though varying in structure, a wide number of inter-jurisdictional governance experiments have been adopted in the United States. A few examples include the Great Lakes Regional Collaboration (http://www.glrc.us/), the Chesapeake Bay Program (http://www.chesapeakebay.net/index.aspx), and California's Natural Community Conservation Planning Act (2001). In fact, at present most natural resources in the United States are managed not only through a number of local, state, and national authorities, but also regional intergovernmental networks offered as venues for agency collaboration and cooperation.

Perhaps unsurprisingly, then, some initiatives by the federal government have been created to serve as a coordinating network for addressing climate change adaptation. The Interagency Climate Change Adaptation Task Force and Council serve as collaborative and coordinating venues for federal agencies on climate change adaptation planning. Other than the CEQ ensuring compliance with planning deadlines, federal agency participants in the Task Force did not have substantial direct oversight or authority over others on adaptation activities. The Task Force served primarily as a venue for Federal agencies to communicate, brainstorm, and develop recommendations for the President on potential federal adaptation strategies (White House Council on Environmental Quality 2010). The Task Force has developed work groups and listening sessions with stakeholders and experts, including state and local authorities, about what might be needed for adaptation policy. Pursuant to these efforts, agencies are developing climate assessments and beginning to formulate adaptation plans (GAO 2013). Similarly, the United States Department of Interior has formed 22 regional landscape conservation cooperatives to help local, state, tribal, regional, national and international authorities to begin to coordinate adaptation planning regarding ecological resources (U.S. Fish and Wildlife Service 2012). Under the Task Force, Council, and Interior initiatives, however, individual agencies develop their own adaptation plans.

4.3 How Prevailing Models Fail to Manage Uncertainty

Unfortunately, the countless attempts to establish inter-jurisdictional networks in the United States have generally made slow and limited progress in overcoming the negative effects of fragmented governance. This is in large part because such experiments, like other past approaches for managing overlapping authority, have paid insufficient attention to the need to cultivate both interagency information sharing and agency learning in the regulatory process. Without these two interrelated features, the existing federal system governing natural resources in the United States does a poor job managing uncertainty.

First, most network-based attempts to reduce the effects of fragmentation have provided at best weak opportunities and incentives for inter-jurisdictional information sharing and collaboration. These regulatory experiments seem to neglect the fact that regulatory inaction is largely a symptom of a bigger problem—the significant barrier to intergovernmental learning created by fragmented governance. Many agencies must cope with longstanding and widespread gaps in the collection and management of scientific information relevant to regulatory decision-making, including data sets that are incomplete, obsolete, or missing information, and difficulties in managing the data that are available (Adler 2005).

This is compounded by the fact that agency officials typically lack access to any analyses or management strategies employed by other agencies, or even those by other regional offices or divisions in the same agency (Karkkainen 2004). Regional collaborations may provide the opportunity to interact with and perhaps learn from a few other regulators on a common resource problem; but such networks rarely set up systematic opportunities for information sharing within a collaboration (let alone more broadly). Consequently, existing regional efforts have too often served as yet another layer of fragmentation to the already disjointed regulatory landscape.

Because much existing information is not readily available, regulators are unable to draw systematically from the experience of other regulators and reduce uncertainty about the efficacy of particular strategies in addressing regulatory goals. Failing to harness the collectively available information and experience leaves regulators isolated and natural resources susceptible to regulatory gaps (Craig 2010). Ultimately, the existing regulatory framework provides little prospect for managing the large-scale effects and considerable uncertainty to follow from climate change. The fact that climate change effects on natural resources until recently have been virtually ignored by this vast array of regulators, including those regional networks specifically tasked with reducing the effects of fragmentation, is telling.

Secondly, most existing natural resource programs in the United States fail to encourage managers and regulators to learn methodically how to manage uncertainty and make resource management progressively more effective at achieving program goals. Regulatory programs are not designed to provide sufficient pressure on resource managers to work with and reduce uncertainty by systematically assessing and adjusting management approaches. In response to numerous critiques of agencies for failing to develop resource management strategies that were enduring and capable of adjusting to new information or changed circumstances, natural resources agencies have increasingly adopted experimental protocols designed to make resource management more adaptive (Camacho 2007). However, even these *adaptive management* innovations have not been structured to require or otherwise pressure managers to learn systematically from their own management experience (Gregory et al. 2006).[6] These experiments also fail to require or provide other incentives for regulators to learn from the activities of similarly situated officials in the same or other agencies. In essence, both conventional fragmented regulation and existing regulatory experiments have failed to cultivate the systemic capacity to manage uncertainty through learning.[7]

[6] Evaluations of collaborative governance experiments have found that scarce information about ecological processes, management strategies, and agency performance contributes greatly to failure by collaborative experiments (Karkkainen 2008).

[7] For a more detailed critique of the capacity of existing regulatory programs in the United States to engage in adaptive learning, see Camacho (2009).

4.4 Integrate Data Gathering and Standardize Adaptive Governance

A governance model can respond to these criticisms and seek to manage uncertainty by leaving decision-making decentralized but creating a network and consistent framework for information generation, gathering and dissemination. This would include creating a shared, easily accessible, adaptive, and interactive infrastructure that promotes inter-jurisdictional information sharing on projected environmental effects (including those from climate change) and potential management strategies (Camacho 2011). A searchable repository would help managers draw from a broader set of environmental data and analyses on the performance of management strategies in diverse contexts (DiMento and Ingram 2005). However, drawing on recent developments in virtual modes of collaboration, an adaptive and collaborative learning infrastructure should move beyond the simple formation of a data clearinghouse to form more interactive and adaptive virtual environment for creating and disseminating information relevant to climate change adaptation (Camacho 2011).

This model would lodge the fairly limited function of information coordination with a central authority (such as the United States Library of Congress or the CEQ) while allowing for resource management decisions to remain with each agency delegated authority over a particular resource. Only requiring information flow between resource managers accommodates many of the core benefits of decentralized governance, neither requiring agency consolidation nor agreement on a particular strategy by all regulators with jurisdiction. As such, making such information available is valuable whether agencies engage in collaborative efforts or in competition.[8] This approach is likely to be particularly useful for problems like climate change adaptation, given the localized character of climate effects but the considerable benefits to individual regulators of information-pooling (Salkin 2010).

Allowing regulators to tap a broader scientific data set and management experience helps curtail uncertainty, impediments to learning, and collective-action problems that result from regulatory fragmentation. Through such a framework, regulators could better learn from the successes and mistakes of other regulators in their management of natural resources. In addition, regulators would serve as sources of external pressure on other regulators to engage in monitoring, reporting, and adjustment of adopted strategies. In this manner, innovations that promote information-pooling and inter-jurisdictional communication could enrich agency decision-making and decrease incentives for regulatory inaction.

The federal Interagency Climate Change Adaptation Task Force has called for the creation of "an online interagency global change information portal/system to

[8] 'The complex scientific underpinnings of many environmental challenges favor allocation of research and information sharing functions to the federal government. Even if inter-jurisdictional competition is viewed as a good, one can embrace allocation of such information-gathering functions to federal actors' (Buzbee 2005, p. 353).

provide 'one-stop shopping' for climate-related information" (White House Council on Environmental Quality 2011a). Despite such recommendations, no comprehensive portal has been launched, though after several years a limited beta version of an interagency portal system has been demonstrated (White House Council on Environmental Quality 2011a, b) and there are disparate government sites that collect selected government climate data (U.S. Global Change Research Program 2013; GAO 2013). Moreover, the Task Force unfortunately has ignored the growing cyber-technology on collaborative learning and has yet to provide an adequate infrastructure to promote such information sharing (Camacho 2011). Encouragingly, the new Presidential Executive Order 13653 (2013) has mandated the development and provision of data and decision-support tools on climate preparedness and resilience, as well as the establishment of a web-based portal for disseminating such information.

A small-scale and incomplete example of a clearinghouse focused on climate change was established by the United States Environmental Protection Agency (EPA) as part of its National Estuaries Program. The National Estuaries Program focuses on promoting intergovernmental collaboration on estuarine management, but it had not created any systematic way for regional estuaries or constituent agencies to share information or learn from others. However, in 2008 the EPA established the Climate Ready Estuaries program, including a pilot 'coastal toolkit' that collected a range of internet links and documents to aid estuaries in identifying vulnerabilities to, and adaptation planning for, climate change (Climate Ready Estuaries n.d.).[9] Nonetheless, this coastal toolkit is very limited in scope and provides few opportunities for collaborative learning (Camacho 2011). Similarly, the United States Geological Survey (2008) founded a National Climate Change and Wildlife Science Center, including eight regional climate science centers. The hope is that these centers will help synthesize climate information, help other agencies develop adaptive management strategies, and facilitate partnerships. The focus of these efforts is not on the assessment of management strategies but rather on developing scientific information on the effects of climate change on resources.

The ACES bill would have improved on existing environmental agency information gathering by requiring periodic vulnerability assessments, plans and strategies on climate change adaptation, and establishment of a National Climate Service in NOAA to coordinate with the United States Geological Survey's National Climate Change and Wildlife Science Center to create a process for developing and providing science and information to assess and address the impacts of climate change on natural resources (ACES 2009). The ACES bill would also have required the National Climate Service to initiate and maintain a clearinghouse that included information on projected climate change, the effects of such climatic changes, and possible adaptation strategies, including hyperlinks to resources on the internet (ACES 2009). Though this information-gathering and dissemination

[9] The pilot program also designates six NEP estuaries for targeted technical support from EPA, and intermittently hosts workshops that assemble officials with jurisdiction over estuaries to discuss adaptation planning (Climate Ready Estuaries 2010).

framework itself might have been fragmented and thus prone to the same regulatory gaps that management agencies are, it nonetheless would have been an improvement on the minimal information infrastructure that currently exists for natural resource governance in the United States.

Nonetheless, a clearinghouse is only as useful as the information that it includes. To truly promote intergovernmental learning on resource management, a clearinghouse would have to include information that today is rarely even generated by resource agencies—systematic data on the past performance of management strategies in accomplishing regulatory goals. Like other clearinghouses, neither the Climate Ready Estuaries toolkit nor the now-defunct ACES bill required the collection and incorporation of systematic evidence on management strategies and programs. Similarly, the National Climate Change and Wildlife Science Center does not appear to contemplate developing and incorporating regular and on-going data on the performance of management strategies and agencies. Though the Interagency Climate Change Adaptation Task Force does advise agencies to share and adaptively manage information for adaptation planning, it ignores the lessons of prior uses of adaptive management and does not require concrete triggers for or otherwise incentivize agencies to engage in continued assessment and adjustment of management decisions (Camacho 2011).

Accordingly, Congress should also adopt an *adaptive governance* framework in which Congress defines clear goals and targets for adaptation, and requires agencies to regularly monitor, assess, and adjust adopted management strategies based on their performance toward such goals and targets. By combining this framework with a robust and coordinated clearinghouse, adaptive governance would proliferate opportunities for information sharing and cultivate learning. This in turn should help reduce the barriers to regulation exacerbated by uncertainty. Furthermore, providing a transparent means for assessing agency progress toward regulatory goals would help promote more effective agency decision-making and serve to make agencies more accountable to legislatures and the public.[10]

4.5 Conclusion

Established models of federalism reasonably contemplate dispersed and overlapping authority as vital to cultivating diversity, innovation, and ultimately more effective regulation. However, conventional natural resource regulation in the United States is poorly designed to promote these potential benefits, leaving resource managers isolated in addressing uncertain regulatory problems and failing to pressure regulators to learn. Prevailing responses for addressing this shortcoming seek either to consolidate or create venues for promoting collaborative decision-making, but these approaches have insufficiently attended to the core challenge of

[10] For a more comprehensive delineation of such an inter-jurisdictional adaptive governance framework, see Camacho (2009, 2011).

reducing and managing uncertainty. In short, natural resource managers and policy-makers not only lack information about future effects and the value of management strategies in particular contexts, they also lack the infrastructure to gain such information.

Unfortunately, few emerging adaptation strategies in the United States adequately seek to alter the existing federal system to provide private and public actors the capacity and incentive to manage the uncertainties of climate change. Even the most robust federal legislative proposals and regulatory actions to date in the United States largely ignore agency learning and accountability. They fail to articulate goals against which agencies would have to assess the performance of adopted plans, hindering the capacity for agencies (and others) to evaluate the effectiveness of adopted strategies. To induce more effective natural resource governance, the United States Congress must establish and refine a shared information framework that creates, collects, and disseminates information on management strategies and the environment. Such a concurrent federal system can reduce the barriers to effective action, cultivate a diverse range of management options that facilitates collective learning, and help reduce uncertainty and manage the effects of climate change.

Biography Alejandro E. Camacho is a Professor of Law at the University of California, Irvine; Director of the UCI Law Center for Land, Environment, and Natural Resources; and a Scholar at the Center for Progressive Reform. J.D. Harvard Law School; LL.M., Georgetown University Law Center; B.A., University of California, Irvine. He writes in environmental law, land use regulation and government organization, with a particular focus on adaptive management and collaborative governance.

References

Adelman DE, Engel KH (2007) Adaptive federalism: the case against reallocating environmental regulatory authority. Minn Law Rev 92:1796–1850

Adelman DE, Engel KH (2009) Adaptive environmental federalism. In: Buzbee WW (ed) Preemption choice: the theory, law and reality of federalism's core questions. Cambridge University Press, Cambridge, pp 277–299

Adler JA (2005) Jurisdictional mismatch in environmental federalism. New York Univ Environ Law J 14:130–178

American Clean Energy and Security Act of 2009

Bardach E (1998) Getting agencies to work together: the practice and theory of managerial craftsmanship. Brookings Institution Press, Washington, DC

Buzbee WW (2003) Recognizing the regulatory commons: a theory of regulatory gaps. Iowa Law Rev 89:1–64

Buzbee WW (2005) The regulatory fragmentation continuum, westway and the challenges of regional growth. J Law Polit 23:323–363

Camacho AE (2007) Can regulation evolve? Lessons from a study in maladaptive management. UCLA Law Rev 55:293–358

Camacho AE (2009) Adapting governance to climate change: managing uncertainty through a learning infrastructure. Emory Law J 59:1–77
Camacho AE (2011) A learning collaboratory: improving federal climate change adaptation planning. BYU Law Rev 2011:1821–1862
Carlson A (2009) Iterative federalism and climate change. Northwestern Univ Law Rev 103: 1097–1160
Climate Ready Estuaries (2010) Climate ready estuaries 2010 progress report. Environmental Protection Agency
Climate Ready Estuaries, Coastal Toolkit (n.d.) Retrieved 28 Feb 2014, from United States Environmental Protection Agency website, http://www.epa.gov/cre/toolkit.html
Craig RK (2008) Climate change, regulatory fragmentation, and water triage. Univ Colorado Law Rev 79:825–927
Craig RK (2010) 'Stationarity is dead' – long live transformation: five principles for climate change adaptation law. Harv Environ Law Rev 34:9–73
DiMento J, Ingram J (2005) Science and environmental decision making: the potential role of environmental impact assessment in the pursuit of appropriate information. Nat Resour J 45: 283–309
Engel KH (2006) Harnessing the benefits of dynamic federalism in environmental law. Emory Law J 56:159–188
Farber DA (2009) Climate adaptation and federalism: mapping the issues. San Diego J Clim Energ Law 1:259–285
Fortney MD (2006) Devolving control over mildly contaminated property: the local cleanup program. Northwestern Univ Law Rev 100:1863–1906
Geltman EG, Skroback AE (1998) Reinventing the EPA to conform with the new American environmentality. Columbia J Environ Law 23:1–56
Gregory R, Ohlson D, Arvai J (2006) Deconstructing adaptive management: criteria for applications to environmental management. Ecol Appl 16:2411–2425
Grumbine RE (1994) What is ecosystem management? Conservat Biol 8:27–38
Intergovernmental Panel on Climate Change (2007) Climate change 2007: impacts, adaptation, and vulnerability, summary for policymakers
Karkkainen B (2004) New governance in legal thought and in the world: some splitting as antidote to overzealous lumping. Minn Law Rev 89:71–497
Karkkainen B (2008) Bottlenecks and baselines: tackling information deficits in environmental regulation. Tex Law Rev 86:1409–1444
National Climate Assessment and Development Advisory Committee (2013) Draft National Climate Assessment
National Climate Change and Wildlife Science Center (2008) Retrieved 2 Aug 2010, from United States Geological Survey website, http://nccw.usgs.gov/documents/NCCWS_factsheet.pdf
Natural Community Conservation Planning Act (2001) California Fish & Game Code, §§ 2800-2840
O'Connell AJ (2006) The architecture of smart intelligence: structuring and overseeing agencies in the post-9/11 world. Calif Law Rev 94:1655–1744
Pew Center on Global Climate Change (2010) Adapting to climate change: a call for federal leadership
Presidential Executive Order 13653 (2013) Preparing the United States for the impacts of climate change. White House, Washington, DC
Ruhl JB (2010) Climate change adaptation and the structural transformation of environmental law. Environ Law 40:363–435
Ruhl JB, Salzman J (2010) Climate change, dead zones, and massive problems in the administrative state: guidelines for whittling away. Calif Law Rev 98:59–120
Salkin PE (2010) Cooperative federalism and climate change: new meaning to 'think globally – act locally'. Environ Law Rep News Anal 40:10562–10571

Schapiro RA (1999) Polyphonic federalism: state constitutions in the federal courts. Calif Law Rev 87:1409–1468
Schapiro RA (2005) Toward a theory of interactive federalism. Iowa Law Rev 91:243–317
Siegel JA (2009) Collaborative decision making on climate change in the federal government. Pace Environ Law Rev 27:257–312
Staudinger MD, Grimm NB, Staudt A, Carter S, Stuart FS, Kareiva P, Ruckelshaus M, Stein BA (2012) Impacts of climate change on biodiversity, ecosystems, and ecosystem services: technical input to the 2013 National Climate Assessment
United States Climate Change Science Program (2008) Synthesis and Assessment Product 4.4: preliminary review of adaptation options for climate-sensitive ecosystems and resources
United States Congressional Budget Office (2005) Uncertainty in analyzing climate change: policy implications
United States Fish and Wildlife Service (2012) National Fish, Wildlife, and Plants Climate Adaptation Strategy
United States General Accounting Office (2003) Great Lakes, an overall strategy and indicators for measuring progress are needed to better achieve restoration goals. Report no. GAO-03-515
United States Global Change Research Program (2013) About the metadata access tool for climate and health, retrieved 20 June 2013 from United States Global Change Research Program website, http://match.globalchange.gov/geoportal/catalog/main/home.page
United States Government Accountability Office (2007) Climate change: agencies should develop guidance for addressing the effects on Federal land and water resources. Report no. GAO-07-863
United States Government Accountability Office (2009) Strategic Federal planning could help government officials make more informed decisions. Report no. GAO-10-113
United States Government Accountability Office (2011) Climate change: improvements needed to clarify national priorities and better align them with Federal funding decisions. Report no. GAO-11-317
United States Government Accountability Office (2013) Climate change: various adaptation efforts are under way at key natural resource management agencies. Report no. GAO-13-253
White House Council on Environmental Quality (2010) Progress report of the Interagency Climate Change Adaptation Task Force: recommended actions in support of a national climate change adaptation strategy
White House Council on Environmental Quality (2011a) Federal actions for a climate resilient nation: progress report of the Interagency Climate Change Adaptation Task Force
White House Council on Environmental Quality (2011b) Instructions for implementing climate change adaptation planning in accordance with Executive Order 13514

Chapter 5
Adaptation Strategies in the Netherlands

Joyeeta Gupta, Judith E.M. Klostermann, Emmy Bergsma, and Pieter Jong

5.1 Introduction

Although climate change has been prominently featured on the global scientific and political agendas since the World Climate Conference in 1979 (WCC 1979), the specific importance of adaptation to climate change has only been underlined about 20 years later. The Netherlands, because it lies largely under sea level, has much to benefit from climate change adaptation. Surprisingly, however, although the Netherlands has been very active in pursuing international climate change politics, the country has not put much effort in politicizing climate change adaptation internationally in this early period and domestically published its National Adaptation Strategy only as late as 2007. This chapter attempts to explain the evolution of Dutch climate change adaptation strategies. It examines adaptation policies in

J. Gupta (✉)
Department of Human Geography, Planning and International Development Studies, Amsterdam Institute for Social Science Research, University of Amsterdam, Room 2.12b, Plantage Muidergracht 14, 1018 Amsterdam, The Netherlands

UNESCO-IHE Institute for Water Education, Delft
e-mail: j.gupta@uva.nl

J.E.M. Klostermann
Water, Climate and Policy at Alterra of Wageningen University and Research Centre, P.O. Box 47, 6700 Wageningen, The Netherlands

E. Bergsma
Department of Political Sciences, Faculty of Social and Behavioural Sciences, University of Amsterdam, Amsterdam Institute for Social Science Research, Oudezijds Achterburgwal 237, 1012 Amsterdam, The Netherlands
e-mail: e.bergsma@uva.nl

P. Jong
Policy and Management, Delft University of Technology, Jaffalaan 5, 2628 Delft, The Netherlands
e-mail: p.jong@tudelft.nl

E. Albrecht et al. (eds.), *Implementing Adaptation Strategies by Legal, Economic and Planning Instruments on Climate Change*, Environmental Protection in the European Union 4, DOI 10.1007/978-3-540-77614-7_5,

four climate-related sectors (water, nature, agriculture and spatial planning) to identify general patterns regarding adaptation strategies in the Netherlands.

When climate change entered international politics in the 1970s and 1980s, international discussions and efforts focussed on counteracting or mitigating the causes of climate change. At the World Climate Conference (WCC 1979), the Hague summit on global environmental issues for Heads of State (Hague Conference 1989) and the Noordwijk Conference on Climate Change (Noordwijk Conference 1989), the emphasis was on realizing the seriousness of the impacts of the climate change problem, and—based on that realization—to promote efforts in the area of greenhouse gas mitigation. Although the first set of reports of the Intergovernmental Panel on Climate Change did indeed include a report on impacts and adaptation measures (IPCC-3 1990), when the United Nations Framework Convention on Climate Change (UNFCCC 1992) was finally adopted in 1992, the emphasis was clearly on mitigation as a global issue and adaptation as a local issue (Bodansky 1993). This emphasis was created for three reasons: to disconnect liability for impacts and related adaptation measures from the emissions at the global level; to make sure adaptation is not treated as a surrogate for mitigation, hence, to promote global action on mitigation; and finally, in recognition of the fact that adaptation measures are context-specific and therefore best constructed and negotiated at local level. However, at the start of the twenty-first century, calls have been increasingly made to "lift the taboo"[1] on climate change adaptation within international politics.

In the Netherlands, similar trends can be found. While there was a flurry of interest in adaptation-related issues in the early 1990s, the emphasis in that decade was on mitigation. It is only in the post-2000 period that there is a gradual re-emergence of the adaptation challenge on the domestic agenda. The following sections elaborate on this in some detail. This chapter is a further analysis based on a content analysis of the relevant policy and legal documents in the Netherlands (Klostermann et al. 2010).

5.2 The Evolution of Climate Policy in the Netherlands

Historically, the Netherlands as a low-lying country has always fought against water—the sea and the river! Since the twelfth century water management authorities have managed Dutch water issues; while this institutional structure was long characterized by high fragmentation, centralization efforts in Dutch water management started in the late eighteenth century under French influence and is currently still on-going; the number of local water boards has, for example, been reduced from about 2,500 in the 1950s to only 26 up to now, and the Delta Act (2011), included in the 2009 Water Act, intends to strengthen national coordination in flood

[1] This expression has been borrowed from Pielke's et al. (2007) commentary in *Nature* in which they argue for rehabilitating adaptation in international climate change politics.

management. This long tradition of being able to manage the water despite being below sea level has created a sense of confidence in the Netherlands that it will be able to cope with the impacts of climate change.

Despite its long history with water management, Dutch climate policy in the 1990s focussed mainly on mitigation. In the first National Environmental Policy Plan of 1988, the government adopted a mitigation target of stabilizing CO_2 emissions by 2000 at 1990 levels (VROM 1989). A year later, the newly elected government enthusiastically revised this target to stabilizing CO_2 emissions by 1994 and reducing them by 3–5 % by 2000. In 1991, the Policy Note on Climate Change (VROM 1991) developed mitigation policies, which were further revised in the second policy note of 1996 (VROM 1996). It should be noted that these ambitious targets were not supported broadly by other ministries and social actors and were not ultimately achieved. In the meanwhile, the Netherlands ratified the 1992 United Nations Framework Convention on Climate Change as well as the 1997 Kyoto Protocol (United Nations 1997) which committed the Netherlands to contribute to an overall goal for developed countries of a −5.2 % reduction of emissions from 1990 levels by 2005. This translated into a −6 % reduction of emissions for the Netherlands. In the run-up to the Copenhagen negotiations on Climate Change, the Netherlands committed itself to reducing its emissions by 30 % in 2020 compared to 1990 levels. In an effort to engage local actors and other ministries, the Netherlands has developed mitigation policies in collaboration with provincial and municipal actors since 1999.

While there have been developments in water management policies in the Netherlands in the beginning of the 1990s (e.g. VenW 1990), climate change adaptation was not seriously considered in any of the early policy documents within this domain. The reasoning was that if there was enough attention paid to mitigation, there would be less attention needed for adaptation. A parallel reasoning was evident at the global level, where apart from listing a series of potential adaptation measures, there were no real commitments with respect to adaptation either in the Climate Convention or in the Kyoto Protocol. However, pressure from the developing countries led to reconsideration of the funding rules for adaptation as applied by the Global Environment Facility; the decision was made to set up an Adaptation Fund from the proceeds of the Clean Development Mechanism under the Kyoto Protocol; as well as to finance the National Adaptation Plans of Action of the Least Developed Countries. A series of meetings of the Conference of the Parties in Nairobi and thereafter focused on the importance of adaptation strategies for developing countries. The Katrina disaster in 2005 in the United States pointed attention to the fact that even the developed countries could be increasingly vulnerable to the impacts of climate change. In the Netherlands, floods in 1995 and water problems of the late 1990s increased the awareness of adaptation needs. This led to the establishment of a Commission on Water Management for the twenty-first century and a series of measures have been taken since then.

A study of the adaptation strategies of the Netherlands is not a simple and straightforward exercise (Klostermann et al. 2009). More than 90 documents provide the basic information regarding the evolution of explicit and implicit adaptation strategies. These can be clustered, although not without some difficulty, in terms of general adaptation strategies and adaptation strategies focusing on four specific sectors—nature, spatial planning, agriculture and water. The choice of four sectors is clearly limited as the National Program for Spatial Adaptation to Climate Change (ARK program; VROM 2006a) identifies nine sectors that are relevant to consider when thinking about climate change adaptation in the Netherlands. However, this chapter focuses on the four sectors that are most strongly related to land use, since land use planning is generally considered to provide the overarching framework for most adaptation strategies in the Netherlands (see VROM 2006a). The following sections thus describe the general adaptation strategies, and the strategies that operate in the four identified sectors.

5.3 General Adaptation Strategies

This section divides adaptation strategies into two phases—the pre-2004 phase and the post-2004 phase.

5.3.1 Pre-2004 Policy Approaches

In the pre-2004 period, three key general measures were taken which have relevance for adaptation. The first is the adoption of the Environmental Management Act of 1993 (EMA 1993), which is a living document and is regularly updated. It did not focus on climate change, but created a number of incentives that can easily be adapted for use and application in an adaptation strategy. These include environmental plans, environmental impact assessments (EIAs), environmental quality standards, environmental permits, reporting and enforcement rules. It also includes environmental subsidies, taxes and provisions for damage compensation. In reaction to the 1995 floods, a law on Compensation of Damage in case of Disasters and Accidents (Disasters Compensation Law 1998) was adopted in 1998 which provides a financial safety net for damage caused by large-scale events, including weather events. The Ministry of Internal Affairs is empowered to activate this law when a large-scale event happens. Since 2000, the Dutch government is obliged to prepare an annual Environmental Balance that assesses which environmental goals have or have not been met and where action should be focused. This assessment is based on the goals set in the regularly updated National Environmental Policy Plans, which was revised for the fourth time in 2001 (VROM 2001). However, these documents do not explicitly include climate-related or adaptation-related goals and strategies.

5.3.2 Post-2004 Approaches

In the post-2004 period, there has been a more rapid focus on climate change and adaptation. While in various (scientific) reports and at different conferences the urgency of taking adaptation measures was established, most Dutch environmental policy at the beginning of this century still did not explicitly include adaptation strategies. Key scientific documents in this period include the Climate Change Report of the House of Representatives (Rooijers et al. 2004) which, although it focuses more on mitigation, does emphasize the need to deal with floods and droughts and to provide adaptation financing to developing countries. In 2005, seven Senators under the leadership of Lemstra submitted a motion that existing spatial planning did not adequately take climate change into account and this motion was unanimously adopted (Lemstra 2005). In 2006, the Scientific Council for Government Policy (WRR 2006) argued persuasively that adaptation should be seen as a 'no-regrets' policy; at the national level the focus should be on the water and allied sectors to improve flood defenses since existing safety norms in the sector were out-of-date. The Council furthermore argued that adaptation measures could best be developed regionally since they are context relevant. Finally, the document noted that there needed to be better links between the water and spatial planning sectors, but that spatial planning alone may not be enough. The document also noted that, since climate change impacts directly affect individuals, it is essential to involve and engage stakeholders who may then be more willing to take action.

In reaction to these reports, the Dutch Government launched two adaptation policies in 2006—the Think Ahead Campaign[2] which focused, inter alia, on the potential extreme weather events and floods that may affect the Netherlands and the role of individuals in dealing with these events; and the Agenda for the Future VROM (2006b) which argues that climate change calls for understanding and revisiting the responsibilities of the different actors and that more responsibility should be passed on to the citizen. Additionally, a collaboration between various governmental actors—ministries, provinces, municipalities and water authorities—and non-governmental organizations adopted the ARK program in 2006 (National Program for Spatial Adaptation to Climate Change 2006–2014; VROM 2006a) which focuses on climate-proofing nine sectors in the Netherlands through spatial planning.

Since that moment on, adaptation issues increasingly received attention from the scientific community and governmental actors. In 2007, the Netherlands Environment Assessment Agency (Mathijssen et al. 2007) organized a conference to reflect on climate change adaptation issues and concluded that there was a need to focus not only on obvious risks in the water sector but also on more latent risks caused by climate change. It focuses specifically on the need to deal with the uncertainty of impacts; uncertain impacts call for taking risky approaches and evaluating them, for

[2] See: http://www.denkvooruit.nl. Accessed 08 November 2013.

generalized rather than specific approaches, for strong leadership, for an understanding of what can be planned in advance and what not, and finally it calls for incorporating multiple problem definitions by building on different stakeholder perspectives.

That same year, scientific work on climate change adaptation was integrated into a research document called the Route Planner (Van Drunen 2006) which identified 96 different measures that can be of relevance to the Netherlands. These measures were classified into different categories, namely: importance, urgency, no-regret measures, additional effects, mitigation effect and complexity (weighted sum of technological, social and institutional complexity). The Route Planner notes that the maximum available options are in the water sector and the lowest in the health sector; and that the biggest challenge is institutional complexity, which often stands in the way of the simplest measures. For example, the policy Room for the River, which tries to provide more space for the river to overflow, faces a number of institutional challenges especially from Spatial Planning Law. The authors call for flexible institutions that can cope with the new kinds of challenges imposed on them. In the same year, the Council for the Ministry of Housing, Spatial Planning and the Environment (VROM Council 2007) argued that uncertainty in climate science calls for structural, offensive and flexible long-term options. Such options should be robust enough to cope with not just the middle scenarios but also the extreme scenarios. Amongst other issues, it mentioned the need for establishing a watchdog to ensure that policies are implemented and to avoid administrative complexity.

Also in 2007, participants in the ARK program adopted the National Adaptation Strategy (VROM 2007a), which focuses on adaptation as primarily a spatial challenge and tries to limit risks within specific compartmentalized areas within specific dike rings. It also aims to use existing ecological processes to deal with climate change. The document focuses on hard measures (technological measures) and calls for the mobilization of large-scale investments.

In the period that followed, several policies were launched that aimed to clarify the different responsibilities for climate change adaptation. In April 2007, for example, the National Risk Strategy (BZK 2007) was adopted. This document identifies climate risks such as floods, droughts and health hazards such as flue pandemics and discusses the significance of each. It argues that crisis management involves differentiated roles for government and citizens. In June 2007, a policy vision "Working Together, Living Together" (AZ 2007) was adopted which focuses on climate proofing the Netherlands through a spatial framework in line with the central vision of all Balkenende cabinets "decentralize where possible, centralize where necessary". And in 2010, a Law on Safety Areas was adopted that allocates various responsibilities to deal with calamities, amongst others those related to water issues.

Table 5.1 provides a chronological listing of measures that have some relevance for adaptation in the Netherlands.

Table 5.1 Chronological general adaptation measures in the Netherlands (compiled from Klostermann et al. 2010)

Year/type	Institution	Focus
1993 Law	Environmental Management Act	Not on climate change; but can be adapted: includes EIA, standards, permits, reporting, enforcement rules; subsidies, taxes, compensation
1998 Law	Compensation of damage in case of disasters and accidents	Safety net for large scale-events
2000 Monitoring	Environmental balance	Takes climate change into account
2001 Policy	Ministries: National Environmental Policy Plan—4	Adaptation inadequately covered
2004 Science	House of Representatives: Climate Change Report	Mitigation; adaptation discusses dealing with floods and droughts and an adaptation fund for developing countries
2005 Motion	House of Representatives: Lemstra Motion adopted	Spatial policy should take climate change into account; FES (Economic Structuration Fund) money should be used also for knowledge infrastructure
2006 Policy	Agenda for the Future	Changes responsibilities for climate change; more responsibility on the citizen
2006 Policy	Government: Think Ahead Campaign	Extreme weather events, floods and the role of individuals
2006 Advice	Scientific Council for Government Policy	Adaptation is a 'no-regrets', regional option; link spatial planning to water sector
2006 Policy	Collaborative program: ARK	Climate-proof Netherlands for nine sectors through spatial planning
2007 Science	The Netherlands Environment Assessment Agency	Focus on obvious and latent risks; policy should deal with uncertainty
2007 Science	Collaborative research: Route Planner	Lists 96 different options for climate-proofing; stakeholder participation
2007 Advice	VROM Council Report	Uncertainty calls for structural, offensive and flexible long-term options; need for watchdog
2007 Policy	Collaborative program: National Adaptation Strategy	Adaptation is a spatial issue; compartmentalize risk; use ecological processes; hard measures; need for large-scale investments
2007 Policy	Cabinet: National risk strategy	Identifies climate risks: floods, droughts, flue pandemic; crisis management involving differentiated roles for government and citizens.
2007 Policy	Cabinet: Working Together, Living together	Climate-proofing through spatial framework
2010 Law	Law on Safety Regions	Allocation of responsibilities to deal with calamities

5.4 Agriculture

The agricultural sector in the Netherlands is one of the oldest sectors and is hugely influenced by developments at the international level within the World Trade Organization, at the European Union level within the Common Agricultural Policy (EC 2009) and global markets.

The agricultural sector does not formally and explicitly discuss climate change and adaptation. The 1993 Environmental Management Act (EMA 1993) includes regulations with respect to environmental impacts of agricultural practices. In the last decade of the twentieth century, the Dutch agricultural sector increasingly had to deal with extreme rainfall causing damage to crops. Several arrangements administered by different governmental bodies were in place to compensate farmers for their losses. This constellation of arrangements was criticised for its differentiating and non-committal character (Duin and Mesu 1995). The 1998 Disasters Compensation Law intended to centralize all damage compensation regulations into one arrangement; however, other regulations also continued to exist for the agricultural sector. For example, several studies to evaluate the possibilities for agrarian insurances were performed (e.g. LTO-Nederland 1999; IRMA 2000) and now, several insurance companies offer rain insurances to farmers. This development seems to shift some responsibility to the farmers by requiring them to insure against extreme rain. Next to these private initiatives, there are also some public arrangements that financially support farmers who have to deal with losses due to rainfall.

In 2004, the government's Agenda for a Living Countryside (LNV 2004) emphasized that non-agricultural policies and laws would have to play a critical part in helping to address climate-related water problems.

In 2005, two research and advisory documents were critical. The Social Economic Council came up with a report (SER 2005) about the opportunities for rural areas and argued that these opportunities can be optimized by linking national with European policies and mobilizing local actors. Climate adaptation is not mentioned in this report. However, the report focuses on the entrepreneurial and adaptive role of the farmer. The Ministry of Agriculture's report 'Choice for Agriculture' (LNV 2005) focused on potential agricultural developments and aims at informing farmers about these potential developments to enhance their adaptive capacity.

In 2006, a Company Premium (LNV 2009) was established and a Fertilizer Law (2006) was adopted. The Company Premium is a follow-up to the changes made to the EU Common Agricultural Policy (EC 2009) and offers income support delinked from production to farmers if they meet some criteria. This premium does not mention climate change adaptation. The Fertilizer law was established also as a follow-up to the non-compliance of the Netherlands to the EU Nitrates Directive (Dienst Regelingen 2008) and provides emission standards and rules on the use of fertilizers. This too does not take climate change into account but will perhaps need modification to do so.

In 2007, the government adopted first a strategy on rural development and then a policy. The strategy (LNV 2007a) describes how financial means from the European Agricultural Fund for Rural Development (EAFRD) will be allocated to local projects that combat biodiversity loss and climate change, and maintain water quality and quantity. The policy links up with the goals in the European Rural Development Policy (European Council 2006) focusing on the competitiveness of the agricultural and forestry sector; improving the environment and the countryside; improving the quality of life in rural areas and encouraging diversification of the rural economy; and building local capacity for employment and diversification through a leader-approach (i.e. a bottom-up approach stimulating the involvement of local actors). Although climate change is not specified, it is expected that these subsidies will help farmers to adapt. In the same year, the Rural Areas Development Act (WILG 2007) was adopted, which although it does not address climate adaptation could easily be adapted to do so. This Act divides responsibilities between the central and provincial governments with the latter held accountable for achieving rural goals. The key tool used in this document is spatial planning. This Act creates a financial investment instrument (Investment Budget Rural Areas (ILG) which provides budgets to provinces. It also changes some rules of the Agricultural Land (Transactions) Act (WAG 1981). This act now empowers provinces to rearrange and redistribute land if there is need to do so based on certain principles; and to reconstruct land areas to reduce chemical pollution, acidification and stench.

In 2010, the subsidy scheme for Rural Area Management revised the existing three subsidy schemes for nature management, agricultural nature management and private management of natural and agricultural ground (Dienst Regelingen, IPO, SNL 2009). It includes some EU funds and hence EU criteria. These measures indicate a growing integration of different measures and laws into a common legal and subsidy system.

Table 5.2 provides a chronological listing of measures that have some relevance for adaptation in the Netherlands.

5.5 Nature

The Netherlands is a densely populated small country; nature reserves are thus concentrated in relatively small areas. Regulations in this area are of relatively recent origin and quite often regulated from Brussels—e.g. The Habitats Directive (European Council 1992) and the Birds Directive (European Council 1979). In 1967 the Nature Conservation Law (NCL 1967) was adopted and provided the framework for action for the coming decades.

The Forestry Act of 1981 aims at protecting lands currently under forests from land use change until and unless there is a significantly important public good to be served. This Act includes reporting, replanting and compensation obligations and a prohibition on felling.

Table 5.2 Chronological implicit and explicit adaptation measures in the agricultural sector in the Netherlands (compiled from Klostermann et al. 2010)

Year/type	Institution	Content
2004 Policy	Agenda for a Living Countryside	Recognizes climate-related water challenges and the role of non-agricultural laws (e.g. NEN) in addressing these. Focuses on decentralization. Spatial planning has a limited role.
2005 Advice	SER: Opportunities for Rural Areas in the Netherlands	Opportunities optimized by linking national with EU policies and mobilizing local actors. Climate adaptation not mentioned and the role of the entrepreneur is emphasized.
2005 Vision	Ministry of Agriculture: The Choice for Agriculture	Describes potential agricultural developments and aims at informing farmers to enhance their adaptive capacity.
2006 Subsidy	Company Premium	Provides income support delinked from production to farmers upon conditions; climate change not addressed.
2006 Law	Fertilizer law	Creates emission norms and use norms for fertilizers; climate change not addressed.
2007 Vision	Dutch Strategy for Rural Development 2007–2013	Describes how the EU EAFRD will be allocated to local projects on biodiversity, climate change and water quality and quantity.
2007 Policy	Dutch Rural Development Policy 2007–2013 (RDP2/POP2)	Linked to EU ERDP goals. Although climate change is not mentioned, subsidies could help farmers adapt.
2007 Law	Rural Areas Development Act (WILG)	Does not address climate adaptation explicitly but could easily be adapted to do so. Provinces are accountable for achieving rural goals. Creates ILG to finance provinces and WAG to empower provinces to rearrange and reconstruct land if needed based on certain principles.
2007 Policy	Agenda for a Living Countryside 2007–2013	Decentralizes responsibilities to provinces. The multi-year programs between water authorities and municipalities incorporate WILG and ILG.
2010 Policy	Subsidy system for Nature and Landscape Management	Subsidizes public and private nature management on agricultural lands.
–	Miscellaneous (public and private) compensation schemes for damage compensation to farmers	Public and private insurances for a contribution in crop damage compensation caused by (extreme) rainfall to farmers, placing more responsibility at the level of farmers.
–	Miscellaneous subsidy schemes on nuts, cattle farms, etc.	There are a number of sector specific subsidies that could perhaps be modified for use to help the sector adapt.

In 1990 the Ministry of Agriculture, Nature Management and Fisheries adopted the Nature Policy Plan (NPP 1990) and introduced the concept of the National Ecological Network (NEN), a concept that aimed to counter the increasing fragmentation and isolation of species in specific pockets of land, by developing corridors for species to move around. These ecological zones are to be created and achieved by 2018 and are to ensure the resilience of the Dutch species, although at the time climate change was not taken into account. This omission has been remedied in a number of studies undertaken since then (Routeplanner 2007; Vos et al. 2007). This concept (NEN) has been integrated into a number of different laws including the WILG.

Five years later in 1995, the NEN was officially adopted in a Spatial Plan for the Rural Area and its related key planning decision (LNV 1995). Several policy documents after that further elaborate on the NEN, and promote the management of nature in relation to agriculture and water. In this context, the impacts of climate change are taken into account. The obligations for nature protection in the rural areas are targeted at provinces and water authorities. The role of land owners in this is emphasized. For example, the concept of a National Climate Buffer, referring to the creation of a natural zone that can absorb climate shocks, is proposed as an extension for the NEN.

In 1998, the Nature Conservation Law of 1967 was amended and focused on protecting areas and landscapes (and not species) through mandating the preparation of nature policy plans with a maximum interval of 8 years, the development of vision statements, designation decisions, preservation goals, management plans, permits and compensation rules, as well as monitoring and enforcement. Also, since 1998, Nature Balances have been made annually that describe the impacts of climate change on nature but do not discuss potential adaptation options. The Nature Exploration documents focus on how different climate scenarios may impact on the achievement of the national target.

Also in 1998 the Flora and Fauna Act (1998) was adopted. This implements the international Convention on endangered species (CITES) and the EU Birds and Habitat Regulations, and aims at protecting endangered species through rules on hunting, trade and ownership, the ‘no-unless’ rule, and a fauna fund to finance these activities. The ‘no-unless’ rule does not allow land use change unless there are no alternatives and the changes are perceived to be in the national interest. This law has not taken climate change explicitly into account.

In 2000, the ministry adopted the Nature for People, People for Nature policy (LNV 2000). It draws attention to raising public attention and support for nature as well as emphasizing that Dutch nature is unique. It uses the climate change problem to emphasize the role of nature in contributing to water management and refers to the concept of Room for the River. It promotes the implementation of multi-level regulations, provides financial incentives and a greening of the Dutch tax system and educational incentives. The concept of the National landscape is expected to help integrate the rural and aesthetic functions of the landscape. This document promotes land acquisition and spatial planning of areas for the National Ecological Network. Also, this policy note introduced the concept of robust ecological corridors as additions to the NEN. While ecological corridors, or climate buffers, have not been officially included in the NEN, 19 national climate buffer projects, supported by payments from the High Water Security Program, have been initiated[3] (Rijkswaterstaat 2007).

In 2007, the Ministry of LNV published a policy document which provides clear rules with respect to interpreting the NEN and explains how the concepts of 'no-unless', 'compensation rule',[4] 'redemarcating the NEN'[5] and 'the NEN balance approach'[6] should be interpreted (LNV 2007b). However, while the NEN is a key policy for nature management in the Netherlands, its financing has been reduced and its policy goals have been softened by the previous government (cabinet Rutte 2010–2012), amongst others pressured by the economic recession. The future status of the NEN has become insecure.

As in other sectors, a coming together of different measures is visible in the Dutch nature sector. Table 5.3 sums up the key policies.

5.6 Water

The Netherlands, as has been mentioned before, lies largely below sea level. It is a delta country with four major rivers (Rhine, Meuse, Scheldt and Ems). Its coast needs to be protected by man-made dunes, dikes and other structures. The water sector is perhaps the most regulated sector in the Netherlands. This section only examines the recent and most relevant regulations and policy decisions that deal with this sector.

In 1990, a New Coastal Defence Policy (VenW 1990) for the Netherlands was adopted. It explained that following the 1953 floods, the dikes and dunes along the North Sea were raised to 'Delta Height'. The protection should ensure that regions would be protected from the extremes of a 1 in 10,000 year storm (Annex II of the Water Act 2009). Given the potential impacts of climate change and following several studies, four options were identified: retreat; selective preservation; preservation; and seaward expansion. In the 1990 Coastal Defence Policy, the choice for dynamic preservation was made (dynamic to allow for some 'natural' movement of the shoreline), but the preservation goal aimed at both combating coastal erosion and dealing with sea level rise, primarily through sand nourishments and replenishment and stone revetments in weak locations. Dyke protection was to be

[3] See also: http://www.klimaatbuffers.nl/english-homepage-2. Accessed 08 November 2013.

[4] If spatial developments are allowed, negative impacts on nature should be mitigated and remaining damage should be compensated.

[5] Allows changing the borders of NEN areas on a small scale when this has a positive effect on quality or quantity. When it happens for other reasons, the 'no-unless' principle applies.

[6] A development approach allowing an integrated approach to NEN areas combining different qualitative or quantitative aims.

Table 5.3 Chronological implicit and explicit nature measures in the nature sector in the Netherlands (compiled from Klostermann et al. 2010)

Year/type	Institution	Content
1967 Law	Nature Conservation Law	Framework for conservation
1981 Law	Forestry Act	Protects forests through reporting, replanting and compensation obligations and a prohibition on felling.
1990 Policy	Nature Policy Plan	National Ecological Network (NEN) to be created by 2018
1995/2001 Policy	Spatial Plan for the Rural Area	Elaborates further on NEN; sees water as an organizing principle; delegates responsibility to decentralized governments and farmers.
1998 Law	Nature Conservation law, amended	Calls for regular nature policy plans, vision statements, designation decisions, preservation goals, management plans, permits and compensation rules, monitoring and enforcement
1998 Policy	Nature explorations and balances	Annual reports on the nature sector by Netherlands Environmental Assessment Agency
1998 Law	Flora and Fauna Act	Protects endangered species through rules on hunting, trade and ownership, the 'no-unless' rule and a fauna fund; does not explicitly take climate change into account
2000 Policy	Nature for People, People for Nature	Land acquisition and spatial protection of NEN; incentives and taxes; education; climate change addressed
2000 Policy	Nature Policy Plan	Promotes robust ecological corridors and climate buffers
2007	Rules of the NEN	Document that interprets key terms—NEN, compensation, no-unless, NEN balance approach

undertaken by maintenance while dune coasts were to be allowed some dynamic movement.

In 1996, a program for testing the flood defences every 5 years was established—to see if these still meet the safety norms. The test conducted in 2006 revealed that 24 % of the barriers did not meet the norms. This led to the establishment of a High Water Security Program in 2006 (Rijkswaterstaat 2007) with 93 measures that needed to be undertaken and would cost 2.3 billion Euros.

In the second half of the 1990s, two near-floods in the Netherlands led to the establishment of a Commission on Water Management for the twenty-first Century. The Commission's 1999 report (Commissie Waterbeheer 2000) concluded that the greatest challenges were in integrating and linking the water sector to spatial planning objectives and developments. The Commission recommended a clarification of responsibilities, greater collaboration between the different concerned actors including scientists, the promotion of no-regrets measures and the need to raise additional resources to deal with the problem. The report recommended that excess water should be retained upstream, in surface water, and, if necessary, in temporary basins.

In 2000, the Third Policy Note on Coasts (VenW 2000) was adopted. It focused on strengthening coastal protection by focusing on the weak parts of the coastal protection chain and dynamic maintenance of the coastal boundaries including through sand replenishment. In 2003, based on an assessment of the weak links in the coastal defence system, the government adopted a program focusing on ten weak links. Appropriate policy is being developed in these regions (VenW 2003).

In 2003, the different administrative and social actors came together to adopt the National Administrative Accord on Water (NAW 2003) and decided to develop policies for areas that fall outside the formal dike protection of the Netherlands, and a policy line to manage the coastal areas with special attention for the parts of the coast that have weaker protection. The NAW approach was evaluated in 2006 and the evaluation concluded, inter alia, that the approaches adopted were very complex and that the financial responsibilities were not always clear. In 2008, the NAW was made up-to-date based on the latest information about climate scenarios and the obligations that flowed from the European Water Framework Directive. That same year, a water test was included in the Spatial Planning Act which calls for testing spatial planning for their impacts on water quality and quantity (RIZA 2003).

In 2006, the third policy note (VROM 2007b) focusing on the Waddenzee was adopted. 90 % of this area is seen as a National Nature Monument and is also covered by the Habitats Directive of the European Union. The document calls for prevention of pollution discharges into the sea, greater cooperation with Germany and Denmark and appropriate policy with respect to gas and fish exploitation.

A 2006 Policy Document (Policy Guideline for Major Rivers: VenW and VROM (2006) replaced a 1997 policy document to focus on room for the river. This document was drawn up in cooperation with social actors. In 2007, a decision was taken to make Room for the River; this was a major shift in mindset from creating hard protective measures to allowing the rivers space to overflow in selected areas if necessary. This includes 40 context-related measures related to the Rhine and the Meuse.

The 2007 evaluation of the Third Policy on Coasts was positive and recommended specific rules regarding areas outside dike protection, weak links in the coastal defence system and sand replenishment. In the same year, a revised law on the water authorities (Water Authorities Modernization Act 2007) was adopted that changes the mandate and management system of these authorities. These water authorities are empowered to make water management plans, water ordinances and charge taxes.

In 2008, a Delta Commission (led by Veerman; Delta Commission 2008) came out with its report focusing on the long-term goals for water management. It concluded that the safety levels for water protection should be increased by a factor of 10, and building in risky places should be based on an evaluation of the costs and benefits. It made recommendations with respect to specific elements of the coastal defence system.

In 2008, the government adopted the National Water Plan (NWP) based on a Water Vision published in 2007 (VenW, VROM and LNV 2009). This plan adopts and integrates the existing programs of coastal protection, Room for the River and river expansion in the Maaswerken, agreements between the state and other actors regarding water shortage and excess, and river basin management flowing from the Water Framework Directive. The NWP recommends taking climate change impacts into account in water policy. It creates a multi-level security approach: in the first layer the focus is on prevention of flooding, in the second layer the focus is on sustainable spatial planning, and in the third layer the focus is on crises management.

Municipal water plans are plans made by municipalities in cooperation with water authorities and social actors and can go beyond their official task of managing the sanitation system to include the broader management issues in relation to water. In 2008, the Act on Municipal Water Tasks (2008) was adopted that amends previous laws and integrates new tasks and although it does not mention climate change, is a result of a recognition of the impacts of climate change at the municipal level. The law allocates responsibilities for sanitation and rainwater within municipal boundaries.

In 2009, the Government adopted a Water Act (2009), which replaces and integrates eight other water laws.[7] The Water Act discusses water shortage, water safety and water quality. It calls for 12 yearly revisions of the norms and six yearly policy revisions. It bundles the existing system of permits. This Water Act integrates past Acts into one consolidated system.

The latest policy development within the Dutch water sector is the adoption of the 2011 Delta Act. This law regulates Dutch flood risk protection and fresh water supply, principally by ensuring sufficient financing for water safety in the Netherlands as well as by creating a "delta commissioner" that coordinates all water safety regulations and oversees the fulfilment of national water safety goals (Table 5.4).

5.7 Spatial Policy

In 2006, the Government adopted the Spatial Policy Note that amended previous documents and presented a national policy for the period until 2020 and discusses the period 2020–2030 as well (VROM et al. 2006). The policy calls for shifts from

[7] Among the laws are the Water Management Act which managed both quality and quantity issues, the Flood Defences Act of 1996, the Groundwater Act, the 1969 Surface Waters Pollution Act, the 1975 Marine Waters Pollution Act, the Act of 14 July 1904 containing provisions on land reclamation and construction of dikes, the Public Works Management Act (sections relating to waterways), the Public Works Act 1900 (sections relating to waterways). The Act on Municipal Water Tasks (2008) has partly been integrated in the Water Act.

Table 5.4 Chronological implicit and explicit adaptation measures in the water sector in the Netherlands (compiled from Klostermann et al. 2010)

Year	Institution	Content
1990 Policy	New coastal defence policy	Choice for dynamic preservation of dikes (maintenance) and dunes (flexible)
1997 Policy	Policy Directive: Room for River	Room for the River concept
1999 Advice	Commission on Water Management for the twenty-first century	The 1999 report requested the Royal Netherlands Meteorological Institute for climate scenarios; greatest challenge linking water to other sectors
2000 Policy	Third Coastal Policy	Focus on weak parts of the coastal defence system
2003 Policy	National Administrative Agreement on Water (NAW)	Collaboration between all governmental actors to deal with water on ten different weak parts
2003 Law	Water test	Included in existing Spatial Planning Law
2006 Law	Tests of flood defences	The last test in 2006 showed that 24 % of the flood defences did not meet the norms and a program with 93 measures has been adopted
2006 Policy	Evaluation of NAW	Collaboration complex, division of responsibilities not clear
2006 Policy	Policy note on Waddenzee	90 % of sea is a nature reserve; measures include international collaboration, limits on fish and gas exploitation, and standards on discharges
2006 Policy	Major Rivers Delta Plan, later renamed as Room for the River	Room for the River: includes 40 context relevant measures for the Rhine and Maas
2006 Policy	High Water Security Program	Improving coastal flood defences for ten weak spots and realizing emergency retention areas
2007 Policy	Evaluation of Third Coastal Policy	Generally effective; more attention to areas outside the dikes
2008 Policy	National Administrative Agreement on Water amended (NAW)	Includes implications for Water Framework Directive and clearer division of responsibilities
2008 Advice	Delta Commission (Veerman)	Water safety levels should be increased by factor 10
2008 Policy	National Water Plan (NWP)	Integrates existing programs for coastal protection, river management and implementation of the WFD
2009 Law	Water Act	Replaces and integrates eight other water laws; climate change not explicitly taken into account; calls for 12 yearly revisions of norms and 6 yearly water policies.
2011 Law	Delta Act	Regulates water safety management at the national level.

planning to development and towards decentralization. It engages to maximize the participation of social actors at multiple levels of governance and thereby to maximize the opportunities for diverse responses. It aims to strengthen the economic competitive position of the country, equitably promote vital cities and villages; and protect important spatial values and the security of the country including water security. Climate change and its impacts are explicitly taken into account and there are efforts to see the impacts also in terms of how they can improve the living environment. The Spatial Planning Note distinguishes between responsibility for running the system and responsibility for achieving goals. The national government is responsible for ensuring the basic quality of the system. The spatial policy for major rivers and the IJsselmeer fall under the responsibility of the state; the spatial policy for the coast, the National Ecological Network, and the national landscapes fall under the responsibility of the state and provincial governments.

A year later, in 2007, an Urgency Program for the Randstad was established to promote 35 projects to enhance the resilience of spatial areas within this economically active region (VenW 2007). In 2008, the Spatial Planning Act of 1965 (SPA 1965) was revised (SPA 2008) to provide new procedures, but this Act does not explicitly take climate change into account. All government levels are empowered to make 'structural visions' and this new term encapsulates a number of different terms used in the past. The visions should integrate and provide direction and bind the authority that has designed them. There are also land use plans to be revised once every 10 years and these plans will be used for giving permits for buildings and demolitions. Where large-scale projects are being planned that do not fit into the nature of the relatively small-scale land use plans, the Environmental Licensing Bill (2010) provides for the possibility of a project license. Finally, the state and the provinces are empowered to make land use plans in case there are national (state) or provincial interests at stake (*inpassingsplannen*). Legal procedures have been simplified so that the response time is reduced. This Spatial Planning Act amends some existing laws: It modifies the Municipal Priority Right (that calls on land owners to give first priority to the municipality when they sell their property) to ensure that these rights are included into the land use plans. The Act includes the Ground Exploitation Act, enabling municipalities to place restrictions on the use of an area.

A number of other laws are also relevant. An Act on Expropriation (1851) allows the state to claim property rights from the owner based on a full compensation to the land owner when there is clear public interest involved; for 'green' reasons such as the National Ecological Network; for 'blue' reasons such as the Room for the River policy or for infrastructure and housing. A Building Decree of 2008 revising a previous Decree of 2003 calls for climate mitigation but not yet climate adaptation to be taken into account in building standards. There are plans to simplify the multiple permits needed for construction purposes into one integrated permit that allows for balancing between different interests.

The Environmental Management Act of 1994 calls for environmental impact assessments for specific types of projects. In addition, a European Directive of 2001 calls for strategic environmental evaluations for strategic projects. These instruments have an impact on spatial policy. Furthermore, a Social Cost Benefit Analysis is compulsory since 2000 for all large projects (Table 5.5).

Table 5.5 Chronological implicit and explicit adaptation measures in the spatial planning sector in the Netherlands (compiled from Klostermann et al. 2010)

Year	Institution	Focus
1851 Law	Law on expropriation	Allows expropriation with full compensation when it is in the public interest.
1965 Law	Spatial Planning Act	Framework for spatial planning in the Netherlands.
2006 Policy	Spatial Policy Note	Amended previous documents and presented a national policy for the period until 2020; shift towards development and decentralization/ participation. Climate change impacts explicit; distinguishes between system responsibility and goal accountability; delegates responsibilities.
2007 Policy	Urgency Plan for the Randstad	35 projects to enhance the resilience of spatial areas.
2008 Law	Spatial Planning Act of 1965 revised	New procedures, but does not explicitly take climate change into account. Tools: 'structural visions', land use plans, project decisions; revises Municipal Priority Right and includes Ground Exploitation Law.
2008 Law	Building Decree (revising a previous Decree of 2003)	Takes mitigation into account, but not adaptation, in building standards.
2010 Law	Environmental Licensing Bill	Provides for the possibility of a license if a project is not in accordance with a land use plan.

5.8 Analysis

This chapter has tried to give a bird's eye view on national adaptation policy in general and in four of the nine sectors seen as important for climate change adaptation in the Netherlands. It shows that there has been an enormous degree of activity in the policymaking sphere in these four sectors in the last few years. The question that rises is: How can this activity be characterised? We identify six key trends in the evolutionary process:

5.8.1 Shift from Sectoral to Integrated Approaches to Adaptation

An evolutionary understanding of the climate change adaptation problem and the challenges it poses to society can be culled from the information provided in the last four sections. While clearly the water sector was most aware of the potential consequences of climate change since the early 1990s (VenW 1990), over time the awareness has reached other sectors and levels. First, importance was accorded to adaptation at the national level and its scope (e.g. Lemstra Motion 2005). Second, the awareness has spilt over to other sectors (e.g. the nature sector—Policy Note Nature for People, 2000; the spatial planning sector—Spatial Policy Note, 2006;

and the agriculture sector, although most policies in this sector deal with climate change implicitly). Third, there is a growing realization of the links between sectors. The lack of linkage, and hence, the need for links between the water and spatial policy sectors grew in significance in the 2000s (Commission on Water Management in the twenty-first century; Lemstra 2005; WRR 2006; National Adaptation Strategy 2007; Policy note Living Together, Working Together, 2007). The Agenda for a Living Countryside (LNV 2004) recognizes that non-agricultural policies are critical for dealing with climate impacts on the countryside. In 2001, the link between nature, agriculture and water was emphasized (Spatial Plan for the Rural Area 2001). In 2000, the role of nature in water management was emphasized (Policy Note Nature for People 2000). These growing links and the diversity of instruments being developed in different sectors is now leading to a fourth phase where integration is key. Here, there is a tendency to move towards simplifying pluralistic and competitive procedures into a comprehensive planning process in an effort to provide forums where multiple objectives and concerns can be integrated into decisions. For example, the Water Act (2009) replaces past laws and integrates a number of issues into one document. Similarly the Spatial Planning Law (2008) also aims to integrate different goals and policy instruments. The Rural Areas Development Policy and Law (WILG 2007) also attempts to integrate different goals for the rural areas and their financing instruments (ILG 2007) and land consolidation instruments (WAG).

An interesting link between the Water Act and the Spatial Planning Act has been established. The spatial aspects of national and provincial water plans are also considered as spatial structural visions of the Spatial Planning Act. This link between the two Acts opens the possibilities to implement the water plans with spatial planning instruments.

5.8.2 *Shift from a No-Priority to a No-Regrets to a Priority Issue*

The move from sectoral through national to integrated also reflects changing perspectives on the adaptation issue. As mentioned earlier, although a substantial part of the Netherlands lies below the sea level, it was not seen as vulnerable; there were high expectations from the global emission reduction strategy which would reduce the need for adaptation. Besides, it was not clear how robust policies could be made to deal with uncertain impacts. This led to a general under-emphasis being given to the adaptation process. However, by 1997, it became increasingly evident (a) that a global emission reduction strategy would at best be a very modest one, (b) that physical impacts of the 1995 floods, the 2003 research on the ten weak spots in the coastal defence system and the 2006 research results that 23 % of the storm surge barriers did not meet national standards, and (c) that although the Netherlands is not seen as vulnerable, it was soon felt that it will be increasingly exposed to the impacts of climate change as the newer reports of the IPCC continued to predict that climate change could have very serious impacts globally.

By the end of the decade, people were referring to adaptation strategy as a 'no-regrets' strategy (Commissie Waterbeheer 2000) and this acquired a high political and scientific allure when it was repeated in the WRR document of 2006 (WRR 2006). However, it was soon realized that adaptation measures would have to go far beyond no-regrets policy to also include important and urgent measures, measures that have a contribution to make to adaptation as well; and more complex integrative measures (Route Planner 2007). Furthermore, the scenarios developed by the Royal Netherlands Meteorological Institute which downscaled global impacts to national and local level provided a framework within which climate change adaptation could take place. In the meanwhile, a philosophy was developing on how to cope with uncertainty; uncertainty was seen as calling for different institutional skills (MNP Conference 2006) and approaches (VROM Council 2007). Adaptation is becoming a national priority, although the current political situation in the Netherlands shows that climate change may be slipping fast from the agenda in the face of the transatlantic recession.

5.8.3 Shift from a Technological to Post-Modern Concept

While technological and rationalistic rule-oriented approaches have been dominant in the past, there is an increasing tendency to innovatively design new principles of management and new instruments to help society cope with a range of new problems and challenges and to meet different goals.

The concept of dynamic protection adopted with respect to the coasts tries to combine the need for national physical security with the need to recognize that coasts are by their very nature dynamic—they move with the ebb and flow of the seas (VenW 2000). While dynamism is essentially applied only to the dune defence system, there are limits to the dynamism and dune replenishment and nourishment are key tools here.

The concept of Room for the River is another such concept that provides rivers the space to grow and contract with seasonal variations (VenW, VROM and LNV 2006). Although this sounds like a simple concept, it has major implications for spatial policy, has to be implemented all along the river banks in different provinces and municipalities—and calls for a series of context relevant institutional measures for effective implementation.

A third post-modern concept is that of the National Ecological Network (LNV 1990) and the related concepts of Robust Ecological Corridors and Climate Buffers (LNV 2000). The National Ecological Network allows for linking up ecological zones all over the country by 2018, while the 13 planned Robust Ecological Corridors allow for larger links between the larger ecological zones. Climate Buffers are expected to enhance the ability of the land to cope with climate change.

The state has in the past mostly focused on rational and efficient measures, but in recent years there appears to be a trend shift in the direction of post-modern concepts: redundancy, flexibility and the recognition of multiple rationalities. All of the three above-mentioned concepts have implications for how people live; they call for recognition that people live with nature and must make space for nature. While this is a theoretically attractive notion, actually implementing it might imply the expropriation of land and will require not only very good quality persuasion, but also remarkable access to resources and a flexible spatial planning system. Possibly some of the difficulties in implementing these have led to a partial return to hard measures in the 2007 National Adaptation Strategy (VROM 2007a).

5.8.4 Shift from a Top-Down Consensus Policy Approach Through Decentralization to a New Balance?

A fourth interesting tendency in the policy process is the move from top-down consensus policy to a more bottom-up approach of engagement of civil society and sub-national authorities. The climate mitigation targets of 1988 and 1989 were not achieved, possibly because of a lack of general support for these targets. The need to engage the population and ensure that policies have public support is seen as critical in the Netherlands in this phase. At the same time, there is an increasing neo-liberal interpretation of the role of state as minimal, of passing on responsibilities to other social actors. The norm of 'individual responsibility' is increasingly seen as a dominant value in Dutch society (Think Ahead Campaign 2005; Agenda for the Future 2006). The motto of "Decentralize where possible, centralize where necessary" and stakeholder participation appear to have become buzz words in the policy discussions. For example, the role of farmers and rural dwellers as entrepreneurs in addressing their own problems is emphasized in a number of documents (SER 2005; LNV 2005; LNV 1995) and subsidies are provided to help them use their own initiative (Rural Development Program 2, LNV et al. 2008; WILG 2007); while there is a Disaster Compensation Law (1998) that aims to compensate individuals in the event of an extreme event, newer initiatives try to ensure that farmers take out their own insurance for such events.

However, the trend towards decentralization of responsibilities to the lower levels is subject to so many strategic visions at the central and provincial level that the question of how to balance and divide responsibilities between different levels is critical. This is especially so in the spatial planning sector. In the nature sector there is a complete clash between the top-down nature of the targets set and the actual physical impossibility to prevent species movements as climate changes, even though the physical boundaries of nature reserves remain static. Furthermore, while it is important to have public support for policies, shifting responsibilities to citizens is an interesting but not always practical suggestion. Although the WRR (2006) claims that individuals will feel more engaged to participate in adaptation measures than in mitigation measures, because it concerns them directly, allocating responsibility to home owners on ground water and storm water is not always practical and the line between state and resident responsibility is difficult to draw (Bergsma et al. 2009). After all, most home owners in cities have absolutely no interest in or knowledge of ground water levels under their houses.

5.8.5 *Different Paradigms in Different Sectors*

A fifth interesting outcome of analyzing the different adaptation policies in the different sectors is that there are vastly different paradigms evident in the different fields. These different paradigms have occurred partly because of the different historical evolutionary processes that these policy fields have undergone. For example, water governance in the Netherlands was traditionally dominated by a Delft University of Technology-oriented approach. This paradigm has been changing under the influence of the trends described above, but still a proper calculation will always be the basis of Dutch water policy. Also, water governance has historically involved the decentralized water management authorities within a common vision of protection from floods and multi-level governance. Cooperation is thus institutionalized. Furthermore, the Dutch have been able to master their environment to such an extent with engineering measures that they are now able to discuss the possibility of social and ecological engineering to provide more space to nature and be more fluid in their protection standards.

The Nature regulations of the European Union appear to be more rigid and static, more top-down and unable to cope with the notion of a fluid natural system where changes in global, local and micro climate can have influences. This is so even though the EU directives are based on bottom-up information; possibly the process of making EU directives is too slow. Multi-level cooperation is far from institutionalized and the notion of space for nature carries a rigid framework of maps with boundaries. The paradigm at work in the nature sector is that the past contains the ideal to which we must strive in the future (in the Netherlands: the nature we had in 1850). Such a paradigm obviously conflicts with the changes climate change may bring. In contrast, the agricultural regulations have focused on providing a framework within which innovation and the market can function, allowing for greater autonomy to the farmer and policy intervention only when a social and/or ecological problem was signalled. The farmers have traditionally coped with climate variability through history. The paradigm in the agricultural sector appears to provide farmers with information inputs and financial incentives and to help them become more adaptive. The spatial planning process is much more densely regulated and has multiple tools and instruments at its disposal—but these can also be experienced as highly constricting when it comes to adaptation to climate change. The paradigm in Dutch spatial planning is to accommodate urbanization processes. Because all the good building locations are already taken, this results in developing unsuitable and marginalized locations, also from the climate change viewpoint. Attempts at making this sector less rigid are evident in the new Spatial Planning Act (2008) and the tools of project decisions. Changing this paradigm will not be easy.

5.8.6 From Adaptation Strategies to Adaptive Capacity

An examination of the sectoral adaptation strategies leads to the following impression. For more or less certain impacts (the sea is expected to rise), there are hard measures being taken like the strengthening of the coastal defence system. However, for the more or less uncertain impacts, the focus is on creating procedures and tools (e.g. the water test), general public awareness and engagement both at community level and sub-national level in order to mobilize people to come up with their own autonomous adaptive solutions. This is clearly the case in the agricultural sector and to some extent in the spatial planning and water sectors (especially with respect to precipitation) but less so in the nature sector. However, increasingly the nature organizations are arguing in favour of dynamic nature management. A critical element of adaptive capacity is trying to ensure that institutional complexity and especially the interplay within and between formal and informal institutions is taken into account. While the interplay between formal institutions is being incrementally revised in the last decade especially in response to the understanding that institutional complexity is perhaps the most complex challenge facing adaptation strategies (Route Planner 2007), the interplay between formal and informal institutions seems crucially important and is at the same time unexplored. More research is needed in the role of informal institutions in adaptation to climate change.

5.9 Conclusions

This Chapter has tried to examine the transition in adaptation policy in the Netherlands over the last 20 years in general and with respect to four sectors. The Netherlands has a long history of coping with water problems. This has led to an accumulation of expertise in this area. There is a saying that God made the world and the Dutch made the Netherlands. With engineering marvels such as the 32 km Afsluitdijk that transformed a North Sea inlet into a freshwater lake, the Neeltje Jans and in more recent years the Maeslant Barrier, floating houses along the Meuse, coupled with a tradition of community management and funding of water works through water management authorities that can be traced back to the Middle Ages, show that the Dutch have a high pedigree when it comes to coping with the vagaries of nature. As a result, nature becomes 99 % managed and the value of the remaining parts becomes contested. A famous Dutch poem says: 'And what remains of nature in this land, a forest that has the size of a hand' (Bloem 1965). Luckily, the paradigm change in the water sector also promises more room for nature.

The above chapter shows that there are six major trends in the development of adaptation policy in the Netherlands. On the one hand, this accumulation of expertise creates confidence in the ability of the Netherlands to be able to rise to any challenge; and on the other hand, one can question whether the Dutch have become over confident. Clearly, climate change is a very complex problem, and the solutions chosen are also complex and pluralistic. The complexity of the entire process raises the hope that society as a whole can be empowered to deal with climate change impacts. However, the fear is that adaptive efforts may be dissipated between different actors and individuals and that the collective action may not amount to more than a sum of the individual acts. The VROM Council warned of this and called for the establishment of a watchdog to monitor the entire process (VROM Council 2007).

Acknowledgements This chapter was partially funded by past research entitled: 'IC12: Institutions for Adaptation: The Capacity and Ability of the Dutch Institutional Framework to Adapt to Climate Change', which is funded by the Netherlands BSIK-Program Climate *changes* Spatial Planning (CcSP).[8]

Biography Joyeeta Gupta is Professor of Environment and Development in the Global South at the Amsterdam Institute for Social Science Research of the University of Amsterdam and at UNESCO-IHE Institute for Water Education in Delft.
Judith E.M. Klostermann is Senior Researcher for Water, Climate and Policy at Alterra of Wageningen University and Research Centre.
Emmy Bergsma is a Ph.D. student in political sciences at the University of Amsterdam, Faculty of Social and Behavioural Sciences, Amsterdam Institute for Social Science Research, department of Political Sciences.
Pieter Jong is a Senior Researcher at the Faculty of Technology, Policy and Management, Delft University of Technology.

References

Act on Municipal Water Tasks (2008) Wet Gemeentelijke Watertaken. Bull Acts Decrees 2007 (276):277
Agricultural Land (Transactions) Act (1981) Wet Agrarisch Grondbeheer (WAG). Bull Acts Decrees 248:374, last revisions: Bull. 2010, 155, Bull. 2008, 180
AZ (2007) Samen werken—samen leven, Policy Program Balkenende IV government, 2007–2011. English text can be found at: http://www.government.nl/Government/Policy_statement
Bergsma EJ, Gupta J, Jong P (2009) IC12: institutions for climate change—case study on individual responsibility in adaptive capacity. Working Document 7, nr. W09/10. Institute for Environmental Studies, Amsterdam
Bloem CJ (1965) Verzamelde Gedichten. Athenaeum—Polak & Van Gennep, Amsterdam
Bodansky D (1993) The United Nations framework convention on climate change: a commentary. Yale J Int Law 18:451–588

[8] http://www.climatechangesspatialplanning.nl

Building Decree (2003) Bulletin of Acts and Decrees, 2002, No. 518; last revision: Bull. 2009, No. 400

BZK (2007) Strategie Nationale Veiligheid. The Hague: ministry of the interior and kingdom relations. Online available at: http://www.minbzk.nl/actueel?ActItmIdt=105235

Commissie Waterbeheer 21e eeuw (2000) Water beleid voor de 21e eeuw—Geef water de ruimte en de aandacht die het verdient. Den Haag: Commissie Waterbeheer 21e eeuw

Declaration of the World Climate Conference (1979) WMO proceedings of the world climate conference: a conference of experts on climate and mankind. Geneva, 12–32 February 1979. WMO - No. 537. ISBN 92-63-10537-5

Decree B (2008) Revision on energy efficiency of buildings. Bull Acts Decrees 2008(325):373

Delta Act (2011) Deltawet waterveiligheid en zoetwatervoorziening (Stb. 2011, 604)

Delta Commission (2008) Samen werken met water—Een land dat leeft, bouwt aan zijn toekomst. Rotterdam: Deltacommissie. Available online at: http://www.deltacommissie.com/doc/2008-09-03%20Advies%20Deltacommissie.pdf

Dienst Regelingen (2008) Brochure Mestbeleid 2008–2009. The Hague: ministry of agriculture, nature and food quality. Accessible online at: http://www.hetlnvloket.nl/portal/page?_pageid=122,1780509&_dad=portal&_schema=PORTAL&p_file_id=30906

Dienst Regelingen, IPO, SNL (2009) Brochure Subsidiestelsel Natuur- en Landschapsbeheer Onderdeel agrarisch natuur- en landschapsbeheer. The Hague: ministry of agriculture, nature and food quality. Available online at: http://www.hetlnvloket.nl/portal/page?_pageid=122,1780509&_dad=portal&_schema=PORTAL&p_file_id=42365

Disasters Compensation Law (1998) Law on compensation of damage In case of disasters and accidents (Wet Tegemoetkoming schade bij rampen en zware ongevallen). Bull Acts Decrees No. 325; last revisions: Bull. 2010, No. 146 and Bull. 2009, No. 617

Duin MJ, Mesu JJ (1995) Wie zal dat betalen? Overheid, schadevergoeding en aansprakelijkheid bij wateroverlast. Bestuurskunde 4(8):361–369

EC (2009) The common agricultural policy explained. Brussels: European Commission, Directorate-General for Agriculture and Rural Development. Available online at: http://ec.europa.eu/agriculture/publi/capexplained/cap_en.pdf

Environmental Licensing Bill (2010) Bull Acts Decrees 2008, No. 496; last revisions: Bull. 2013, 144

Environmental Management Act of 1993 (1993) Bull Acts Decrees, No. 693; 1994, No. 80; last revisions: Bull. 2010, No. 187 and Government Gazette 2010, No. 20455

European Council (1979) Council Directive 79/409/EEC on the conservation of wild birds. Available online at: http://europa.eu/legislation_summaries/environment/nature_and_biodiversity/l28046_en.htm

European Council (1992) Council Directive 92/43/EEC on the conservation of natural habitats and of wild fauna and flora (Habitats Directive). Available online at: http://eur-lex.europa.eu/LexUriServ/LexUriServ.do?uri=CELEX:31992L0043:EN:NOT

European Council (2006) Council Decision (2006/144/EC) on Community strategic guidelines for rural development (programming period 2007 to 2013). Available online at: http://eur-lex.europa.eu/LexUriServ/LexUriServ.do?uri=OJ:L:2006:055:0020:0029:EN:PDF

Flora and Fauna Act (1998) Flora en Faunawet. Bull Acts Decrees, No. 402, 1999, No. 264; last revisions: Bull. 2008, 4 and Government Gazette 2010, No. 20455

Hague Declaration on Climate Change (1989) Adopted by 22 Heads of State, The Netherlands

IPCC (1990) Climate change: the IPCC impacts assessment. UNEP and WMO, Australian Government Publishing Service, Canberra

IRMA (2000) Risicofinanciering van oogstschade door extreme weersomstandigheden. Wageningen Universiteit & Research Centrum, Wageningen

Klostermann J, Gupta J, Jong P, Bergsma E (2009) Institutions for climate adaptation: An inventory of Institutions in the Netherlands that are relevant for climate change—working document 4. Institute for Environmental Studies, Amsterdam

Klostermann J, Gupta J, Jong P, Bergsma E (2010) Institutions for climate adaptation: Applying the adaptive capacity wheel on the background document of the content analysis—working document 5. Institute for Environmental Studies, Amsterdam

Law on Safety Regions, (2010) Wet veiligheidsregio's. Bull Acts Decrees 2010(145):146

Lemstra W (2005) Motie-Lemstra (CDA) c.s. inzake lange termijninvesteringsstrate-gieën op het gebied van ruimtelijk-economisch beleid in Nederland (EK, XXI-C). The Hague: The Senate

LNV (1990) Natuurbeleidsplan. Governmental decision. House of representatives, year 1989–1990, 21149, no. 2–3

LNV (1995) Structuurschema Groene Ruimte: het landelijk gebied de moeite waard. Deel 4: Planologische Kernbeslissing. The Hague: Ministry of Agriculture, Nature and Food Quality

LNV (2000) Natuur voor mensen, mensen voor natuur. Nota Natuur, Bos en Landschap in de 21e eeuw. The Hague: Ministry of Agriculture, Nature and Food Quality. Available online at: http://www.minlnv.nl/portal/page?_pageid=116,1640360&_dad=portal&_schema=PORTAL&p_file_id=41642

LNV (2004) Agenda voor een Vitaal Platteland—Visie: Inspelen op veranderingen. The Hague: Ministry of Agriculture, Nature and Food Quality. Available online at: http://www.minlnv.nl/portal/page?_pageid=116,1640321&_dad=portal&_schema=PORTAL&p_file_id=13747

LNV (2005) Kiezen voor Landbouw—Een visie op de toekomst van de Nederlandse agrarische sector. The Hague: Ministry of Agriculture, Nature and Food Quality. Available online at: http://www.minlnv.nl/portal/page?_pageid=116,1640321&_dad=portal&_schema=PORTAL&p_file_id=13085

LNV (2007a) Meerjarenprogramma Vitaal Platteland 2007–2013. The Hague: Ministry of Agriculture, Nature and Food Quality. Available online at: http://www.minlnv.nl/portal/page?_pageid=116,1640321&_dad=portal&_schema=PORTAL&p_file_id=14197

LNV (2007b) Spelregels EHS. The Hague: Ministry of Agriculture, Nature and Food Quality. Available online at: http://www.minlnv.nl/portal/page?_pageid=116,1640321&_dad=portal&_schema=PORTAL&p_file_id=20630

LNV (2009) Brochure bedrijfstoeslag 2009. The Hague: Ministry of Agriculture, Nature and Food Quality. Online available at: http://www.hetlnvloket.nl/portal/page?_pageid=122,1780509&_dad=portal&_schema=PORTAL&p_file_id=18868

LNV, VROM, V&W and the Provinces (2008) Plattelandsontwikkelingssprogramma 2007–2013 voor Nederland (POP2)—Version of 6 June 2008, The Hague. Available online at: http://www.regiebureau-pop.nl/nl/info/4/

LTO-Nederland (1999) Weer zelfverzekerd? Verkennende studie naar de mogelijkheden voor risico-afdekking tegen schade door extreme weersomstandigheden. Dutch Federation of Agriculture and Horticulture, The Hague

Mathijssen J, Petersen A, Besseling P, Rahman A, Don H (Eds.) (2007) Omgaan met Onzekerheid in Beleid; Report on the Conference Dealing with Uncertainty in Policy-making, held on 16 and 17 May 2006 in The Hague. Bilthoven: The Netherlands Environmental Assessment Agency. Available online at: http://www.rivm.nl/bibliotheek/rapporten/550032003.pdf

Nature Conservation Law (1967) Natuurbeschermingswet. Bull Acts Decrees 572

Nature Conservation Law (1998) Natuurbeschermingswet. Bull Acts Decrees, No. 403; 1999, No. 15; last revisions: Bull. 2010, No. 135, 142, Government Gazette 2010, 20455

Noordwijk Conference on Climate Change (1989) The Noordwijk Ministerial Conference held in November 1989 in Noordwijk, The Netherlands

Pielke R, Prins G Jr, Rayner S, Sarewitz D (2007) Lifting the taboo on adaptation. Nature 445: 597–598

Rijkswaterstaat (2007) Brochure Veilig achter duin en dijk—Hoogwaterbeschermingsprogramma. Lelystad: Programmabureau HWBP. Available online at: http://www.helpdeskwater.nl/waterkeren/hwbp

RIZA (2003) Handreiking Watertoets 2. The Hague/Lelystad: RIZA. Available online at: http://www.av-consulting.nl/activiteiten/watertoets/handreiking2.pdf

Rooijers FJ, Keizer I de, Slingerland S, Faber J, Wit RCN, Verbeek J, van Dor-land R, van Ulden AP, Hutjes RWA, Kabat P, van Ierland EC (2004) Klimaatverandering, klimaatbeleid. Inzicht in keuzes voor de Tweede Kamer. Main report and background report. Delft: CE report no. 3.748.1
SER (2005) Advies: Kansen voor het platteland. The Hague: SER publication number 12. Available online at: http://www.ser.nl/nl/publicaties/adviezen/2000-2007/2005/b24110.aspx
Spatial Planning Act (Wet op de Ruimtelijke Ordening) (1965) Bull Acts Decrees 1962, 286
Spatial Planning Act (2008) Wet ruimtelijke ordening. Bull Acts Decrees, 2006, No. 566, 2008, No. 227; last revisions: Bull. 2010, No. 135 and Government Gazette 2010, No. 20455
United Nations (1992) United Nations Framework Convention On Climate Change. Available online at: http://unfccc.int/resource/docs/convkp/conveng.pdf
United Nations (1997) Kyoto protocol to the United Nations framework convention on climate change. Kyoto
Van Drunen M (red.) (2006) Naar een klimaatbestendig Nederland—Samenvatting Routeplanner. Joint publication of the national research programs Climate change Spatial Planning, Living with Water, Habiforum and CURNET. Available online at: http://www.klimaatonderzoeknederland.nl/gfx_content/documents/ark/Samenvatting%20routeplanner.pdf
VenW (1990) Kustverdediging na 1990. First coastal policy. The Hague: ministry of transport, public works and water management
VenW (2000) Derde Kustnota: tradities, trends en Toekomst. Third coastal policy. The Hague: Ministry of Transport, Public Works and Water Management
VenW (2003) Procesplan Zwakke Schakels In De Nederlandse Kust. The Hague: Ministry of Transport, Public Works and Water Management. Available online at: http://www.verkeerenwaterstaat.nl/Images/Procesplan%20Zwakke%20Schakels_tcm195-101251.pdf
VenW (2007) Randstad urgent. The Hague: ministry of transport, public works and water management. Available online at: www.vrom.nl/get.asp?file=docs/publicaties/w947.pdf.
VenW, VROM and LNV (2006) Brochure PKB Ruimte voor de Rivier - Investeren in veiligheid en vitaliteit van het rivierengebied. The Hague: Ministry of Transport, Public Works. Available online at: http://www.ruimtevoorderivier.nl/files/Files/brochures/EMAB%20PBK%20Nederlands.pdf
VenW, VROM and LNV (2009) Nationaal waterplan 2009–2015. The Hague: ministry of transport, public works. Available online at: http://www.verkeerenwaterstaat.nl/Images/NWP_tcm195-270588.pdf
Vos CC, Opdam PFM, Nabuurs GJMM, Bugter RJF, Epe MJ (2007) Klimaatverandering en ruimtelijke adaptatie natuur: wat we (niet) weten. Brinkman Climate Change, The Hague
VROM (1989) Nationaal Milieubeleidsplan; Kiezen of verliezen. House of representatives, year 1988–1989, 21 137, no. 1–2. The Hague
VROM (1991) Nota Klimaatverandering. Ministry of Housing, Spatial Planning and the Environment, The Hague
VROM (1996) Vervolgnota Klimaatverandering. Ministry of Housing, Spatial Planning and the Environment, The Hague
VROM (2001) Nationaal Milieubeleidsplan 4. The Hague: ministry of housing, spatial planning and the environment. Available online at: http://www.vrom.nl/pagina.html?id=2706&sp=2&dn=1076
VenW and VROM (2006) Beleidslijn grote rivieren. The Hague: ministry of transport, public works. Available online at: http://www.verkeerenwaterstaat.nl/Images/Beleidslijn%20grote%20rivieren%20handreiking_tcm195-162030.pdf
VROM (2006a) Nationaal programma Adaptatie ruimte en klimaat. The Hague: ministry of housing, spatial planning and the environment. Available online at: http://www.vrom.nl/pagina.html?id=2706&sp=2&dn=6241
VROM (2006b) Toekomstagenda Milieu: schoon, slim, sterk. The Hague: ministry of housing, spatial planning and the environment. Available online at: http://www.vrom.nl/pagina.html?id=2706&sp=2&dn=6139

VROM (2007a) Maak ruimte voor klimaat! Beleidsnotitie. The Hague: ministry of housing, spatial planning and the environment. Available online at: http://www.vrom.nl/pagina.html?id=2706&sp=2&dn=7503

VROM (2007b) Ontwikkeling van de Wadden voor natuur en mens - Deel 4 van de planologische kernbeslissing Derde Nota Waddenzee. Text after acceptance of the parliament. The Hague: Ministry of Housing, Spatial Planning and the Environment. Available online at: http://www.vrom.nl/pagina.html?id=2706&sp=2&dn=7040

VROM, LNV, VenW and EZ (2006) Nota Ruimte—Ruimte voor ontwikkeling. The Hague: Ministry of Housing, Spatial Planning and the Environment. Available online at: http://www.vrom.nl/notaruimteonline/

National Accord on Water (2003) Nationaal Bestuursakkoord Water. The Hague: Het Rijk, IPO, VNG and de Unie van Waterschappen. Available online at: http://www.helpdeskwater.nl/wetgeving-beleid/nationaal/?ActItmIdt=1280

VROM Council (2007) De hype voorbij—Klimaatverandering als structureel ruimtelijk vraagstuk. The Hague, VROM council, Advice 060. Available online at: http://www.vromraad.nl/Download/37238_DeHypeVoorbij2.pdf

Water Act (2009) Waterwet. Bull Acts Decrees, No. 107, 549; last revisions: Bull. 2010, 135, Government Gazette 2010, 20455

Water Authorities Modernization Act (2007) Wijziging van de waterschapswet. Bull Acts Decrees, 2007, No. 208, 581; last revisions: Bull. 2010, 199

WILG (2007) Wet Inrichting Landelijk Gebied. Bulletin of Acts and Decrees, 2006, No. 666, 677; last revisions: Bull. 2009, 18

WRR (2006) Klimaatstrategie - tussen ambitie en realisme. The Hague: WRR report no. 74. Available online at: http://www.wrr.nl/content.jsp?objectid=3619

Chapter 6
Liability for Damage Caused by Climate Change: A Way to Internalize the Costs of Adaptation?

Ingmar Piroch

6.1 Introduction

The effects of climate change are a growing risk factor for companies. Apart from direct (e.g. extreme whether) and indirect consequences (e.g. emission ceiling, damage to a company's reputation) climate change is likely to increasingly become the subject of court proceedings, as demonstrated internationally (see the overview by Verheyen and Lührs 2009, pp. 73–79 and 129–138; see also Maag 2009, pp. 185–213; Thorpe 2008, pp. 79–105; Grossman 2003, pp. 1–61; for a summary of pending international court proceeding see: http://www.climatelaw.org). Companies in the U.S. already consider the risk of ligitation ranking second on the list of climate relevant risks (KPMG 2008, p. 37; Chatzinerantzis and Herz 2010, p. 594 speak clearly of a 'juridification of climate change').

In Germany the risk of liability for climate change-related damage is rarely dealt with. However, litigation dealing with climate change is going to rise in the foreseeable future (Chatzinerantzis and Herz 2010, p. 594). The German economy, which strongly focuses on export, as well as energy providers who are increasingly acting internationally, should not underestimate the influence of court proceedings on the standing and reputation of companies (as in Hoffmann 2007, pp. 45–46), may it be at the stock market, in their ability to obtain credit or due diligence. This article attempts to identify possible risks of liability under German Law (Sect. 6.2)[1]

[1] Claims for compensation against the government (see BGH, decision of Dez. 10th 1987—III ZR 220/86—forest damage due to air pollution) and claims made by investors or shareholders against a company are not considered, for the latter see e.g. Verheyen et al. (2008); for responsibility for ecological damage based on public law see Knopp and Wiegleb (2009).

I. Piroch (✉)
Centre for Law and Administration (ZfRV), Department of Constitutional, Administrative and Environmental Law, Brandenburg University of Technology (BTU) Cottbus-Senftenberg, Erich-Weinert-Str. 1, 03046 Cottbus, Germany
e-mail: piroch@tu-cottbus.de

E. Albrecht et al. (eds.), *Implementing Adaptation Strategies by Legal, Economic and Planning Instruments on Climate Change*, Environmental Protection in the European Union 4, DOI 10.1007/978-3-540-77614-7_6,

as well as take a look at the future regarding internalization of the adaptation costs for climate change (Sect. 6.3).

6.2 Liability for Individual Damage Caused by Climate Change

6.2.1 Causes of Action: Overview and Limits

A civil action for damages caused by climate change may be based on contract, property or tort law.

For the following reasons the first two will not be considered further: A claim in contract, such as by an end-customer under a utility supply contract with an energy supplier requires a breach of contract. However, there is no general duty on the supplier to provide the services without producing greenhouse gases under German Law (Chatzinerantzis and Herz 2010, p. 595). A cause of action could only arise where the company has violated its own rules of corporate social responsibility (CSR) and these voluntary commitments are part of the contract, i.e. the end-customer has only entered into the contract because of the companies' sustainability strategy, which explicitly includes avoiding the emission of greenhouse gases.

A claim under property law could be based on Art. 906 para 2 BGB (German Civil Code), dealing with causes of action in nuisance. Nuisance provides causes of action to claim compensation or to stop an interference when the enjoyment of a property by its owner is impaired as a result of interference eminating from neighbouring land. The property which is the source of interference need not be immediately adjacent, but must be geographically proximate to the affected property (BGH [German Federal Court of Justice], decision of Jun. 18th 1958—V ZR 49/57). However, in spite of a possible geographical proximity to the source of an interference, damage caused by climate change is likely to be caused by geographically remote facilities, as the emissions are likely to have an effect over longer distances. As a result, claims based on Art. 906 para 2 BGB would not arise due to the lack of a proximity.[2]

Therefore claims can only arise out of the law of torts (Delict), in particular out of the fault-based liability of Art. 823 BGB and strict liability based on the Environmental Liability Act (UmweltHG), which will be examined more closely in the following.

[2] Similar to Art. 906 para 2 BGB a claim for compensation based on Art. 14 para 2 BImSchG [Federal Immission Control Act] also fails due to the missing neighbouring relation between damaging and damaged party, see BGH, decision of Dez. 10th 1987—III ZR 220/86; Jarass (2010), Art. 14 margin number 23.

6.2.2 Law of Torts (Delict)

The law of torts is a means of compensating damage arising out of acts which interfere with legally protected rights (von Bar 1998, margin number 1). In case of climate change-related litigation, legally protected rights that can be impaired would be esp. property and health. According to Art. 823 para 1 BGB the infringement of these rights must be unlawful and negligent (fault-based liability). In relation to unlawfulness and fault, the central question is whether compliance with public law[3] means that the emitter of greenhouse gases is acting lawful, or at least not negligently ('permit defence'). There is no general restriction of liability simply by complying with public regulations (BGH, decision of Sep. 18th 1984—VI ZR 223/82—Kupolofen case). Complying with the benchmarks of the Emmission Control Legislation is merely an indication that the polluter has done everything reasonable to avoid infringement of protected rights. But these public regulations are only minimum standards. Obligations under civil law are based on the judge-made law of the general duty to take care (*Verkehrssicherungspflichten*), defining the limitations of liability for potentially dangerous behavior (von Bar 1998, margin number 104). These duties may exceed public regulations in individual cases. Therefore, unlawful and negligent acts are possible in civil law in spite of compliance with the benchmarks of the Emission Control Legislation (see Chatzinerantzis and Herz 2010, p. 595).

Liability based on the Environmental Liability Act differs from the requirements of Art. 823 BGB, as liability may arise regardless of lawfulness or unlawfulness of the act and does not require fault where personal injury or damage to property occurs due to atmospheric environmental impact. But this is only the case where the damage was caused by a facility listed in Annex 1 UmweltHG,[4] because strict liability in accordance with this act is based on the assumption that the operation of dangerous facilities justifies liability irrespective of fault.

In both cases of liability only *absolute* rights are protected, but the individual's assets as such as well as the community assets such as the climate are not directly protected (Chatzinerantzis and Herz 2010, p. 596). However, precondition for liability is the causation, or rather attributability of the damage to a certain act. German Law distinguishes here between two forms of causality, the causal link between an emission and the infringement of protected rights on the one hand (*haftungsbegründende Kausalität*) and the causality of an act of infringement for the occurrence of a damage on the other hand (*haftungsausfüllende Kausalität*). The latter can be compared to the remoteness of the damage concept in common law. As the central questions of causation for the liability of damage caused by climate change relate to the proximate facts (of a case), this will be examined closely below.

---

[3] As, for example, the emission levels of the Technical Instructions on Air Quality Control (TA Luft) or the BAT (best available techniques) reference documents according to the Directive 2010/75/EU (IED).

[4] This includes e.g. large combustion plants.

6.2.2.1 Causation

Within causation it has to be examined whether a specific act is the cause of infringement of the legally protected right. Applied to the liability for damage caused by climate change, the emission of greenhouse gases has to be the cause of climate change and its consequences, which, in turn, must cause the infringement of the legally protected right (such as property or health). For damage caused via an environmental path (e.g. air) this can be further distinguished between influential causation and causation of violation. As the emission of greenhouse gases is a classic environmental influence or rather environmental change (esp. as defined in Art. 3 para 1 UmweltHG), where it is sufficient that gases disipate in the air (Salje and Peter 2005, Art. 1, 3 margin numbers 84–86 and 113–116), an influential causation can be easily reasoned for.

6.2.2.2 Causation of Violation

Far more difficult is the question of whether the environmental change is the cause of an individual violation of a legally protected right (causation of violation). The liability for damage caused by climate change that is examined here shows a complex (and controversial) chain of causation between act and result, the latter being climate change. Most natural scientists consider greenhouse gases to be the cause for global warming. The causation of climate change by humans, which is necessary for individual liability, is not only acknowledged in politics and natural sciences, but has also been confirmed by some courts.[5] In this respect, we shall assume that it is possible to prove causation between climate change and the emission of greenhouse gases. But this alone is not enough. We would have to prove that a certain damaging environmental incident is a specific consequence of climate change.

This could be difficult in the case of environmental phenomena that only change in their frequency or intensity, such as thunderstorms, torrential rain or hail. As they have always occured, we can hardly refute the supposition that a thunderstorm is to be attributed to natural causes. Even though natural phenomena increase in intensity due to climate change, a causal connection is difficult to prove, as the increase can hardly be put into precise numbers which are the condition for causation. For environmental changes that would not have occurred as they have without climate change the decision would have to be different. Contrary to quantitative or qualitative changes they are 'new' phenomena, which are in most cases caused by

[5] The scientific explanation that the rise of temperatures worldwide can be attributed to the emission of greenhouse gases has already been legally accepted by the New Zealand Environment Court in his decision of September 2002, Environmental Defence Society v. Auckland Regional Council and Contact Energy Limited. With the decision of April 2nd 2007—Massachusetts et al. v. EPA, the U.S. Supreme Court accepts that climate change is man-made as a scientific fact, cf. Scouteris (2008), p. 144 f.

climate change, like sea level rise, desertification, rise of the snow line and melting of glaciers (Chatzinerantzis and Herz 2010, pp. 596–597).

If a causal connection between the emission of greenhouse gases and climate change, as well as between climate change and 'new' environmental phenomena can be established,[6] the next step would be to examine whether this chain of causation can be attributed to only one emitter out of several.

6.2.2.3 Cumulative Causation for Summation and Remote Damage

According to the commonly used equivalence theory (*conditio sine qua non* doctrine), every incident that cannot be dismissed without the result in its precise form failing to exist is causal (von Bar 2000, margin number 413). But concerning damage caused by climate change we deal not only with one emitter, but with an unlimited number of emitters. This situation, which emerges quite frequently within cases of environmental liability, is discussed under the umbrella term of summarized immissions or cumulative damage respectively (for individual case groups see Spindler, in: Feldhaus 2004, Art. 14 margin number 127). Here the emissions produced by the individual emitter are not enough to cause the complete damage in its actual form. But if individual emissions are interacting in such a way that, when one source of emission is omitted, a necessary condition for a precise impairment is omitted as well, we find a progressive increase of damage and cumulative causation of each individual emission (Spindler 2008, pp. 321–322). The aim is to ensure that liability cannot be avoided by pointing to other emitters of greenhouse gases. Condition to prove cumulative causation is that a certain amount of the violation of legally protected rights can be definitely attributed to the isolated contribution of one emitter.

6.2.2.4 Problems in Determining an Individual Contribution

If, for instance, emissions of two different facilities cause damage to the paintwork of a vehicle parked in their immediate neighbourhood, we can establish a linear causal chain between the environmental influences and the damage by means of the equivalence theory (BGH, decision of Sep. 18th 1984—VI ZR 223/82—Kupolofen case). This also applies if the emission of just one of the facilities would not have caused the damage and it only occurred because of the interaction with the emissions of the second facility (cumulative causation). Such a liability presupposes that a certain part of the damage can be ascribed to the isolated contribution of one certain emitter (Spindler 2008, p. 324).

[6] Some inconsistencies in the work of the IPCC, which have become public, show the uncertainty of climate research and illustrate that it is vulnerable and difficult to proof, see Evers et al. (2010), pp. 140–149.

For damage caused by climate change a causal connection between the emission of greenhouse gases and the particular environmental phenomenon can be found in a general way, but an individualization necessary for liability, in the sense that a linear causal chain of a single source of emission leading to the occurrence of a damage can be established, is difficult to prove. Even if it could be established, the precise portion of causation can only be estimated. And due to the large number of emitters, whose emissions intermingle and cause damages as a whole or in conjunction with other emissions which can occur worldwide, it is difficult to attribute the damage to one emitter. This can only be done in general and by predication of probability, but not with certainty.

Additionally, we have to consider time as a factor of liability for damage caused by climate change, which has so far not been taken into account. Here, the question arises of whether emissions from the nineteenth century are equally relevant for liability reasons as, for example, those of today's air traffic. Within the legal context we have to resolve whether liability applies to the time of emission or the time of immission. If the time of immission—the moment of the damage occurring—is the relevant reference point, then the emission of greenhouse gases dating back much further can be relevant for liability reasons, as the effects of climate change extend over a long period of time. Would only the time of emission be taken into account, today's emitters could counter that their emissions have not become effective yet.

But within the examination of causation, aspects, such as forseeability or breach of duty, have to be considered as well (theory of adequate causation or scope of the rule theory respectively). Liability, including liability for summarized imissions, can only be claimed for damage the emitter had to be able to assume and therefore avoid (Spindler 2008, p. 324). As the effect of greenhouse gases to climate are only known for a comparatively short time, emissions that date back a number of decades can not be relevant for liability reasons.

The reference to missing effectiveness of today's emissions due to the long-lasting climate processes once again illustrates the basic problem of liability for damage caused by climate change. From the natural scientist's point of view it is certainly possible to assign an intensifying effect on climate change to today's emissions. But it is not possible to substantiate damage with ample certainty to the contribution of a certain emission source. The processes leading to climate change are too complex. Numerous large and small scale emission sources have released greenhouse gases that have mixed and influenced each other in their effect. The only thing that seems certain, is that these emissions lead collectively to the climate change that can today be observed (Chatzinerantzis and Herz 2010, p. 597).

6.2.2.5 Joint Liability

If the damage caused by climate change cannot be attributed to a certain emission source of greenhouse gases, the question arises of whether it is dispensable to prove the contribution of a single emitter if an adequately large circle of opposing parties is presented. Initially this reasoning seems to be plausible: emitters are combined to a unit on the grounds of the interaction of their emissions and the precise damage is attributed to this unit of emitters (BGH, decision of Feb. 13th 1976—V ZR 55/74, for a case where vibrations caused by two quarries lead to damage on a building, but it is impossible to exactly define the contribution of the different sources to the damage). At this point the question arises, on which criteria the circle of opposing parties shall be defined, as not all large and small scale-emitters worldwide can be made liable. The most obvious way to act would be to hold large scale emitters liable (in the case of Kivalina vs. Exxon the claim was filed against 24 oil companies such as Exxon, BP and Shell, electricity utilities, and a coal company, U.S. District Court for the Northern District of California—CV-08-01138-SBA). But the total of emissions caused by small-scale emitters (car traffic, households etc.) can be held against this, as they are significant in the calculation of the total pollution. The differences between damage caused by climate change and the case decided on by the BGH (as mentioned above) can be seen as the damaging parties were determinable. The concepts of 'market share liability' and 'industry wide liability' developed by American jurisdiction also presuppose at least a definable number of potential emitters, even if their contributions to the damage caused cannot be verified individually (Fischer 1981, pp. 458–459). Therefore statements of probability about the proportional responsibility for the damage concerned would also play a role here, if only on a larger scale.

6.2.2.6 Pro Rata Liability

The German liability scheme, which is focused on clear causation relations, reaches its limits when environmental damages in question are caused by long-range effects without an individual chain of causation (Spindler 2008, p. 324; Chatzinerantzis and Herz 2010, p. 597). On the one hand it is designed to grant full compensation for damage which, on the other hand, is only granted if the infringement of a legally protected right has certainly been caused by a specific act (all-or-nothing-principle).

A pro rata liability would offer an alternative to the all-or-nothing-principle. Attempts for a proportional liability can be found within the causality between an act of infringement and the damage (Art. 287 ZPO [Code of Civil Procedure]), provided that the infringement can be individualized as mentioned above. But pro rata liability in general does not exist in applicable law and cannot be reasoned for with analogy (cf. Spindler 2008, pp. 309–344), making an amendment necessary. Pro rata liability would have to be designed in such a way that, contrary to the existing way of having to definitely prove causation, a liability based on the probability of causing a damage is introduced. The introduction of such a liability concept is afflicted with difficulties, which shall be only outlined. First of all potential emitters are confronted with the problem of carrying joint liability for all possible other sources as well. Therefore they can hardly avoid liability through their own behaviour, which leads to a lack of incentive to take precautionary measures for environmental protection. From the damaged party's point of view, the interest in enforcing small compensation claims is remote. Furthermore they are confronted with high costs to determine the proportional contribution of the emitter. These problems multiply when a large number of emitters and damaged parties is involved, as is the case with damages of climate change (cf. Spindler 2008, pp. 325–326). If a climate liability law was introduced, the question would arise in the opposite direction: whether positive effects of climate change could be balanced by enabling large-scale emitters to demand profit gains from farms or the tourism industries that have profited from climate change (see Chatzinerantzis and Herz 2010, p. 597).

6.3 Summary and Conclusion

Liability for damage caused by climate change is, ultimately, a question of evaluation. The emissions of many have produced climate change. But considering the thought of compensation for damage or the polluter-pays-principle, it only seems appropriate to pass on the costs to those who have contributed to the development of climate change on a large scale and, at the same time, have notably benefited from their emissions.[7]

But German liability law cannot answer this question of evaluation. Instead, it rather reaches its methodic limits assessing damage caused by climate change. German liability law is based on the all-or-nothing-priciple. If, as is the case with damage caused by climate change, the general public, including the damaged party, is to be held liable, any attempt of constructing liability based on the all-or-nothing-principle has to fail as no certain injuring act is the basis for the damage, but a risk produced by the general public. And it is not primarily the reproof of arbitrariness that stands against the liability of an individual for the damage caused by climate change (so Chatzinerantzis and Herz 2010, p. 598) but the fact that a legal foundation for a probability liability (or a *pro rata* liability) does not exist. Therefore, the decision of who is to be made liable for possible damage caused by climate change lies with the legislator.

Whether it is right to introduce *pro rata* liability or a special 'climate liability' seems arguable. Not all types of environmental damage can be met by liability mechanisms and a liable party cannot be found for every environmental damage (von Bar 2000, margin number 1). Liability does not only mean compensation for damage but the potential liable party has to have the opportunity to prevent a liability by his own behavior. But this presupposes foreseeable and avoidable causal connections. When they are not given, as is the case with damage caused by climate change, the consequence is that there is no liability.

[7] This question of compensation is politically even more important on an international level, as the poorest countries have to bear a great amount of the consequences of climate change, which in turn is caused mainly by the industrial nations.

But the government has other means to internalize the costs for climate change. Damage caused by climate change can be balanced by collective measures such as compensation funds or compulsory levies for the large-scale emitters (Wagner 1990, pp. 106, 143–144 and 202). Emission trade is offering a trend-setting instrument as the large scale emitters are already combined into a community based on legal criteria. Emission allowances give the possibility to proportionally connect the individual emission to the damage caused by climate change. The government could then auction off the allowances, thus collecting the 'fees' from the emitters that in turn are to be used to compensate for the damage caused by climate change. This procedure is already possible within the European Law (Art. 10 para 3 lit. a Directive 2003/87/EC—greenhouse gas emission allowance trading scheme as amended by Directive 2009/29/EC, OJ L 140 June 5 2009, p. 63, see also von der Heide 2007, pp. 264–302). Given the economic and budgetary crisis, the question is if and how the member states implement this (for the relation of economy and ecology see Knopp and Piroch 2009, pp. 409–413).

Acknowledgment Special thanks go to Ms Janet Varga for her care in translating the text.

Biography Since 2007 Academic Assistant at the Centre for Law and Administration (ZfRV) at the Brandenburg University of Technology Cottbus–Senftenberg.

Academic work is dealing with various fields of environmental law (e.g. air pollution control, waste control, nature conservation, soil conservation, land use)

References

Chatzinerantzis A, Herz B (2010) Climate change litigation – Der Klimawandel im Spiegel des Haftungsrechts [Climate change litigation – climate change in the mirror of liability law]. NJOZ (Neue Juristische Online Zeitschrift) 11:594–597

Evers M, Stampf O, Traufetter G (2010) Die Wolkenschieber [The cloud pushers]. In: Spiegel no. 13 of 29 March 2010, pp 140–149

Feldhaus G (2004) BImSchG commentary, 16th edn. CF Müller, Heidelberg

Fischer JM (1981) The availability of private remedies for acid rain damage. Ecol Law Quart 9 (3):429–488

Grossman DA (2003) Warming up to a not-so-radical idea: tort-based climate change litigation. Colum J Environ Law 28(1):1–61

Hoffmann J (2007) Herausforderung Klimaschutz [Challenge climate protection]. Nomos, Baden Baden

Jarass HD (2010) BImSchG commentary, 8th edn. CH Beck, München

Knopp L, Piroch I (2009) Umweltschutz und Wirtschaftskrise [Environment protection and economic crisis]. ZUR (Zeitschrift für Umweltrecht) 9:409–413

Knopp L, Wiegleb G (eds) (2009) Der Biodiversitätsschaden des Umweltschadensgesetzes [The biodiversity damage of the environmental damage act]. Springer, Heidelberg

KPMG (2008) Climate changes your business. http://www.kpmg.de/docs/Climate_change_risk_report.pdf

Maag KB (2009) Climate change litigation: drawing lines to avoid strict, joint, and several liability. Georgetown Law J 98(1):185–213

Salje P, Peter J (2005) UmweltHG commentary, 2nd edn. CH Beck, München

Scouteris BC (2008) Klimawandel, US standing-Doktrin und 'regulation by litigation' [Climate change, U.S. standing-doctrine and 'regulation by litigation']. EurUP (Zeitschrift für Europäisches Umwelt- und Planungsrecht) 3:144–147

Spindler G (2008) Kausalität im Zivil- und Wirtschaftsrecht [Causation in civil and commercial law]. AcP (Archiv für die civilistische Praxis) 208:283–344

Thorpe A (2008) Tort-based climate change litigation and the political question doctrine. J Land Use Environ Law 24(1):79–105

Verheyen R, Lührs M (2009) Klimaschutz durch Gerichte in den USA – 1. Teil: Öffentliches Recht, – 2. Teil: Zivilrecht [Climate protection due to courts in the U.S. – first part: public law, – second part: civil law]. ZUR (Zeitschrift für Umweltrecht) 2:73–79, 3:129–138

Verheyen R, Breham J, Bacri C, Queinnec Y, Pellegrino E (2008) A brief legal opinion: minimum benchmarks for reporting of companies on (climate) risks under European Law – an analysis of Italian, French and German Law. http://www.climate-mainstreaming.net/autoeu08.pdf

von Bar C (1998) The common European Law of torts, vol 1. CH Beck, München

von Bar C (2000) The common European Law of torts, vol 2. CH Beck, München

von der Heide A-K (2007) Zurechnung von Schadstofffernwirkungen [Attribution of long range effects of harmful substances]. Nomos, Baden Baden

Wagner G (1990) Kollektives Umwelthaftungsrecht auf genossenschaftlicher Grundlage [Collective environmental liability on a cooperative basis]. Duncker & Humblot, Berlin

Chapter 7
Strategy Development and Risk Management in the Context of Emission Rights Trading

Magdalena Mißler-Behr and Sana Mehicic

7.1 Introduction

On February 16, 2005, the legal framework for emission reduction according to the Kyoto Protocol and the Burden Sharing Agreement came into force. The Kyoto Protocol includes three main mechanisms for emission reduction, namely Emission Trading, Joint Implementation and Clean Development (Kyoto Protocol 1997). The target is to achieve, by 2012, an emission reduction of greenhouse gases of at least 5 % compared to the basis year 1990. The European Community (since 01.01.1992 European Union) member states are in obligation to reduce their emission output by 8 % (Konzak and Heßler 2006). Complementarily, the Burden Sharing Agreement regulates the contribution quota of every European Union member state to emission reduction until 2012 (Elspas and Stewing 2006). Therefore, in 2005, with introduction of the first trading period (2005–2007), the European Community established the use of the Emission Trading mechanism. Thus, the European Community implemented, as one of the first regions world-wide, one of the three environment protection mechanisms of the Kyoto Protocol. As a result, the companies were compelled to trade for a right which had previously been free of charge, namely the right to pollute the air. Already in the first trading period several million emission allowances have been traded. This first emission trading period involved 1,200 companies and 12,000 plants across the 25 European Community member states (Sandhoevel 2006). The companies concerned can be found especially in carbon-intensive industries, such as energy industry and refinery industry. Those industries

M. Mißler-Behr (✉)
Department for Planning and Innovation Management, Brandenburg University of Technology (BTU) Cottbus-Senftenberg, P.O. Box 10 13 44, 03013 Cottbus, Germany
e-mail: magdalena.missler-behr@b-tu.de

S. Mehicic
Brandenburg University of Technology Cottbus-Senftenberg, P.O. Box 10 13 44, 03013 Cottbus, Germany
e-mail: ls_planung@b-tu.de

E. Albrecht et al. (eds.), *Implementing Adaptation Strategies by Legal, Economic and Planning Instruments on Climate Change*, Environmental Protection in the European Union 4, DOI 10.1007/978-3-540-77614-7_7, © Springer-Verlag Berlin Heidelberg 2014

obtained 78 % of the emission allowances. The production companies, such as coking plant, steel industry, cement industry, glass industry, ceramic industry, cellulose industry, paper industry, got the remaining 22 % of the allowances. In the focus of the companies' emission reduction were plants related to energy transformation (Salje 2006).

The yearly reports indicate that in most enterprises internal and external processes for application of Emission Trading have been established (DEHSt 2006). But it is also a fact that a number of companies is finishing the first trading period with a particularly high or particularly low number of emission allowances on their official account. Both positions are inefficient and cost-intensive. For the second trading period (2008–2012), the emission allowance allocation volume has been reduced. Further on, the allowance allocation is now done based on the benchmark, except for lignite power plants (brown carbon plants) (BMU 2007). This forces companies to buy supplementary allowances for old plants in the second trading period. Also, the sanction costs for emission reduction not achieved, respectively for the number of required emission allowances not achieved, have been increased from 40 Euro per ton to 100 Euro per ton (Requate and Graichen 2005). As a consequence of those changes, the second emission trading period is more restricted. In order to achieve the required number of the allowances and to meet the restricted requirements of the second trading period, companies should establish strategy development in the context of emission reduction.

A company's strategy development should enable analyzing different mechanisms and supporting the company in the process of the mechanisms selection. But, what are the appropriate approaches for strategy development in the context of emission reduction? In order to work out a company's suitable strategy development in the content of emission reduction, this paper discusses especially two methods, namely Strategic Environmental Assessment, and Risk Management. To this end, it is to be discussed how those two approaches can contribute to the company's strategy development in the context of emission reduction. The first chapter gives a short overview of the principles and mechanisms for emission reduction provided by the Kyoto Protocol. Based on this, Strategic Environmental Assessment and Risk Management will be described and discussed concerning their contribution to a company's strategy development.

7.1.1 Principles for Emission Reduction

To enable the emission reduction, the Kyoto Protocol offers three different mechanisms, as mentioned above. Those are based on two principles, namely the cap-and-trade and the baseline-and-credit principles (Betz 2005). Cap-and-trade refers to the aggregate cap, here the number of the emission allowances that are allocated by the state to the company. If the allowances number does not cover the company's emission output, the company has the possibility to trade needed emission allowances and *vice versa*. At the end of every year of a trading period each emitter should have at least as many emission allowances as correspond to its emissions into the atmosphere (Requate and Graichen 2005). Otherwise the emitter has to pay sanctions. Of the three mechanisms provided by the Kyoto Protocol, emission trading is the one based on the cap-and-trade principle. In contrast to the baseline-and-credit mechanisms, emission trading is obligatory for companies and is especially referring to the reduction of carbon dioxide emission (Betz 2005). It is therefore this mechanism we shall concentrate our further considerations on.

The second principle (baseline-and-credit principle) allows a company to make emission reduction credits according to its own baselines. The Kyoto Protocol mechanisms based on this principle are Joint Implementation (JI) and Clean Development Mechanism (CDM). Those mechanisms can be used optionally. Within the JI, projects can be implemented by European Community companies in countries listed in Annex-I of the United Nations Framework Convention on Climate Change (UNFCCC) (referred to as 'Annex-I states'); Within the CDM, projects can be implemented in developing countries (Non Annex-I states) (Konzak and Heßler 2006). In contrast to CDM implementation, JI project implementation requires from both companies' states an existing reduction commitment under the Kyoto Protocol. The promotion of these projects should lead to an emission reduction of greenhouse gases in comparison to the lack of the specific project (Betz 2005). The emission allowances obtained by those projects can be used for the fulfilment of the company's own emission reduction obligations as well as for trading. The rules for the second trading period (2008–2012) allow for the fulfilment of 22 % of emission reduction obligations through allowances obtained by the JI and CDM projects (Zuteilungsgesetz 2012—ZuG 2012).

7.1.2 Emission Trading Mechanism

The Emission Trading Mechanism was the most frequently used and is the obligatory mechanism. It is therefore briefly outlined in the following by way of illustration. The Emission Trading Mechanism itself does not directly reduce a company's emission output. In fact, it makes environmental protection a business and motivates the companies to invest in emission-reducing technologies (Gerlach 2006). The companies are allocated a certain amount of emission allowances from a national institute, in Germany: German Emissions Trading Authority (DEHSt) at the Federal Environment Agency. This amount is smaller than what the companies need to keep the current business running without modernization by use of emission-reducing technology. A company which can reduce emission output with low investment costs has the motivation to provide modernization due to the fact that emission allowances can be sold. As emission allowances are allocated free of charge, the company makes profit vending them. A company which cannot reduce the emission output with low investment costs can buy additional emission allowances (Betz 2005). However, as the market price for allowances increases and investment costs for emission-reducing technology get lower than trading allowances on the market, the company should prefer replacing strong emission-output-producing machines by better ones. Due to the company's dependency on the allowances offer volume and on the related price on the trading market, in the long run an investment in emission-reducing technologies may be cost-efficient (Schleich et al. 2006).

So, the company can buy additional emission allowances if the number of allowances originally obtained does not cover the company's actual emission output in the atmosphere, that is short position or shortage. Otherwise, emission allowances can be sold if the company has a surplus of allowances, i.e. long position, surplus (Betz 2005). In addition, a company has the possibility of banking and borrowing the allowances allocated to them (Requate and Graichen 2005). In case a company has a surplus of allowances it can turn to banking. This means that the company can transfer the surplus of allowances for use in the next trading year of the same trading period. In contrast to banking, borrowing of emission allowances is a transfer of the company's allowances from the next year to the present year of the same trading period. Both banking and borrowing cannot be applied between two different trading periods, as for example banking between 2007 and 2008 or borrowing between 2008 and 2007 (Betz 2005).

7.2 Strategy Development

In the above, we described different mechanisms for emission reduction and for achieving the required number of emission allowances at the end of a trading period. From the company's point of view the question arises which of those mechanisms is the suitable one to be used in the next trading period. The elaboration of this strategic question should be tackled by means of strategy development. Strategy development is '...the determination of the basic long-term goals and objectives of an enterprise, and the adoption of courses of action ... necessary for carrying out these goals' (Chandler 1962; Mintzberg 1994; Welge and Al-Laham 2001). Thus, strategy development enables the company to find an appropriate decision regarding a special issue by studying the company's objectives, feasibilities and feasibility's consequences to achieve those objectives. The result of the strategy development is then a selected and described strategy. Thus, the strategy is a company's action that enables accomplishment of the company's objectives. Additionally, in the context of the emission reduction, the strategy is a suitable action that enables achieving the required number of emission allowances at the end of the trading period. Therefore, different emission-reducing mechanisms are the instruments that form the company's strategy. A substantial requirement on the strategy development in this context is to offer methodical support in order to work out the instruments that should be used.

To accomplish the above requirement, strategy development includes decision-making stages which describe the procedure of the strategy development. Therefore, the strategic decision-making stages are: identifying of objective of strategic action, identifying of alternative ways to achieve objective, selecting between alternatives, fine-tuning of selected alternatives, announcing of draft plan and finally, implementation and monitoring of strategic action (João 2005). Apart from the decision-making stages for strategy development, the consideration of the environmental information and legal framework is necessary. To consider these environmental aspects the methodical support of the Strategic Environmental Assessment (SEA) for the company's strategy development will be described and discussed.

By the decision-making under certainty all substantial current and future information are known (Rosenkranz and Mißler-Behr 2005). So the company can take, from its point of view, the best decision which emission-reducing instruments should be applied. But the point is that, in general, such certainty is not given. In particular, the decision-making with regard to emission-reducing instruments is a new working area and therefore related to various uncertainties (Stewing and Grit 2006b). Those uncertainties arise internally within a company and can, for that reason, partly be influenced. But a number of them are external uncertainties and cannot be directly influenced by the company, as for example market prices and political developments. Both of them complicate the decision-making concerning the selection of emission-reducing instruments. Moreover, they increase the existing enterprise risks. Therefore, a profound analysis and prognosis of the development of the uncertainties with regard to emission-reducing instruments is essential for a company's strategy development. Based on those evaluations, the enterprises have to develop a suitable system to handle those uncertainties (Stewing and Grit 2006b). The handling of uncertainties is a major task of Risk Management. Therefore, the tasks and methods of Risk Management are addressed according to their contribution to the company's strategy development.

Before the contribution of Strategic Environmental Assessment and Risk Management to strategy development is described, we shall briefly attend to the process of the strategy development and its timeframe.

7.2.1 Strategy Process and Timeframe

For starting the process of strategy development, the official state deadlines of the emission trading system can be used as the first point of reference. Within the strategy development especially the following deadlines have to be considered: application for allocation, issuance and starting of the trading period (Greinacher 2004). The first deadline is the company's application for allocation of emission allowances. For this the company has to report an official statement on emission allowances according to its planned emission output for the next trading period. The application for allocation for the next trading period is to be submitted by March 31 of the final year of the preceding trading period. The next relevant point in time is when the state presents the information on how many emission allowances will be in fact allocated to the company. Related allowances for the current trading period are provided to the company on a yearly basis, on February 28, by transferring the allowances to the applicant's account in the emission trading register. The third deadline is the start of the trading period.

As a consequence, the strategy development should be placed before the first deadline, namely application for allocation of emission allowances. The information of the number of emission allowances obtained for the last period and the in-house evaluation of future emissions, based on the production planning for next year, should be used as background information for the strategy development. After receiving feedback on the application and information about the quantity of emission allowances obtained (second deadline), the strategy has to be adjusted, if necessary. In addition, strategy adjustment has to be carried out throughout the entire trading period.

7.2.2 Strategic Environmental Assessment

In the environmental context, Strategic Environmental Assessment is described in the European Community Directive 2001/42/EC, also known as the SEA Directive (SEA Directive 2001). It had to be implemented by the European Community member states by mid-2004. According to the SEA Directive, before a state's programs are published, the public must be consulted and an environmental report has to be generated. The environmental report shall include environmental effects of the program as well as reasonable alternatives. Two principles can be discerned as the essential basis of Strategic Environmental Assessment. The first principle refers to the identification of feasible alternatives and their comparison. The second principle includes improving the program according to the environmental effects (João 2005). In this context Strategic Environmental Assessment is '...the process of evaluating the environmental impacts of proposed policies, plans or programs, in order to inform decision-making' (João 2005). To enable the implementation of those principles and thus improving the political decision-finding, Strategic Environmental Assessment affects political strategic decision-making with the following specifications: *reshaping objectives, identifying targets and indicators, predicting and evaluating, enhancing impacts of chosen alternatives, reporting, implementing guidelines as well as monitoring environmental impacts* (João 2005). In the following there is a brief description of those requirements.

Reshaping the objectives refers to the adjustment of the program's objectives according to the environmental effects. Based on the first requirement, *identifying targets and indicators* refers to the identification of the program's alternative and alternative related objectives as well as measurements according to the environmental effects. The third specification, *predicting and evaluation,* assists to evaluate and to predict the impact of the alternatives according to their environmental effects. After the impact of the alternatives on the environment has been analyzed, the *enhancing impact of the chosen alternatives* requires the development of reasonable alternatives e.g. by combining the best aspects of existing alternatives (João 2005). Once the most reasonable alternative has been made out, the fifth requirement, *reporting* refers to the elaboration of the environmental report mentioned above (Albrecht 2005). Finally the *establishing of environment implementation guidelines* assists the definition of the action in order to achieve the objectives of the reasonable alternative. *Monitoring* in this context means controlling the implementation of the alternative and observing the development of its environmental effects. The decision about the monitoring measures is left to the decision of the Member States (Albrecht 2005).

7.2.3 Risk Management

The application of Risk Management for companies is given through legal regulations (Wolf and Runzheimer 2003). Independently of that, companies apply Risk Management for strategic decision-making, as for example for a new product launch. As described at the beginning of Chap. 2 above, from the company's point of view, risk is uncertainty as to the development of the company's decisions' consequences. The uncertainty is characterised through too little or too much information, inconsistent information or inexact information (Rosenkranz and Mißler-Behr 2005). To handle this uncertainty, different qualitative and quantitative methods support Risk Management (Rosenkranz and Mißler-Behr 2005; Schmitz and Wehrweim 2006; Brühwiler 2007). It is common to support strategic issues with qualitative methods in the first place. This may include risk raster, scenario analysis but also different creative techniques (Rosenkranz and Mißler-Behr 2005; Brühwiler 2007). Based on the above, the risks also affect numerical values in different departments as for instance in the Planning department or Accounting department. Therefore, different quantitative tools can describe the effects of risk (Rosenkranz and Mißler-Behr 2005). At this point, the tools of data analysis and decision supporting methods maintain risk assessment and impact.

A detailed description of the Risk Management tasks can be found in DIN EN ISO 14971 as follows: *finding of risk strategy, risk identification, risk assessment and risk controlling*. Finding of risk strategy concerns a definition of the company's risk behavior. Consequently, the formulation of the company's risk attitude is addressed. Hence, the definition of the risk strategy is a part of the company's strategy, for instance the company's risk diffusion and risk concentration attitude towards decision-making (Rosenkranz and Mißler-Behr 2005). To enable the risk strategy definition, different risk types and companies' preferred risk handling have to be defined with this task. Risk types refer to risks with similar impact, for example, the risks that a company prefers to carry out and risks that a company chooses to accept or to limit (Schmitz and Wehrweim 2006).

As a second task, the *risk identification* concerns the identification and description of different upcoming risks related to decision-making (Rosenkranz and Mißler-Behr 2005). Here, different decision alternatives and related upcoming risks are described. Direct single risks as well as indirect risks of the decision alternatives have to be incorporated in this risk identification. Moreover as the risk identification is the basis for realistic risk assessment, a precise risk characterisation is necessary. Different aspects can be included in risk characterisation. This depends strongly on the upcoming risks (Schmitz and Wehrweim 2006). Possible characterisation aspects are: affected company's department, time of formation, risk frequency, time needed to treat the risk or for instance short-run or long-run risk. Further on, to allow early identification of occurring risks, the impact of related risks on the company's performance has to be identified. For this, key performance indicators are in use as qualitative and quantitative information. The key performance indicators concretize the objectives such as sales volume, cost reduction, budget and other. As the identified risks affect those key performance indicators by changing the indicator's values, the risk effects on the company's objectives are visible. Assistance in the selection and definition of the measurements is provided for example by the Balanced Scorecard concept (Kaplan and Norton 1997).

As a third task, the *risk assessment* concerns the evaluation of identified risks. Hereby, causes and dependencies between risks are considered. The form of risk evaluation depends on the existing data. Risk evaluation can thus be done in the form of a qualitative assessment. For instance, a risk could be considered important or not important, of high or low priority (Schmitz and Wehrweim 2006). Risk evaluation can also be done quantitatively, e.g. by working out risk frequencies and monetary values. Additionally, risk evaluation can be implemented by using complex evaluation methods, especially when risks are precisely defined and quantitative data are available. For this, different statistical methods can facilitate the risk assessment. The statistical methods include the quantitative evaluation of data and different risk measurements (Rosenkranz and Mißler-Behr 2005). For example, expected value, standard deviation or confidence intervals are frequently used for risk assessment. In the following, there is a short illustration of those methods. Expected value can, for instance, indicate the future allowance price. Based on the expected value, the standard deviation describes how strong the old price varies around the expected value (expected future price). Thus, the probability of whether the elaborated allowance price is realistic can be worked out. If, for instance, there is an 80 % standard deviation regarding the expected value (here: expected price), the elaborated allowance price is unrealistic. The company is then confronted with a high risk and thus has to pay more than foreseen in the budget. Also the use of the confidence interval supports risk assessment. The confidence interval gives the probability distribution of one value between the average value and an extreme value. For illustration, the existing data could indicate that the probability of paying a high price for the allowances is between 70 % and 82 % at the end of the trading period. The possibility that low prices will be paid is unlikely. The discussed methods are probability-based methods. In this context, today's information and communication technologies meet the expectations of a detailed probability distribution analysis. The detailed determination of the value probability distribution as well as the simulation of the values coherences is feasible. To provide this kind of evaluation the availability of quantitative data is necessary. Further on, the selection of the suitable statistical methods is essential for the success of the risk evaluation. Thus the selections criteria should be well prepared (Brühwiler 2007). Here, it should be differentiated between the organizational criteria and methodical criteria. The organizational criteria may be the available resources. The methodical criteria are for instance available information and qualitative data as well as a method's required data input.

The fourth task, namely *risk controlling*, monitors the risk development as well as the implementation of above described Risk Management tasks (Burger and Buchhart 2002). The risk development is applied by controlling the as-is and to-be key performance indicators. Once the deviance between as-is and to-be is specified, the causes as well as the handling of the deviance should be addressed. Further on, the implementation of all Risk Management tasks, such as finding risk strategy, risk identification and risk assessment, should be monitored. In case decision-based information is changing, the effects of this information have to be evaluated.

After this overview about the Strategic Environmental Assessment and Risk Management, in the following we focus on their contribution to a company's strategy development.

7.2.4 Strategic Environmental Assessment and Risk Management Within Strategy Development

Strategic Environmental Assessment is a political approach, implemented by the European Community member states for the states' use. According to this, the specifications of the SEA Directive refer especially to the consideration of environmental issues for political decision-making (Kläne and Albrecht 2005). But SEA can also support a company's decision-making (Marshall and Fischer 2004). In the context of the Emission Trading, because of the involvement of environmental issues in decision-making, a transformation of the SEA requirements to the company's strategy development is conceivable and recommended. Such transfer into the company's strategy development can be carried out as follows: The requirement *reshaping of objective* should refer to the first stage of a company's decision-making, namely identification of objective of strategic action. This way a company should introduce the environmental legal framework and environmental objectives into its strategy development. Working out the company's objectives, not only the production planning, business volume, reduction costs but also the planned emission reduction and planned investment for emission-reducing technology should be considered. Based on that, the second stage of the company's decision-making, namely identifying of alternative ways to achieve objective, is supported by the requirement *identifying of targets and indicators*. In the context of a company's decision-making, this requirement refers to including and combining different emission-reducing instruments to achieve objectives described above. Thus, the reconsideration regarding the objectives and measurements of different emission-reducing instruments should be elaborated. In the third stage of a company's decision-making (choice between alternatives), the SEA requirements *predicting and evaluating* as well as *enhancing the impacts* can be applied. By means of those requirements the emission-reducing instruments should be evaluated with regard to their achieving emission output and related costs. In addition, the reasonable alternative, which may also be a combination of emission-reducing instruments, is to be preferred. This alternative should indicate the best emission output achievement due to the reasonable investment. The last stage of the company's decision-making is specifically announcing a draft plan and implementation and monitoring of strategic action, which can be supported by the last two requirements of the SEA, i.e. *reporting and establishing of environment implementation guidelines*. Considering those requirements in the company's decision-making, the action plan for implementation of the emission-reducing instrument should be elaborated and communicated. Additionally, *monitoring* the effects of the used emission-reducing instruments enables early reaction to any undesirable development of the instrument. Summing up, it can be said that the Strategic Environmental Assessment gives conceptual support in order to include the company's related environmental aspects regarding emission reduction.

The implementation of Risk Management for the companies is given on the one hand through legal regulation, and on the other hand the companies implement the Risk Management to assure the fulfilment of strategic decisions (Rosenkranz and Mißler-Behr 2005). Since finding a company's strategy for the next trading period is strategy decision-making related with uncertainties, Risk Management contributes to the identification and evaluation of upcoming risks. The integration of Risk Management tasks *finding of risk strategy* at the first stage of a company's decision-making enables the elaboration of a company's risk strategy. As the risk strategy is a preferred company's behavior in respect of handling risks, it should be considered when defining a company's objective. Thus, the development and selection of emission-reducing instruments depends on the company's risk affection (Lafeld et al. 2006). At the second stage of a company's decision-making, the integration of the Risk Management task *risk identification* enables finding risks related to an alternative. Hence, different upcoming risks of emission-reducing instruments can be identified and described. The third stage of a company's decision-making, namely selecting of the alternatives, is supported by the Risk Management task *risk assessment*. So, different emission-reducing instruments-related risks can be evaluated. Due to the fact that the alternatives can be compared, the evaluation of the risks related to an alternative is the basis for fine-tuning the chosen alternative. The last two stages of a company's decision-making, i.e. announcement of a draft plan and implementing and monitoring strategic action, are supported by the Risk Management task *risk controlling* of the risk development. Consequently, not only the strategy implementation, but also the risks-related development of the strategic action should be monitored.

Summing up, it can be said that on the one hand Strategic Environmental Assessment points out the requirements to be considered in strategy development and on the other hand Risk Management provides a methodical pool for qualitative and quantitative support. The methods support also the identification of alternative related risks, the risk evaluation and the risk controlling. The Fig. 7.1 shows the

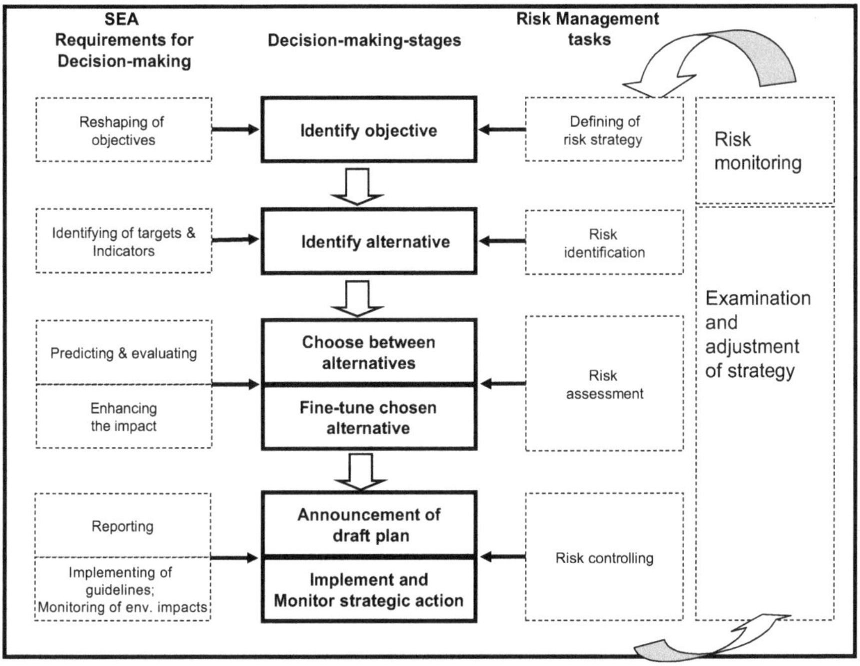

Fig. 7.1 Strategy development with SEA and risk management

contribution of Strategic Environmental Assessment and Risk Management to strategy development.

In order to illustrate the company's strategy development, we will discuss the stages of application.

7.2.4.1 Identifying of Objectives

Strategy development includes as a first stage identifying of objective of strategic action. From the company's internal point of view those can be budget, product planning, short- or long-term cost reductions plans (Enzensberger and Hermes 2004). According to the requirements transferred from Strategic Environmental Assessment, compliance with the emission obligation should be one of the strategic objectives. Therefore, an investment budget must be foreseen for emission allowances and emission output reduction in order for the company to be able to handle a shortage or surplus position. Based on this objective different risk types have to be analyzed with a view to which risk could be borne by the company and which risk should be avoided, limited or handed over. Different risk types can be construed here, as for instance market risks (demand and offer of allowances on the market), legal framework risks (changing allocation regulation, development of the sanctions costs) as well as reputation risks (non-compliance with emission output also causes reputation risk) (Stewing and Grit 2006a). Based on this, the risk diffusion degree should be settled. For example, the market risk can be limited by using a combination of emission-trading instruments; the reputation risk could be avoided by achieving the set emission reduction requirements; and legal framework risks could be kept low by contracting experts for legal questions related to emission trading.

7.2.4.2 Identifying of Alternatives

The second stage, identifying alternative ways to achieve objective, includes the finding of possible solutions that achieve the objectives defined above, especially handling of shortage or surplus position. For this, according to the requirements transferred from Strategic Environmental Assessment, objectives and measurements of different emission-reducing instruments should be described. Additionally, the Risk Management enables the identification of alternative related risks from the company's point of view.

Different instruments handle a shortage or surplus of emission allowance. For instance, the company can reduce emissions through investment in the emission-reducing machines (Spangardt and Meyer 2005). Therefore, the plants of the company have to prove accordingly the reduction of the emission output. This can be achieved for example by changing the fuel type, supplying of biomass and spare fuels, improving of heat recovery or improving of isolation (Bockamp et al. 2005; Stewing and Grit 2006a). Additionally, the replacement of the old machines by the new ones should be considered. If the shortage still cannot be covered, the company could buy additional allowances on the market or borrow them from the next year of the same trading period (Spangardt and Meyer 2005; Konzak and Heßler 2006). The borrowing should be applied for the short-term production fluctuations; otherwise the short position could be transferred to the end of the trading period. But also with the implementation of JI projects, CMD and national projects the company can reduce the shortage. The advantage of those optional instruments is the additional risk diffusion, due to the fact that emission allowances are obtained independently of the emission trading market. Also supplementary customer acquisition as well as the possibility of opening new market access can be a result of international projects. To implement those projects, the enterprises should be informed about project-related regulations. For example, the implementation of JI projects is possible if the enterprise's country fulfils the following participation criteria: ratification of the Kyoto Protocol, determination of the assigned amount, establishment of a national system for the estimation of the greenhouse gas emissions and emission reduction, establishment of a computer-made national register, punctual setting-up of annual inventories and submission of additional information about the assigned amount. Also the JI project host country must accomplish the same criteria. Thus, current information about the legal regulation for the implementation of the JI projects, CMD projects and national project has to be collected. Also paying sanctions can be considered as a possible instrument. Paying sanctions is an option if the sanction costs are lower than the investment needed to obtain the necessary number of emission allowances. Still, the sanctions contain, besides paying the monetary value and delivering the missing allowances in the next trading year, also a public naming of the companies that did not fulfil the required number of allowances. The costs caused through the damage of reputation as well as the environmental development are not considered by selecting this instrument. Due to the fact that the emission allowance prices are lower than the sanction costs, and the reputation damage is not known, paying sanctions is not a preferred instrument (Requate and Graichen 2005). To enable instrument characterisation, different key performance indicators should be added as the instrument's performance measurements. Especially the measurements that describe the emission-reducing instruments, as for instance number of allowances, reduced emission-output, allowance price, price development, organizational costs, sanctions costs can be used as performance indicators (Betz et al. 2003).

Further on, the identified instruments should be described with related upcoming risks. We concentrate especially on the risk of obtaining additional allowances by the Emission Trading Instrument. A trading of allowances is linked to the planning uncertainty due to possible price fluctuations. The price fluctuations may have different reasons: the allocated number of the emission allowances (Spangardt and Meyer 2005) or a high economic growth causes increasing production, especially for the energy-intensive industry (Marci 2005). This leads to an increasing energy use in the production and to rising emissions output. Consequently, the company's demand of emission allowances increases, which leads to an increase of the allowances prices. But also weather and environmental changes affect the allowance price. For instance, in extremely dry summers the demand for air conditioning increases. Air conditioning is energy-intensive, therefore the companies produce more energy and, consequently emission output increases. The demand for the allowances rises and the prices for allowances increase. As a result, obtaining the needed number of allowances, for the price originally planned in the budget, is not certain; this constitutes a risk for the company. Further on, it is not known if the required quantity will be offered on the market. So, the company is faced with the volume risk (Spangardt and Meyer 2005). Closely related to the volume risk are inappropriate production planning as well as unpredictable production downtimes and peaks. Those can increase the amount of the allowances that have to be purchased on the trading market. Besides the price and volume risks, the project circumstances as well as the political environment of JI projects and CMD projects may lead to non-performance of the planned emission allowances quantity (Spangardt and Meyer 2005). Consequently, the additional amounts of allowances have to be acquired on the emission trading market. In the Fig. 7.2, the above discussed risks are characterised with risk type, risk frequency, point of risk formation and effect of the risks. The risk type describes the time period of the risk duration. By way of example, a price risk occurs at the moment when a company trades for the allowances, so it is a short-run risk. The frequency describes the risk occurrence. So, the frequency of the price risk is determinated by the company's demand, or by how often the company trades.

Risk	Risk type – Duration?	Frequency – How often?	Point of formation – When coming?	Effect of the risk – What effects?
Price risk – does the company get the allowances for the desired price ?	Short-run	At demand	At demand	Higher costs than planned in budget
Development of certificate price at the trading markets – how do environment uncertainties effect the allowances price ?	Short-run	At demand	At demand	No predictability; uncertainty
Volume risk – does the company get the needed amount of allowances?	Short-run	At demand	At demand	No adequate allowances amount Pay sanctions Loss of production
Production downtimes or unexpected peak loads – Development of productions demand?	Short-run	Repeated in trading period	Increased/Decreased customer demand	To few allowances or to many allowances
Project risk – Are the project objectives achieved within the given costs? Can the planned allowances be obtained with this project?	Long-run	One per Project in the trading period	During project or at the end of project	External purchase of allowances is necessary High resource cost Resource cost saving in other company's areas

Fig. 7.2 Risk characterisation raster

The point of the risk formation gives information about the point in time when the risk is likely to appear. Accordingly, the price risk is relevant for a company that has current demand. Finally, the effect of the risk describes the risk's consequences. A trading of the allowances on the market may result in higher costs for emission reduction than originally planned in the budget.

The risk characteristics may also contain additional and different description aspects. But it is important that the risk identification and description is elaborated for each emission-reducing instrument. This way the risk comparison of the instruments, according to the company's risk behavior defined in the first stage of strategy development, is possible.

7.2.4.3 Selecting Between Alternatives

Based on the above elaborated information, the final comparison of the instrument is run on the stage selecting between alternatives and fine-tuning of selected alternatives. Here, according to the transferred requirements of Strategic Environmental Assessment, the alternatives have to be predicted and evaluated. As a result, the reasonable alternative should be selected. For this purpose, Risk Management applies the scenario analysis. The scenario analysis sketches the possible future development of the defined scenarios. The future development of scenarios thus depicted serves as decision basis (Welge and Al-Laham 2001). By the means of the scenario analysis, in the first place different emission-reducing instruments can be systematically combined and analyzed in scenarios. The substantial elements of the instruments are related and compared. Based on this comparison, as a second step, the future development of different emission-reducing instruments combination (the scenarios) can be evaluated. Several different future pictures can be sketched in this way. Consequently, the substantial steps of the scenario selection are: pre-selection, evaluation and fine-tuning (Mißler-Behr 2001).

At the preselection the scenarios can be described through qualitative and quantitative elements (Kahn and Wiener 1968). The emission-reducing instruments can be described with to-do activities, key performance indicators and identified risks, see Fig. 7.3. Moreover, with the selection criteria, like the risk diffusion degree, different scenarios can be sketched. Especially the risk diffusion, and with it the combination of presented instruments, is recommended in order to cover shortage. To prevent a shortage of emission allowances in the second trading period, a first scenario may be defined as using Emission Trading (buying and borrowing) as one instrument only. The second scenario may be defined as using Emission Trading with borrowing options as well as the adjustment of machines. A third scenario may include Emission Trading with borrowing options, adjustment of machines and realization of the projects. As three different instruments are in use, the last scenario has the highest risk diffusion.

Based on the preselection made, the next step focuses on the evaluation of those scenarios. In current practice, the information on the scenarios is rarely presented in a quantitative data form. But a suitable economical evaluation of those scenarios requires quantified information about future production and related emission output. It is therefore essential to work with a reliable and comprehensive data basis. Such data basis can be created by empirical investigations in the enterprise. In case that the data are not available, different simulation models can be applied, such as the Flow-Sheeting System (Stewing and Grit 2006a). On the basis of such simulation models the planned production and pertinent emission output can be computed. The scenarios can then be evaluated with the help of the information determined in the simulation. In particular, the marginal reducing costs, i.e. marginal cost for emission-reducing for one ton of carbon dioxide, have to be elaborated (Stewing and Grit 2006a). Consequently, the scenarios with lower long-term marginal reducing costs than expected value of allowance price have to be considered for the fine-tuning. Additionally to marginal reducing costs, also average reduction costs and total reduction costs per scenario should be considered (Stewing and Grit 2006a).

Finally in the fine-tuning, due to the results of the evaluation phase, the decision can be made as to which scenario is preferred for the application. The final decision-making can be provided in a view of key performance indicators elaborated in the first stage of decision-making. In the context of emission trading, the company budget, forecasted production plan, cost reduction plans as well as forecasted allowance price can be used as suitable measurements (Enzensberger and Hermes 2004).

After making the decision on the strategy to be implemented, the strategic action plan for the implementation as well as for the related monitoring process should be elaborated. The monitoring should be applied, especially for the strategic action controlling, risk controlling and emission-reducing effects controlling. Moreover,

PRE-SELECTION						EVALUATION			FINE-TUNING
Instruments		Actions needed	Performance indicators	Related Risks	Risk type	Scenario I	Scenario II	Scenario III	Decision based indicators
Emission Trading	**Vending of allowances**	Apply for membership on trading platform Vending of needless allowances	Number of allowances Selling price, Organizational costs	Price risk Volume risk Project risk Production downtimes Development of allowances price at the trading markets	Short-run Short-run Long-run Short-run Short-run				Company budget Product planning
	Buying of allowances	Apply for membership on trading platform Acquisition of allowances vol.	Price, Organizational costs			Yes	Yes	Yes	Short- or long-term cost reductions
	Banking/ Borrowing	Taking from forthcoming trading period / saving in actual trading period	Production demand development, Organizational costs	Demand development others			Yes	Yes	Development of allowances price
Investment	Adjusting of machines	Review of machines	Emission output Investment costs	Technology maturity others	Long-run		Yes	Yes	
	Buying of machines with lower emissions	Review of machines Review of offer on new machines	Emission output Investment costs	Technology maturity others	Long-run				
	Realization of projects	Collecting the Information respective regulations Partner selection	Investment costs Resources costs Transaction costs	Resources costs Transaction costs Application risk	Long-run			Yes	
Pay sanctions		No action for emission reduction	Sanctions costs	Reputation damage	Long-run				

Fig. 7.3 Scenario building

the process of strategy development and strategy adjustment should be established as a continual process. In the case of changes in the environmental legal framework or of new internal and external company's effects on strategy implementation, the exiting strategy should be reviewed and modified as needed.

7.3 Conclusion and Recommendations

In order to meet the changed volume of allowances in the second trading period as well as the related changes in the legal framework, the company's strategy development in the context of emission trading should be improved. In this context we addressed strategy development and strategy process as well as the contribution of the Risk Management and Strategic Environmental Assessment to strategy development. We came to the conclusion that both Risk Management and Strategic Environmental Assessment contribute to improving the company's strategy development with regard to achieving the required number of emission allowances during one emission trading period. The Strategic Environmental Assessment is in the first place a political approach that affects political decision-making within the consideration of the environment aspects. But, as environmental aspects become more important for companies and companies are obliged to manage their emission output, it is recommended to establish a company's strategy development in the context of emission reduction and to involve these environmental aspects also in the company's strategy development. Those environmental aspects should be included as early as possible in the company's strategy development. Consequently the requirements of Strategic Environmental Assessment can be transferred into the company's strategy development. Doing so assures that the environmental information and legal framework are considered at an early stage within the company's strategy development. Apart from the Strategic Environmental Assessment, Risk Management also supports the company's strategy development by finding of risk strategy, risk identification, risk assessment as well as risk controlling. Therefore, qualitative and quantitative methods support Risk Management tasks. Based on the company's risk behavior, the company's risk strategy should be defined. This risk strategy then serves as an initial point in decision-making. For instance in order to meet the risk diffusion behavior, different emission-reducing instruments should be combined. In order to find the right combination, the risk identification and assessment should be implemented for all emission-reducing instruments. Thus, the combinations can be compared and the decision on the preferred one can be taken.

In summing up it can be said that both Strategic Environmental Assessment and Risk Management support a company's strategy development with forming of general company's strategy development framework. The Strategic Environmental Assessment provides a conceptual approach for consideration of the environmental information, without proposing concrete evaluation methods. On this point the Risk Management supports the Strategic Environmental Assessment with qualitative and quantitative evaluation methods. As environmental development and changes in environmental legal framework even in the short future are not known, some risks are not perceived until their occurrence. In order to handle the uncertainties, the company's strategy development should be established as a continuous process.

Biography Magdalena Mißler-Behr is Head of the Department for Planning and Innovation Management at Brandenburg University of Technology (BTU) Cottbus-Senftenberg.
Sana Mehicic was a member of Graduate School in Risk Management from 2007 to 2009 at BTU.

References

Albrecht E (2005) Legal context of the SEA directive – links with other legislation and key procedures. In: Schmidt M, João E, Albrecht E (eds) Implementing strategic environmental assessment. Springer, Berlin, pp 31–52

Betz R (2005) Flexible instruments in climate change – a guide for businesses. Ministry of the Environment, Climate Protection and the Energy Sector Baden-Württemberg (ed) Karlsruhe

Betz R, Schleich J, Wartmann S (2003) Flexible instruments in emissions trading. Joint implementation. Clean development mechanisms. A guide for businesses. Ministry of the Environment, Climate Protection and the Energy Sector Baden-Württemberg (ed) Stuttgart

BMU – Federal Ministry for the Environment Natrure Conservation and Nuclear Safety (2007) Emission trading in the second Period 2008–2012. Simple – effective – efficient. http://www.umweltrechtsreport.de/news/hg_emissionshandel2008_2012.pdf. Accessed 25 Feb 2014

Bockamp S, Folke C, Rode H, Jacobi A (2005) Technology for reducing CO2-emissions. In: Lucht M, Spangardt G (eds) Emissions trading. Economic principles, legal regulations and technical solutions for climate protection. Springer, Berlin, pp 153–175

Brühwiler B (2007) Risk management as a management task. Taking into account the latest international standards. Haupt, Bern

Burger A, Buchhart A (2002) Risk controlling. Oldenbourg, München

Chandler A (1962) Strategy and structure – chapters in the history of American business. MIT Press, Cambridge, MA

DEHSt– Deutsche Emissionshandelsstelle Emissionshandel (German Emissions Trading Authority) (2006) CO_2-Emissionen des Jahres 2005. Evaluation of the actual emissions of the emissions trading sector in 2005 in Germany. http://www.umweltbundesamt.de/sites/default/files/medien/publikation/long/3551.pdf. Accessed 25 Feb 2014

Elspas M, Stewing C (2006) System and essential terms of emissions trading. In: Elspas M, Salje P, Stewing C (eds) Emissions trading. A practical trading books. Carl Heymanns, Köln, pp 1–13

Enzensberger N, Hermes H (2004) Strategic planning issues in the context of emissions trading. In: Energiewirtschaftliche Tagesfragen, No 9, pp 587–591

Gerlach A (2006) Sustainability entrepreneurship in the context of emissions trading. In: Antes R, Hansjürgens B, Letmathe P (eds) Emission trading and business. Physica, Heidelberg

Greinacher D (2004) Emissions trading is on its way: required actions for companies. Surf Min World Min 56(5):2–10

João E (2005) Key principles of SEA. In: Schmidt M, João E, Albrecht E (eds) Implementing strategic environmental assessment. Springer, Berlin, pp 3–13

Kahn H, Wiener H (1968) Voraussagen der Wissenschaft bis zum Jahre 2000, 4th edn. Molden, Wien

Kaplan R, Norton D (1997) Balanced scorecard – implement strategies successfully. Schäffer-Poeschel, Stuttgart

Kläne C, Albrecht E (2005) Purpose and background of the European SEA directive. In: Schmidt M, João E, Albrecht E (eds) Implementing strategic environmental assessment. Springer, Berlin, pp 15–31

Konzak O, Heßler P (2006) Emissionshandel in der Praxis. C.F. Müller, Heidelberg

Kyoto Protocol to the United Nations Framework Convention on Climate Change, adopted December 11th, 1997, UN document number FCCC/CP/1997/L.7/Add.1. In: International Legal Materials 1998, pp 32 ff. Date of entry into force: February 16th, 2005

Lafeld S, Hüwener M, Sandhövel A (2006) Risk management in the emissions trading market from a business perspective. In: Elspas M, Salje P, Stewing C (eds) Emissions trading. A practical trading book. Carl Heymanns, Köln, pp 481–491

Marci M (2005) Trading. In: Lucht M, Spangardt G (eds) Emissions trading. Economic principles, legal regulations and technical solutions for climate protection. Springer, Berlin, pp 117–136

Marshall R, Fischer T (2004) Best practice use of SEA – industry, energy and sustainable development. In: Schmidt M, João E, Albrecht E (eds) Implementing strategic environmental assessment. Springer, Berlin, pp 673–691

Mintzberg H (1994) The rise and fall of strategic planning. Free Press, New York, NY

Mißler-Behr M (2001) Fuzzy-based controlling tools. Development of fuzzy approaches. Gabler, Wiesbaden

Requate T, Graichen P (2005) The rocky road from theory to the practice of emissions trading: the EU directive on CO_2-emission trading and its national implementation. Perspektiven der Wirtschaftspolitik 6(1):41–56

Rosenkranz F, Mißler-Behr M (2005) Identify and manage business risks – introduction to quantitative planning. Springer, Berlin

Salje P (2006) Affected of emissions trading. In: Elspas M, Salje P, Stewing C (eds) Emissions trading. A practical trading book. Carl Heymanns, Köln, pp 67–83

Sandhoevel A (2006) Foreword. In: Antes R, Hansjürgens B, Letmathe P (eds) Emission trading and business. Physica, Heidelberg

Schleich J, Erhart K-M, Hoppe C, Seifert S (2006) Generous allocation and a ban on banking-implications of a simulation game for EU emissions trading. In: Antes R, Hansjürgens B, Letmathe P (eds) Emission trading and business. Physica, Heidelberg

Schmitz T, Wehrweim M (2006) Risk management: basic theory and practice. Verlag W. Kohlhammer, Stuttgart

SEA Directive (2001) Directive 2001/42/EC of the European Parliament and of the Council of 27th June 2001 on the Assessment of the Effects of Certain Plans and Programs on the Environment. In: Official Journal of the European Communities L 197, 21.7.2001, pp 30–37. Date of entry into force: July 21st, 2001

Spangardt G, Meyer J (2005) Risk management in the emission trading. In: Lucht M, Spangardt G (eds) Emissions trading. Economic principles, legal regulations and technical solutions for climate protection. Springer, Berlin, pp 219–232

Stewing C, Grit W (2006a) Measures for emission reduction. In: Elspas M, Salje P, Stewing C (eds) Emissions trading. A practical trading book. Carl Heymanns, Köln, pp 332–369

Stewing C, Grit W (2006b) Development of a risk management. In: Elspas M, Salje P, Stewing C (eds) Emissions trading. A practical trading book. Carl Heymanns, Köln, pp 370–374

Welge M, Al-Laham A (2001) Strategic management – principles – processes – implementation, 3rd edn. Gabler, Wiesbaden

Wolf K, Runzheimer B (2003) Risk management and KonTraG. Design and implementation, 4th edn. Gabler, Wiesbaden

Chapter 8
The Possibilities and Potential Advantages of the Life Cycle Assessment in the Framework of Climate Change Mitigation

Marek Gawor

8.1 Introduction

The term 'climate change' has attracted the public attention and discussion over the past 20 years. The United Nations Framework Convention on Climate Change has defined the climate change as 'a change of climate which is attributed directly or indirectly to human activity that alters the composition of the global atmosphere and which is in addition to natural climate variability observed over comparable time periods' (United Nations 1992). The US Environmental Protection Agency set up the starting date for these human-dependent changes on a large scale at the beginning of the Industrial Revolution around the year 1750 and following increase of the greenhouse gases concentration in the atmosphere. Subsequently, the US EPA has identified observable changes related to this phenomenon, namely:

- Atmosphere changes;
- Temperature changes;
- Precipitation and storm changes;
- Sea level changes;
- Ocean acidification;
- Land cover and land use change (US EPA 2012).

As one can see, the consequences of climate change are very broad and affect various compartments of the natural and human environment. In recent years, however, the term has been simplified and associated with the increase of the mean temperature on earth. The indicator chosen by scientists and politicians to measure the climate change rate and the industrial and political measures is the carbon dioxide (or CO_2-equivalent for other greenhouse gases) emission. It has

M. Gawor (✉)
Faculty for Environmental Science and Process Engineering, Department for Waste Management, Brandenburg University of Technology Cottbus-Senftenberg, Lehrstuhl Abfallwirtschaft, P.O. Box 101344, 03013 Cottbus, Germany
e-mail: Marek.Gawor@B-TU.de

E. Albrecht et al. (eds.), *Implementing Adaptation Strategies by Legal, Economic and Planning Instruments on Climate Change*, Environmental Protection in the European Union 4, DOI 10.1007/978-3-540-77614-7_8,

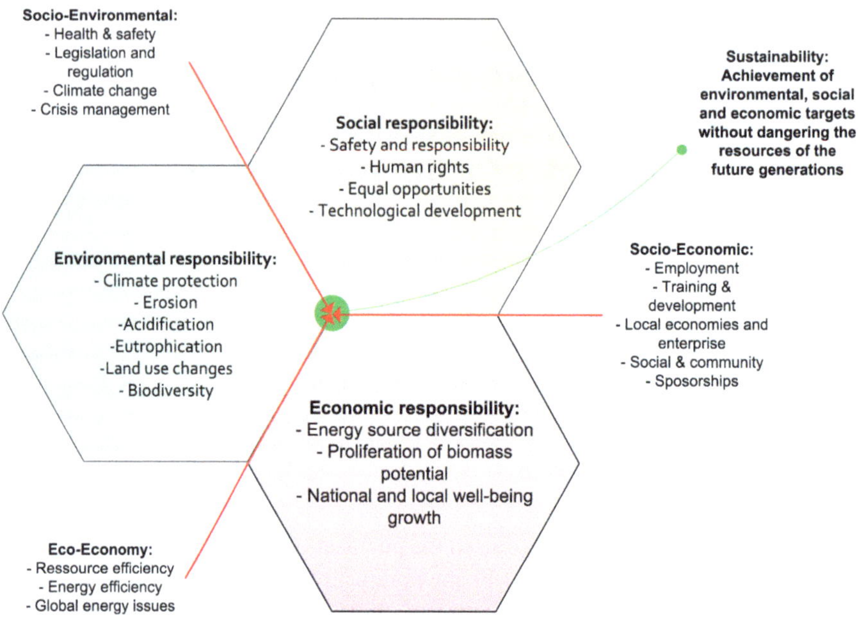

Fig. 8.1 Elements of the sustainable approach (partially adopted from Verify Technologies Limited (2010), after Thrän and Gawor (2012)

been introduced in the intergovernmental agreements such as the Kyoto Protocol, European level standards (EU Renewable Energy Directive) or the national legislation (Renewable Fuel Standards issued by the US EPA and other national plans aiming at the GHG reductions).

Another strategy for combating the environmental changes in general and climate change in particular is the application of voluntarily standards and certification schemes. These non-binding measures may be seen as softer ways to protect the environment than laws and directives.

Above-mentioned legal and non-legal environmental measures relate to the more holistic concept of sustainability, namely balancing social, economic and environmental spheres of life (Fig. 8.1). Several tools have been developed in order to facilitate the development of the sustainable products, services or political measures. One of the most widely applied decision supporting tools is the Life Cycle Assessment (LCA).

The LCA studies of the environmental damages caused by the product or service cover all environmental aspects of the product life cycle, from the raw material acquisition to the final disposal of the product. Furthermore, this technique enables assessing the environmental impacts in the 'environmental' part of the sustainability question, such as climate protection through the minimization of the GHG emissions, erosion, acidification and eutrophication. Thus, the LCA technique has become a practical basis for the current decision-making, i.e. in Germany when creating and updating the Renewable Energy Sources Act. Therefore, understanding of its implications for the final results as well as the various methods of performing the LCA will become a crucial part of the climate change abatement activities.

This chapter will summarize and present the methodologies and tools for performing the LCA studies useful for the assessment and mitigation of the climate changes. The first sections will describe the methodologies of the LCA according to the ISO standards as well as national and international laws on the example of the EU directives and the German Renewable Sources Act. Secondly, the core element of the LCA will be presented, namely Life Cycle Impact Assessment (LCIA) according to the different methodologies and approaches.

The next section will present some case studies of various LCA performed for the products such as gas turbines or the bioenergy production. The most important issues of the current discussion on the indirect land use changes and uncertainties of the delivered results will be given as well. Subsequently, the possible use of the modelling results will be discussed in the framework of the CDM measures relating to the Kyoto protocol. Finally, conclusions and the future developments in this field of the research will be discussed. As the outcome, the comprehensive overview of the possible LCA application in the climate change abatement will be presented.

8.2 LCA Methodology: ISO Standards, EU Laws and Directives, Labels and Certification

8.2.1 Introduction

The origins of the LCA technique may be traced back to the post-war period, when it became clear that the use of resources and energy carriers may create significant damages to the environment. The first forerunners of the methodology, however, focused more on the savings that could be made in the production processes and were initiated by the commercial players such as the Coca Cola Company. The first LCA studies, bearing the name Resource Environmental Profile Analyses (REPAs), appeared in the late 1960s and the beginning of the 1970s. They investigated the difference between various packaging materials or the energy requirements of the production of various goods. The interest in such research strongly increased after the energy crises of the 1970s and oil shortages, which forced the industrial players to put their efforts to producing more energetically efficient (Horn 2009).

As Horn indicates, the subsequent developments of the LCA have led to the creation of the systematic, inclusive and analytical approaches to the environmental impact assessment. One has to mention that although the principles of the Life Cycle Assessment have been applied for more than 50 years, the method is considered to have still a potential for improvement. The main advantage (and at the same time problem area) of the LCA is the flexibility of use—it may be applied for any product or service using varying assumptions and value-based choices, thus creating results which practically shall not be compared with each other. The difficulties of performing the comparative LCA studies is especially well visible in case of modelling bioenergy pathways, where interests of scientific, political and industrial lobbying groups are colliding with each other. Therefore, the activities of the international organizations or the transnational political bodies such as the EU aiming at the creation of the well-defined framework for performing the LCAs is becoming even more important than before.

As for the internationally accepted LCA standards, Elcock (2007) identifies those created and/or being under development by the International Organization for Standardization (ISO), Society for Environmental Toxicology and Chemistry (SETAC) and the United Nations Environmental Program (UNEP). Since the ISO norms are the oldest and the most widely accepted standards, the next sections will shortly describe what mandatory and optional elements of the LCA studies should look like according to those standards. Furthermore, it will be shown how the principles of such international frameworks were adopted and implemented in the European legal measures for the minimisation of the GHG emissions. Subsequently, the main tools used by the LCA practitioners will be described, followed by the description of the LCIA methods, including advantages and disadvantages of the specific approaches.

8.2.2 ISO Standards Shaping the LCA Process

The International Organization for Standardization (ISO) is the most widely known international body that creates and publishes standards for almost any technical and engineering branch worldwide. So far, the organization consists of 162 national standards bodies from all regions of the world; furthermore, its portfolio comprises of more than 18,000 standards for business, government and society with practical tools for all three dimensions of sustainable development: economic, environmental and societal (ISO 2010a). For the LCA studies, the most important group of standards belong to the ISO 14000 family, relating to all kinds of environmental management and organized according to the Plan-Do-Check-Act logics (Fig. 8.2). However, the LCA activities originally located in the planning phase of the product development, may be used also in three other phases of the cycle, namely as the basis for the environmental labelling ('do'), as a measure for checking the environmental performance of the products ('check') and to implement the corrective action, as well as to develop new policies or standards. The application of the LCA on all these managerial levels is very well visible in the area of the renewable energies, such as biofuel certification schemes.

All environmental standards created by the ISO organization are being subject of the continuous development and update process, depending on current market demands. For instance, the upcoming new standards ISO 14067 and 14069 will provide guidelines on the calculation of products' carbon footprints (ibid). Since the

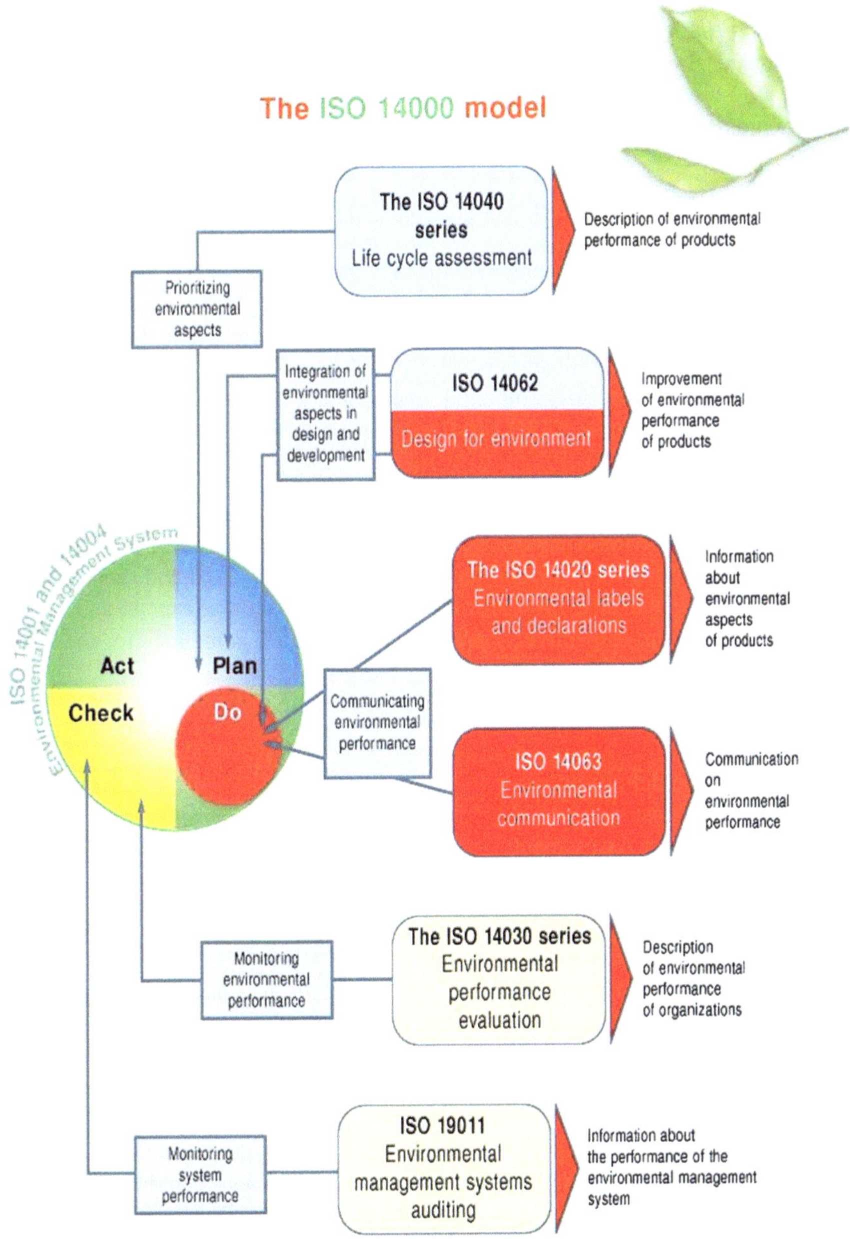

Fig. 8.2 The most important ISO 14000 family standards (Gawor 2005; ISO 2007)

carbon footprints give an insight into the amounts of greenhouse gases emitted during the products' life-cycle, these standards will have a high potential of application in climate mitigation actions. However, the existing standards, especially those belonging to the 14040 family, also provide a sound basis for a GHG mitigation assessment in conjunction with the LCA perspective. Currently, those standards include:

- ISO 14040:2006: Environmental management—Life cycle assessment—Principles and framework;
- ISO 14044:2006: Environmental management—Life cycle assessment—Requirements and guidelines;
- ISO/TR 14047:2003: Environmental management—Life cycle impact assessment—Examples of application of ISO 14042;
- ISO/TR 14049:2000 Environmental management—Life cycle assessment—Examples of application of ISO 14041 to goal and scope definition and inventory analysis (ISO 2010b).

The ISO 14040 and 14044 standards are the core elements of the LCA studies for the practitioners. The first standard describes the basic principles of the LCA; the latter one (substituting the previous standards 14041 for the goal and scope definition and inventory analysis, 14042 for the life cycle impact assessment, and 14043 for the interpretation, respectively) present more detailed guidelines for the LCA process.

In general, ISO standards distinguish between four main phases of the LCA study, namely:

- Goal and scope definition;
- Inventory analysis;
- Life Cycle Impact Assessment;
- Interpretation.

In the goal and scope definition the general framework of the study needs to be determined. The LCAs may be created for various purposes. Thus, the quality requirements may be different for a study performed for internal company purposes ('screening LCA') and for a comparative assertion of various products, whose results will be available for the public. In the ideal case, the LCA encompasses the entire life-cycle pathway of the product, including the data about raw material acquisition, production, use, and waste treatment (Fig. 8.3). However, the time and financial constraints may require a simplification of the study and narrowing the boundaries of the system under the study. Such simplifications shall be specified in the goal and scope definition. Furthermore, the choice of the functional unit should be defined in this phase of the LCA. The functional unit specifies what kind of function the system performs (i.e. making 1 cup of coffee or transporting 1 t of material over a distance of 1 km). The systems and products should be compared on the basis of the same functional unit.

After specifying the framework of the study, the inventory analysis shall take place. Here, the input data is collected, including all materials, energy and waste

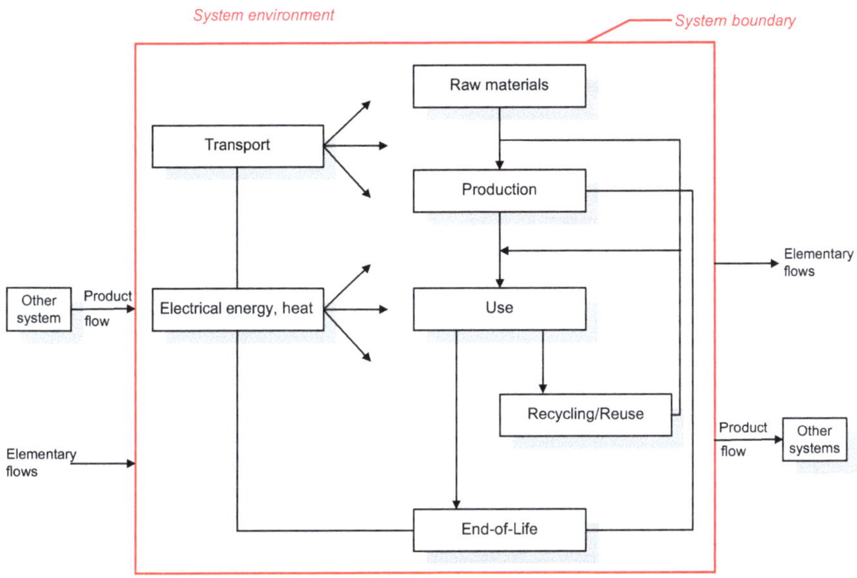

Fig. 8.3 The elements of the LCA product system according to the ISO 14040 family standards. Adopted from ISO (1997, 2006a, b)

streams being part of the product system within specified system boundaries. This element of the LCA is usually very time-consuming, and includes the usage of tools such as questionnaires or literature research. In case of bioenergy modelling, the inventory analysis methodology may greatly influence the final results, especially in cases where no scientific consensus exists, for example in the methane reductions from the use of residues materials from anaerobic digestion. The use of these residues as a fertilizer reduces the need for artificial fertilizers; however, the values of these reductions and subsequent GHG mitigation may be different on various fields and under various climatic conditions.

In the Life Cycle Impact Assessment, the collected LCI data are converted and assigned into groups of indicators relevant for specific environmental compartments or similar type of impacts according to environmental mechanisms (Fig. 8.4). Thus, assessment of the product's impact may be done in a holistic manner, including the effects on i.e. human health, ecosystem quality or resources. In this phase, the impacts may be sorted according to the value-based choice. Some LCIA methods allow even a creation of single-score results for all products. Since, however, such value-based choices may not be decoupled with some uncertainty of the results, the ISO standards declare those measures as optional ones.

The mandatory LCIA elements are as follows:

- Selection and definition of impact categories—identifying relevant environmental impact categories (e.g. acidification, eutrophication, terrestrial and marine eco-toxicity, global warming potential, etc.);

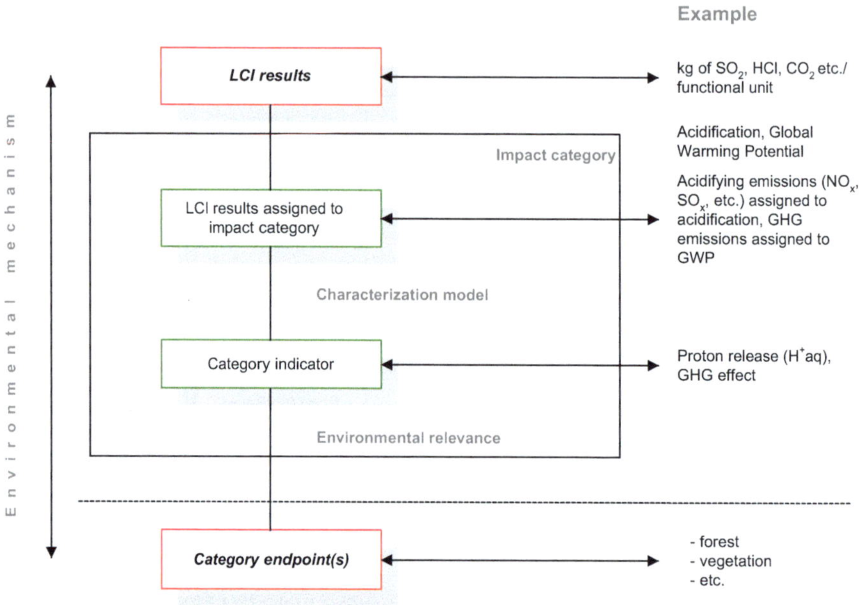

Fig. 8.4 Sample scheme of the calculation of the category endpoints in the LCIA phase. Adopted from ISO (2006a)

- Classification: assigning LCI results to the impact categories (e.g. classifying CO_2 emissions to the global warming effect);
- Characterization—modelling LCI impacts within impact categories using science-based conversion factors (i.e. modelling of potential impact of CO_2 and methane emissions on the global warming);
- Normalization—expressing potential impacts in ways that can be compared (e.g. comparison of the impacts on global warming coming from carbon dioxide and methane emissions; the normalization can be made on the basis of the given area and time period, i.e. GHG emissions in Germany in 2005);
- Grouping—sorting and possibly ranking the indicators (i.e. sorting the indicators by location: local, regional, and global or the impact category: human health, ecosystem quality, etc.);
- Weighting—converting and possibly aggregating indicator results across impact categories using numerical factors based on value-choices and a personal attitude to environmental problems;
- Evaluating and reporting LCIA results (data quality analysis)—gaining a better understanding of the reliability of LCIA results (ISO 1997, 2006a, b; Gawor 2005).

The last LCA step—interpretation—in fact occurs at every previous stage of the LCA. The collected and/or obtained data are being analysed and interpreted to give an impression about the impact of the analysed system on the environment and the plausibility of the collected data. Theoretically the LCA does not require the use of any specific software. However, the complexity of modern products and processes, as well as numerous LCIA methods available for the practitioner make the LCA studies almost impossible to be made without LCA-specific tools. The next section will shortly present the two most widely applied programs, namely SimaPro and Umberto, along with an example of the LCIA method integrated in both tools.

8.2.3 Methods and Software Facilitating the LCIA

The production of even simple goods such as a pen or a computer mouse may involve several hundred materials and processes; in case of more complex products (TV sets, washing machines or industrial gas turbines), the number of system elements easily exceeds several hundred thousand. One of the biggest challenges of the LCA was to allow decision-makers to model their products in a simple, transparent and effective way on the one hand and simultaneously satisfy the scientific requirements set up by the ISO norms on the other hand. The solutions to this—at first glance impossible—task are provided by the specialized software packages, such as SimaPro, GaBi, GEMIS or Umberto. This non-exhaustive list may be easily extended since several institutions have created their own software models for performing the LCA (compare Gloria 2010). However, only few of them are recognized worldwide, creating a sound basis for comprehensive and reliable studies. Certainly SimaPro and Umberto belong to this group. Their strengths may be summarized as follows:

- Easy-to handle graphical user interface;
- Inclusion of several databases with several categories of materials, processes, and waste management options, which enable even inexperienced users to perform simple screening LCAs within minutes;
- The possibility to create and compare various scenarios on the basis of a single product;
- Inclusion of ready-to-use LCIA methods. These methods classify environmental burden using various impact categories as well as scientific and value-based factors. Thus, environmental burden may be shown using various perspectives;
- Enormous saving of time and financial resources when performing the LCA with the provided data;
- The possibility of creating own modules, materials, processes or valuation systems, as well as adaptation—if necessary—of the datasets provided with the software.

The disadvantages of such software have to be mentioned as well:

- The provided modules/materials are often inefficiently described; thus it is impossible to track back the most important assumptions made by database developers;

Table 8.1 LCIA methods suitable for the climate change assessments

No.	Name and developer	Details—climate change assessment	Examples of availability
1.	IPCC (Intergovernmental Panel on Climate Change)	The method specifies the climate change factors with the time frame of the 20, 100, and 500 years, relating the effects of the greenhouse gases to the effect caused by 1 kg of the CO_2.	SimaPro, Umberto, implemented within several LCIA methods and assessments
2.	Eco-Indicator 99	Damage-oriented approach classifying environmental effects into categories of human health, ecosystem quality and resources; climate change and ozone layer impact categories present; global warming according to the IPCC factors; older version (EI 95) included also summer and winter smog categories.	SimaPro (Eco-Indicator 95, Eco-Indicator 99), Umberto (Eco-Indicator 99)
3.	BEES (Building for Environmental and Economic Sustainability); National Institute of Standards and Technologies (NIST), United States	LCA tool for building and construction materials; GWP factors adopted from the TRACI method; also smog, ozone depletion, indoor air pollution and criteria pollutants (outdoor air pollution such as particulate matter) taken into consideration.	Adopted in SimaPro, available as separate software
4.	CML 1992/2001; Center of Environmental Science of Leiden University, the Netherlands	General LCIA method, including ozone layer depletion, photochemical oxidation, Global Warming Potential 100a (adopted from the IPCC).	SimaPro, Umberto
5.	Ecological Footprint	Includes the CO_2 emissions from the use of the fossil fuels.	SimaPro
6.	EDIP (Environmental Design of Industrial Products) 97 and 2003	Global Warming based on the IPCC factors in 100a time frame; stratospheric ozone depletion potentials and photochemical ozone creation potentials (stemming from the human and vegetation sources) implemented.	SimaPro

Some LCIA methods, which are not listed here, combine the modelling factors using the other LCIA methods (i.e. Impact 2002+) (Goedkoop et al. 2008; Ministry of Housing, Spatial Planning and the Environment 2000; Gawor 2005, 2010; Goedkoop and Spriensma 2001; Lippiat 2007)

- Information provided by the databases may be outdated, leading to a situation where environmental impacts assessed by the software are over- or underestimated;
- Simplifications and errors of calculations are not always clearly stated or the information is hidden for the inexperienced user behind many submenu levels.

Thus, the LCA software may be seen as a very powerful tool which simultaneously creates some risks when applied without taking into consideration its limitations. In the climate change considerations, such software greatly facilitates the assessment of the climate change measures by providing the ready LCIA methods in the form of the predefined valuation systems. Methods that may be seen as the most appropriate and suitable for this task are for instance IPCC GWP or the Eco-Indicator 99. Table 8.1 shortly summarizes key features of these methods.

Figure 8.5 presents a simplified calculation procedure applied in the Eco-Indicator 99 within the framework of the LCA. The first step is the collection of data in the inventory phase of the LCA, which includes all relevant environmental data concerning the extraction of ores, several conversion and processing steps together with material and energy inputs at each stage. Eventually transportation and disposal processes are taken into consideration. The second step is based on the resource, land use and fate analysis of the significant elements and compounds, such as heavy metals, CO_2 emissions, land use, extraction of minerals, etc. Thus, it may be determined which element/material occurring within a life cycle affects specific environmental compartments. Subsequently, exposure and effect analysis helps to determine whether the absolute values of these impacts are significant for the environment or not (since i.e. extraction of 1 kg of platinum may have a different effect on the resource availability than extraction of 1 kg of sand). Subsequently, the damages to the specific environmental compartments are expressed on the basis of real damages, such as local effects on the species or the occurrence of the respiratory problems affecting humans. These specific impacts on the environment are then collected into the more general categories, such as damages to human health, ecosystem quality and resources. Finally, after the normalization and weighting steps, the single indicator result may be presented as the final output of the Eco-Indicator 99 methodology.

An interesting option that is offered within the Eco-Indicator methodology uses three basic cultural archetypes of individual perspectives and attitudes towards environmental problems. The 'Individualist' approach accepts only proven environmental effects, believing that technology may help to avoid many problems, and sets a short-time perspective of considering the environmental effects. On the other side, the 'Egalitarian' LCA practitioner takes into consideration long-term environmental effects, even if the scientific evidence is not well proven, believing that the environmental problems may lead to the catastrophe. The 'Hierarchist approach' may be seen as a trade-off between these two radical approaches, believing that problems might be avoided if a proper policy is implemented. Hierarchist-LCA practitioners include the environmental effects on the basis of

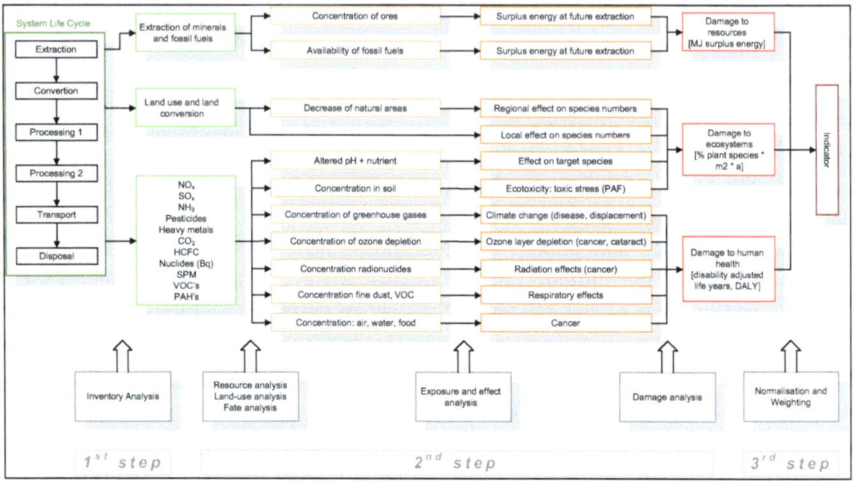

Fig. 8.5 LCIA according to the eco-indicator 99 (Ministry of Housing, Spatial Planning and the Environment 2000; Gawor 2005; Goedkoop and Spriensma 2001)

the scientific consensus with a balance between short- and long-term environmental effects (Ministry of Housing, Spatial Planning and the Environment 2000).

The choice of a specific approach may have a visible influence on the LCA modelling results (Gawor 2005, 2010). The variety of LCA methods and the lack of scientific consensus with respect to modelling approaches, as well as large results variation depending on the methodology chosen is currently being a huge hurdle for political decision makers. The next section will describe shortly the implemented legal framework relating to the above-mentioned methodologies and aiming at the mitigation of the climate change phenomenon on the basis of consistent and generally applicable methodologies.

8.3 Climate Change Assessment in Legal Acts and Voluntary Agreements

8.3.1 Introduction

This section will present key laws and standards implementing the LCA rules and simultaneously facilitating the introduction of climate change measures on the local scale of the single enterprise and interregional scales of national and international plans. The appearance of the LCA in the industry—although already present in the bigger enterprises as a part of voluntary commitment towards environmental protection—has been officially adopted in the European Union by the Energy-using Products (EuP) Directive in 2005.

On the other hand, national and international measures to reduce GHG emissions were related to the Kyoto Protocol dating back to the year 1998. The commitment to achieving the ambitious goals of the Protocol on the European level has required the creation of appropriate European laws and directives. An excellent example of GHG reductions is the increasing use of renewable energies in the energy sector, which allows saving considerable amounts of greenhouse gases when used for the electricity generation, heat provision or as a fuel in the automotive sector. An example of a transnational law facilitating the use of renewable energies might be the Renewable Energy directive, as well as national laws and action plans enforcing the use of renewable energies. One of the most famous and effective national laws is the German Renewable Energy Sources Act (EEG), which has led to historical change of the energy market in Germany through the guaranteed feed-in tariffs and focused support on desired technologies and substrates (such as combined heat and power systems or the manure from the agricultural farms). Since all above-mentioned laws may be seen as successful measures for the climate change mitigation, directly or indirectly relating to the LCA technique, their short description will be presented in the following sections.

8.3.2 Energy-using Products Directive[1]

The Directive 2005/32/EC of the European Parliament and the Council from July 6, 2005 introduced a legal framework for the setting of an ecodesign requirement for industrial products that use any kind of energy—fossil fuels, electricity or renewable energy sources. Simultaneously, the directive requirements are applicable both to products that are manufactured in the EU and imported. This definition excluded transportation and small-production systems (below 200,000 units of sale within the EU internal market). The EuP Directive introduced the ecodesign requirements in accordance with the LCA approach, stating that:

- The entire LCA shall be covered, starting from the raw material selection and use, manufacturing, packaging, transport, distribution, installation, maintenance, use and the end-of-life;
- Aspects taken into consideration in the LCA shall include consumption of materials, energy and resources, as well as emissions to air, water and soil, and pollution through physical effects such as noise, vibrations, radiation and electromagnetic fields. Furthermore, the expected generation of waste materials (especially hazardous waste) and the waste minimization options (such as reuse, recycling, recovery of the energy and materials) shall be considered.

Information about specific environmental performance of products as well as the way of impact minimization shall be clearly stated to the consumers, allowing them

[1] Unless otherwise stated, information in this section was stemming from European Union (2005)

to choose the products with better environmental characteristics. Simultaneously, this information shall provide hints about the installation, use and maintenance of the product, helping to minimize the environmental impacts, as well as options for the end-of-life disposal. Thus, with the help of this basic information, the large-scale GHG emissions may be avoided. Examples of such consumer behaviour triggered by the product's manual are refrigerators with an integrated freezing compartment. Most of the manuals suggest defreezing the products by taking them out of the freezing compartment and putting them in the cooling compartment one day before preparing the meal. Thus, the energy used otherwise for the cooling may be saved, simultaneously contributing to the climate change abatement.

The EuP Directive sets up general and basic requirements without specific and quantified requirements on the GHG emissions or any other impacts. Although one may ask a question about the effectiveness of such legal framework, it may be seen as a step in the good direction. As in the case of the renewable energies, the high diversity of the industrial products does not allow to create very specific requirements within such framework directive. However, it may be expected, that such requirements will be set in the future for all products offered on the EU market.

8.3.3 *Renewable Energy Directive (RED Directive)*[2]

The EuP Directive has created a general framework for the ecodesign methodology of the individual products. Its impact on the fulfilment of GHG reductions on the European level may be, however, predicted in a very restricted way. The fulfilment of the ambitious GHG targets required issuing the regulation specific for the industry branch, which creates a significant part of the climate change impacts, namely the energy generation. The Directive 2009/28/EC came into force on April 23, 2009 with the main focus on the promotion of the renewable use of energy from renewable sources. The main targets of the Directive were the following:

- GHG emissions reduction with the help of renewable energy sources (which should have the priority and guaranteed access to the energy grid), energy savings and increasing efficiency (by using i.e. combined heat and power, CHP systems) in the energy sector;
- Promotion of the security of the energy supply, technological innovation and providing regional development incentives (including export prospects) and employment opportunities;
- Encouraging decentralized energy production, thus supporting regional development, shortening the transport distances and reducing energy transmission losses as well as creating a local added value;

[2] Unless otherwise stated, the information in this section was taken from European Union (2009).

- '20-20-20' target: in the year 2020, 20 % improvement of the efficiency shall be achieved, with the 20 % of the renewable energy part in the overall energy share (10 % for the transport sector), with 20 % GHG emission reductions. These values represent generic levels for the entire EU energy market mix. The country-specific targets are to be achieved with the help of so-called national renewable energy action plans. The directive established procedures of including the creation of the guarantees of origin for electricity, heating and cooling from renewable energy sources;
- Offering the agricultural sector additional income possibilities through supporting the use of manure, slurry and agricultural residues as an input in the biogas plants;
- Setting up a framework for the development of sustainability criteria for bioenergy sources;
- Creating a basis for LCA studies focusing on renewable energy sources.

The last two aspects, mainly sustainability and the LCA characteristics of the bioenergy sources, are the key elements that have to be understood and correctly applied in order to fulfil GHG targets on the European scale. As for the LCA methodology, the Directive 2009/28/EC provides the practitioners clear rules for the calculation purposes, such as i.e. the energy contents of biofuels, normalisation rules, default-values for biofuels/LCA processes, GHG emissions factors and, finally yet importantly, the minimum GHG savings for biofuels and bioliquids together with their fossil comparators. The GHG savings of the biofuels represent the key figure from the Directive, since the minimum value is set up to the 35 % in comparison to fossil comparator. If the GHG saving is smaller than a predefined value, biofuel may not be treated as sustainable and thus cannot be taken into consideration in the overall fulfilment target of the EU member country. From the years 2017 and 2018, the minimum required GHG saving will amount to 50 % and 60 %, respectively.

8.3.4 *Renewable Energy Sources Act in Germany and Its Role in the Development of the Green Energy Market on the National and International Scale*[3]

The Renewable Energy Sources Act has been implemented in the year 2000, while its most important novels came into force in the years 2004, 2008 and 2011. Its main features may be characterized as follows:

[3] Unless otherwise stated, the information in this section is based on the English translation of the Renewable Energy Sources Act and its general description (Federal Ministry for the Environment and Nuclear Safety 2010a, b).

- It guarantees the priority of the connection to the grid, purchase and transmission and payment for the electricity from the installations using renewable energy sources, such as i.e. hydropower, tidal power, wind energy, solar, geothermal and wind energy, energy from biomass including biogas, landfill gas and sewage treatment gas, as well as biowastes of municipal and industrial origin;
- This electricity is being paid based on constant feed-in tariffs, set up in advance for a period of 20 years;
- These fees are based on real costs of electricity (depending i.e. on the type of substrate input in case of biomass installations or the costs of installing other renewable energy equipment, as in case of photovoltaic systems), size of the installation (smaller installations having greater marginal costs obtain higher payment), year of commissioning (thus creating the incentive to invest into renewable energies earlier rather than waiting for the decrease of the installation costs), type of the technology applied as well as the grade of technological development (i.e. additional bonus is given for future and promising technologies such as Organic Rankine Cycle, ORC, or the generation of combined heat and power, CHP, which leads to the higher overall conversion efficiencies).

Section 27 of the EEG provides details of the feed-in tariffs for the biomass sources. The EEG payments for 1 kWh of generated electricity range from 11.67 cents (installations up to 150 kW) to 9.18 Euro-cents (installations between 150 and 500 kW), 8.25 Euro-cents (installations between 500 kW and 5 MW) and 7.79 Euro-cents (installations between 5 and 20 MW), respectively.[4] Furthermore, it is possible to co-fire the biomass with non-renewable energy sources, but the installation operator has to provide proof of the amounts of the biomass used (subsequently, only the part of the 'green' energy may obtain the tariff payment). Additionally, the EEG act enables additional payments for:

(a) using the innovative technologies such as thermochemical gasification, fuel cells, ORC, Stirling engines, technologies of conversion of the straw and other stalk biomass (without any other substrates added), etc. (2 Euro-cents per 1 kWh);
(b) using energy crops and manure; the energy crops that might be used are listed on the so-called positive list (examples: biomass growth from meadow and pastures, fodder crops, palm oil, forest waste wood); the negative lists, on the contrary, present the substrates that may not be categorised as the energy crops (i.e. rejected vegetables, bioethanol, wood shavings and sawdust). The bonus amounts from 6 cents for the first 500 kW output to 4 Euro-cents for 1 kWh for the output between 500 kW and 5 MW. In the latter case, burning wood is paid with 2.5 Euro-cents per kWh (the first 500 kW output). Additionally, it is possible to obtain 4 Euro-cents per 1 kWh (output up to the 150 kW) and

[4] These and subsequent figures are valid for the installations commissioned in the years 2008–2011. It can be expected that the payments will decrease in the succeeding EEG novels due to the decreasing overall costs of the bioenergy generation.

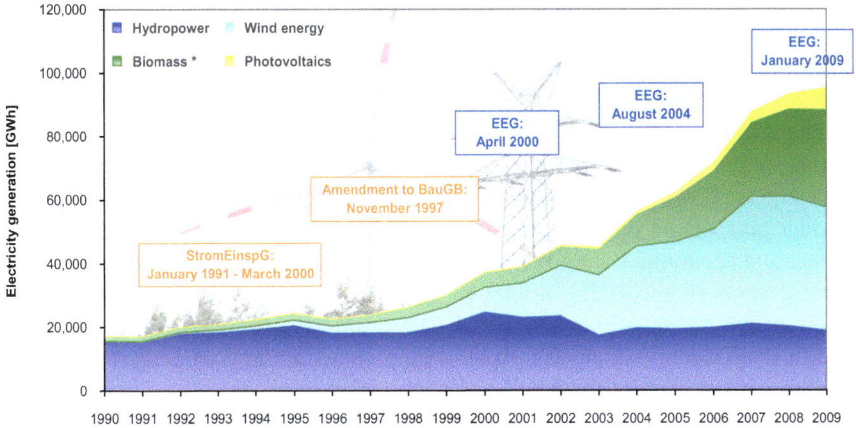

Fig. 8.6 Development of electricity generation from renewable energy sources in Germany 1990–2009 (Federal Ministry for the Environment and Nuclear Safety 2010a, b)

1 Euro-cent (output between 150 and 500 kW), if the amount of manure in the both cases exceed at least 30 % of the input substrate. Furthermore, another 2 cent bonus is possible for the installation generating electricity as part of the landscape management (up to the 500 kW output);

(c) using the combined heat and power systems (3 Euro-cents per 1 kWh); in case of the CHP, the EEG act defines the positive and negative lists of eligible installations as well.

As one can see, the EEG regulations provide a rather complicated system of the remuneration of renewable energy. Many of those bonuses are combinable with each other; thus, the EEG act has influenced strongly the shape and behaviour of the renewable market in Germany. The market players (i.e. the biomass plants generating energy or the farmers planning their biogas installation, as well as industrial companies providing the necessary equipment) orient themselves according to the current EEG bonuses. I.e. nobody commissions the 150 kW plants, but rather 190 kW in order to get the full payment (it is necessary to plan the installation in such a way that after including the idle time for the repairs and maintenance the full possible EEG payment for electricity generated in the installations up to 150 kW output may be used). The same rule applies for the 600 kW plants (instead of 500 kW).

On the one hand, the EEG has allowed Germany to boost the renewable energy market (compare Fig. 8.6). On the other hand, after several changes it became hardly understandable even for the experts in the green energy branch, as well as it created the incentives for the illogical energy use (i.e. using waste heat for the stall heating in order to obtain CHP bonus). Therefore, the German government has started the discussion leading to the novel of the EEG in 2012. The Act was partially simplified with respect to the available bonuses, the values of remuneration have been lowered in many cases due to a decrease in costs (the effect of scale and learning curve in the industry). However, the main direction of the renewable energy support remained unchanged.

The EEG Act in addition to Renewable Energy Directive set up relatively specific requirements for the achievement of the climate change mitigation goals. The targets, however, may not be achieved without taking into consideration a wider perspective of the sustainable development, including not only environmental and economic issues, but also social aspects and implications of the recent developments in the GHG mitigation actions. The first attempts to clarify the requirements on the sustainability have been already identified for the biofuels. The next sections will present those requirements as well as further voluntary measures in the form of the labels and certification schemes.

8.3.5 Sustainability Requirements for the Biofuels and Bioenergy and Future Energy Generation Systems as the Tool of the Climate Change Mitigation

The member states of the European Union as well as other industrialized countries are promoting renewable energies to avoid or minimize the negative impacts of energy generation on the environment. However, the political targets set are rather ambitious, and thus hard to achieve with the help of the domestic biomass resources. Furthermore, in the era of the global economy it may not be prohibited (and at the same time such prohibition would be highly undesired by the developed countries) to import the biomass resources i.e. from the developing countries. Such biomass imports somehow contradict the original assumptions on the advantages of the bioenergy (being locally harvested and used), but it seems that the international biomass trade will grow strongly in the future. The biomass production, however, does not have to be sustainable in the third countries; on the contrary, it may lead to even higher environmental damages, which may not be overbalanced by the GHG reductions through the bioenergy production. In order to avoid such a situation, developing countries are currently putting efforts to create generally accepted sustainability standards. These standards shall ensure that imported bioenergy carriers (such as pellets or palm oil) will help combating the climate change on the global level without creating the so-called leakage effects, namely reducing the GHG emissions in one country with their simultaneous increase at another point of the globe, which makes the idea of bioenergy useless. The sustainability rules set up i.e. by the European Commission focus on the following problem areas:

- Protection of the areas with highly biodiverse ecosystems;
- Protection of the areas with high carbon stocks while maintaining the balance of carbon stock in other areas;
- Promoting sustainable agricultural production and forest management;

- Monitoring the GHG performance of bioenergy carriers with the help of the LCA technique, starting from the source and finishing at the final fuel along the entire process chain. Simultaneously, the proper allocation methodology shall be used for the division of emissions between the co-products;
- Reducing energy consumption and increasing the efficiency of the energy production (European Commission 2010b).

These EU sustainability requirements are relatively general; Fritsche et al. (2006) identified the additional problem areas related to sustainability considerations, namely:

- Land use, land availability and land-use conflicts, including clarification of land ownership, avoiding the negative impacts from bioenergy-driven changes in land use, and securing the priority of food supply over bioenergy generation;
- Minimization and combating of the soil erosion and degradation;
- Securing the supply of drinking water, especially in the arid and semi-arid regions, minimization of water use by i.e. optimizing the agricultural systems and re-use of the treated waste water while avoiding water contamination;
- Minimizing the socio-economic problems by improving labour conditions (such as appropriate wages, prohibiting illegal overtime and child labour, etc.) while simultaneously protecting the workers' rights and ensuring a fair share of the income and its correct distribution. The human health impacts have to be also avoided and minimized (by education in the field of health and safety, proper use of pesticides, etc.).

Many of those aspects are difficult to quantify in the official lawmaking processes; on the other hand, the market offers a variety of products, and only some of them are the subject of regulations. For many consumers, however, it is important to buy products which are provided without violation of sustainability rules. At this point of the process, the voluntary labels and certification systems offer a connection between products, legal requirement and market expectations, additionally with the benefits for the improving companies' image in the field of Corporate Social Responsibility. In the area of the bioenergy and biomass trading, the main labelling and certification systems may be assigned to one of the following types:

- Forest certification systems (with Forest Stewardship Council (FSC) and Program for the Endorsement of Forest Certification (PEFC) being active as the main umbrella forest certification organizations);
- Biomass energy crops certification systems (such as Roundtable for Sustainable Palm Oil (RSPO) or Roundtable on Sustainable Biofuels (RSB);
- Certification systems used in the power generation sector (with various labels indicating the environmental performance of the products offered, such as Green Gold Label and Electrabel/Laboretec);
- Certification systems used in the emission trading (Vis et al. 2008; van Dam 2010).

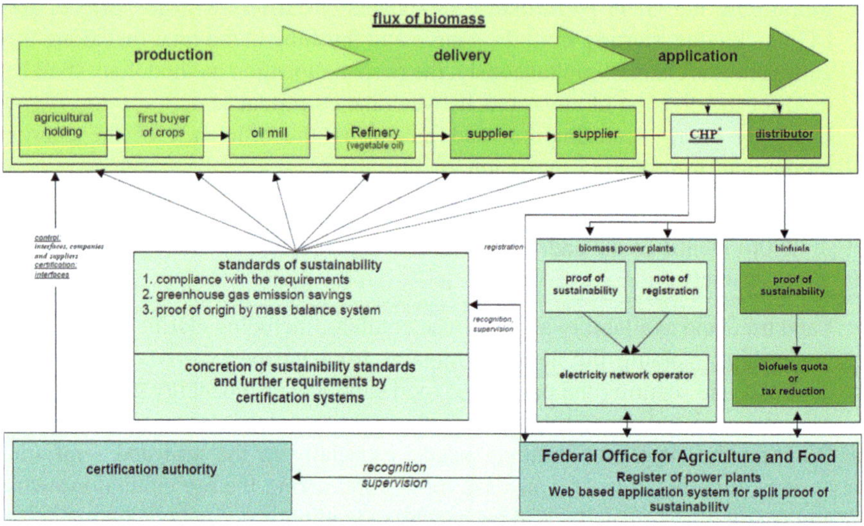

Fig. 8.7 The certification system of the bioenergy production in Germany (Schnau 2010)

Although the certification systems are the subject of the political and scientific discussion with respect to the requirements, monitoring and methodological tools, it seems that the certification of i.e. bioenergy fuels will eventually become binding at least in the European Union. Some indications on how such certifications may look like may be observed on the example of i.e. the German Biofuel Sustainability Act (Fig. 8.7). As one can see, certification encompasses all life-cycle phases of the biomass flux, including production, delivery and application. The GHG reduction potential is one of the three most important standards of sustainability that have to be fulfilled. Thus, a constant link between the existing LCA methodology and the binding legal acts is established.

8.4 LCA Projects and Their Contribution to the Mitigation of Climate Change

8.4.1 Introduction

Previous sections presented the theoretical and legal basis of the LCA and its implementation. The next section will demonstrate how the LCA/LCIA methodology may be used in practice to evaluate and minimize the human impact on the climate. Two types of LCA projects will be discussed: industry and product-oriented looking at the example of the Siemens gas turbines, as well as LCA focusing on the policy-making on the national level, using the example of the Deutsches Biomasseforschungszentrum (research centre focusing on the research in the field of energy-related biomass use) and its research on the various bioenergy pathways.

8.4.2 Comparative LCA of the Gas Turbines

Siemens Power Generation, a department of the Siemens Energy Division, with its headquarter in Erlangen, Germany and locations all over the world, belongs to the most important global players in the field of systemic solutions of energy generation, including various types of power plants (gas, steam in various configurations, from the single components to the turnkey plants) and renewable solutions (Siemens 2010). As part of its environmental activities, the company had initiated a project aiming at performing the LCA of the entire reference power plant. This section will present an overview of the partial results of the project, namely the comparative analysis of the two gas turbines offered in the past by Siemens—SGT-1000 F and V64.3. Figure 8.8 gives an impression about the scale and construction of the Siemens gas turbines.

The data for the modelling were collected during the master thesis research, using mainly Siemens internal information about the product (SAP export sheets, as well as the information from the company's portal on the environmental issues—so-called WISE). Furthermore, additional information about the material composition was taken from the internet databases on the materials and alloys used in the industry. Simultaneously, the data was adopted for use in the SimaPro modelling software, in order to find the equivalent materials in the databases and thus to represent possibly detailed impacts of the turbines on the environment, starting from the raw material acquisition to the final disposal. The LCIA modelling has been done according to the Eco-Indicator 99 methodology, using the hierarchic normalisation/weighting sets (Gawor 2005).

In general, more than 20 subassemblies have been identified with several hundreds of materials used. The characterization results (Fig. 8.9) show that the highest impact on the environment is caused by turbine parts constituting the major part of the product, namely main gear, rotor, combustor, rotor and blades as well as the outer housing. Depending on the impact category, the percentages of the impact contribution of those subassemblies vary. In case of the carcinogens, for example, the main gear, cooling air compressor as well as motor create the highest damage to the environment. For the acidification/eutrophication, higher impact is assigned to the combustor. Such differences are usually the result of the application of various materials in the production, which may be determined only after detailed analysis of the LCI results.

The normalization results for the gas turbine SGT 1000 F show that most of the environmental impact is created in the categories of respiratory inorganics, fossil fuels and minerals (Fig. 8.10). Such results are specific for products with very high amounts of the metals in the overall composition (compare Gawor 2010). The aggregation of the results into damage categories (Fig. 8.11) reveals that the

Fig. 8.8 Example of Siemens gas turbine. Here: SGT5-8000H (Copyright: Siemens AG 2007)

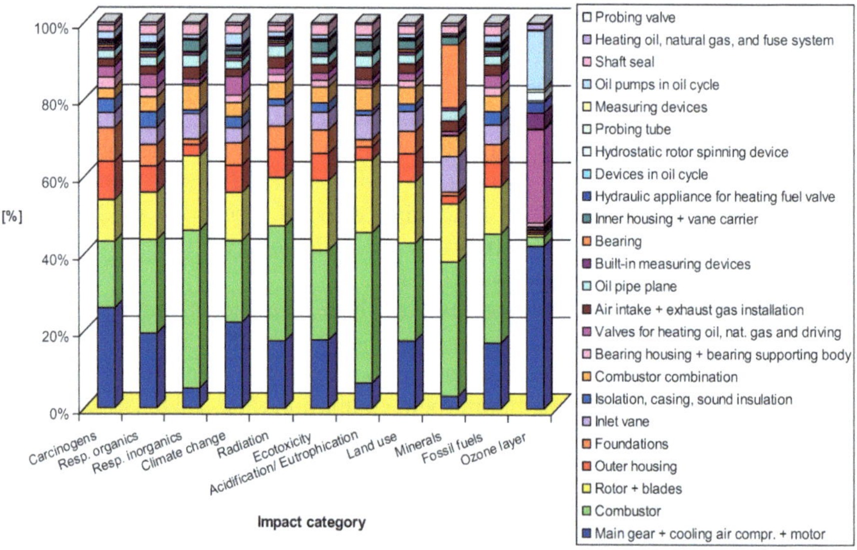

Fig. 8.9 Characterisation results for the gas turbine SGT-1000 F (Gawor 2005)

production of the gas turbine materials influences especially human health and resources, having a minor impact on the ecosystem quality. The damages to the ecosystem quality are not posing a main problem, since the material composition of the gas turbine does not contain any substances significantly hazardous for the marine or terrestrial ecosystems. However, resources are being used and the

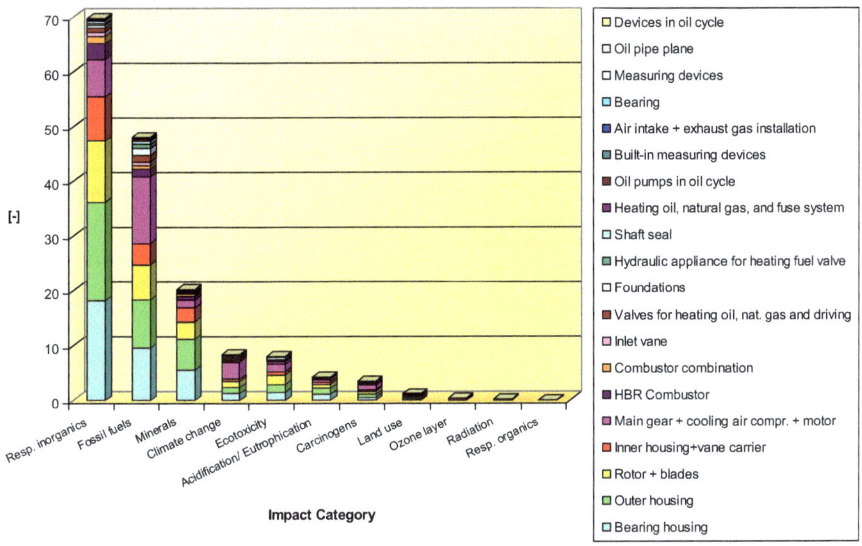

Fig. 8.10 Normalisation results, gas turbine V64.3 (Gawor 2005)

production of (raw) materials influences human health through i.e. the exhaust gas production.

However, LCA results of the single product do not comprise the full picture. One must compare possible options in order to get a more comprehensive view of the product's environmental impact. Such comparisons have to be done very cautiously, using the same assumptions, functional units, etc. (ISO 2006a, b). In case of Siemens products, the input data was collected using the same sources and methodology, thus the comparison does not contradict the LCA rules. Figures 8.12 and 8.13 present the comparison between both turbines analysed in the LCA study. It is clearly visible that SGT-1000 F, being the newer product in the portfolio of the company, has a much lower impact on the environment. In the case of ozone layer, for example, its impact is approximately 70 % smaller than the impact of the V64.3. The differences in the results are mainly due to lowering the components weight and applying the materials, which have smaller environmental burden while maintaining the product's functionality and durability (i.e. using low-alloyed steel instead of high-alloyed materials).

Another insight into the environmental performance is provided in Fig. 8.13; here, the end-of-life options were analyzed. The best scenario is the recycling, and the municipal-type disposal, like landfilling, may be positioned between the recycling and the incineration options. For all end-of-life scenarios, the environmental characteristics of the SGT-1000 F turbine indicated the progress towards production of the less-polluting and more environmentally-friendly materials.

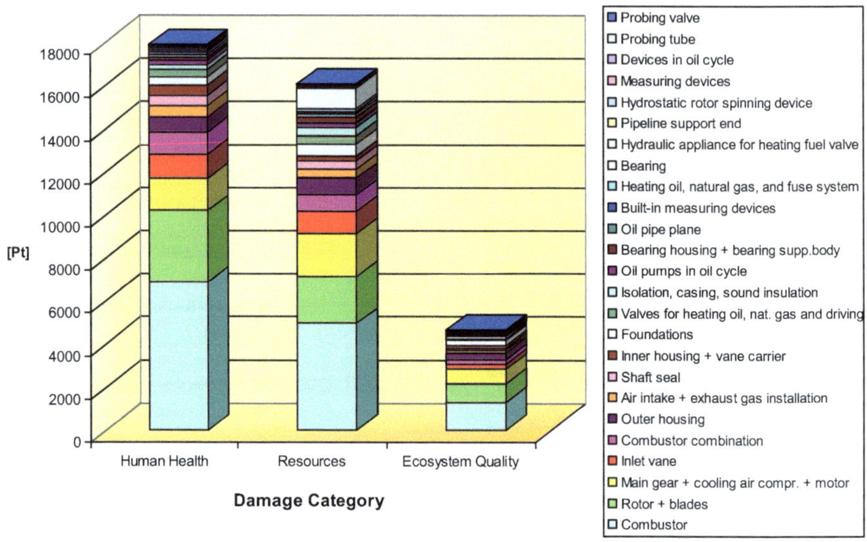

Fig. 8.11 Weighting, gas turbine SGT-1000 F (Gawor 2005)

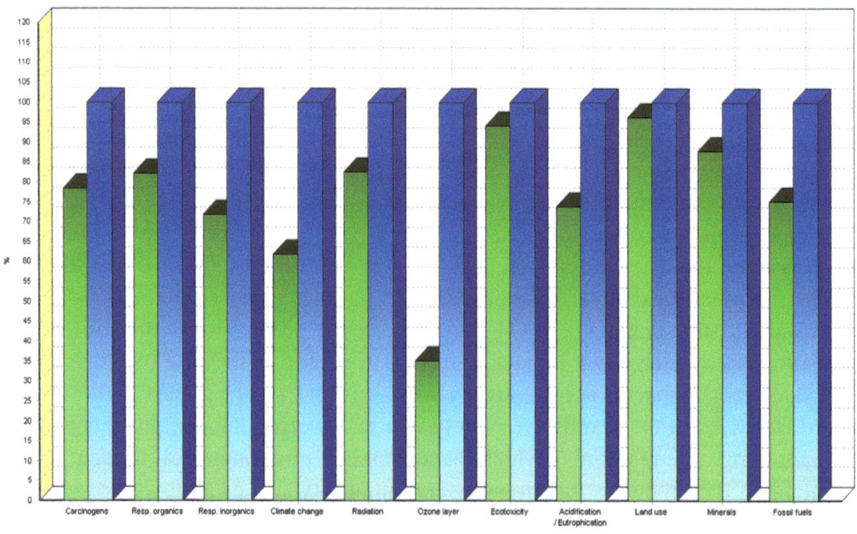

Fig. 8.12 Comparison of the environmental impacts of the gas turbines SGT-1000 F (*green colour*) and V64.3 (*blue colour*) (Gawor 2005)

8.4.3 LCA of the Biomass Energy Sources

The above-described study presents an example of the single industrial product that does not belong to the mass-consumer product group. The LCA, however, may also

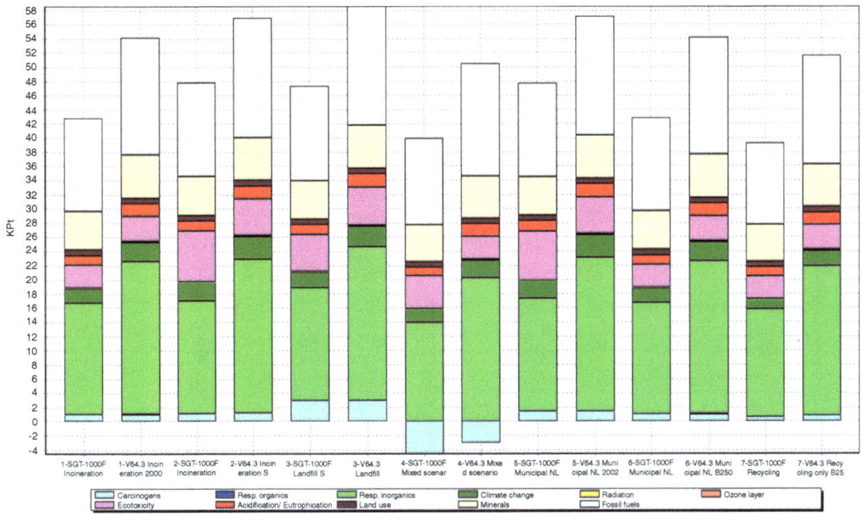

Fig. 8.13 Various end-of-life options for the Siemens gas turbines (Gawor 2005)

be used for goods and services that, through the effect of the scale, may have a much higher impact on climate change on the global scale. An example for such products may be the bioenergy carriers, especially those using biomass as substrate for their production.

Research on biomass as a regenerative energy source belongs to the main interest areas of the Deutsches Biomasseforschungszentrum (DBFZ) located in Leipzig, Germany. This research institution investigates the possibilities of sustainable provision and use of biomass for the energy sector, including the technological, ecological, economic, social and energy-management aspects along the entire utilisation chain, from the production of substrates to the conversion and utilization of energy (Deutsches Biomasseforschungszentrum (DBFZ) 2010). An example of such research may be a study made for the Agency for Renewable Resources (FNR) focusing on the economical and ecological analysis of natural gas substitutes from renewable resources (Müller-Langer et al. 2009).

The authors of the study analysed the environmental and economical impacts of the 11 various plant configurations, starting from the 2.5 MW plant through the 5, 10, 22, 75, 77 and 293 MW plants, respectively. The plants availability ranged from technical solutions already available on the market (2.5 MW, 5 MW) as well as those being planned to be technically realizable in the future (10 MW or more). Furthermore, impacts on the use of the several substrates were analyzed, including manure, forest residues, short coppice plants, etc. Furthermore, two main pathways of the biomethane production were taken into consideration, namely upgrading raw biogas from the agricultural substrates, and producing biomethane from so-called

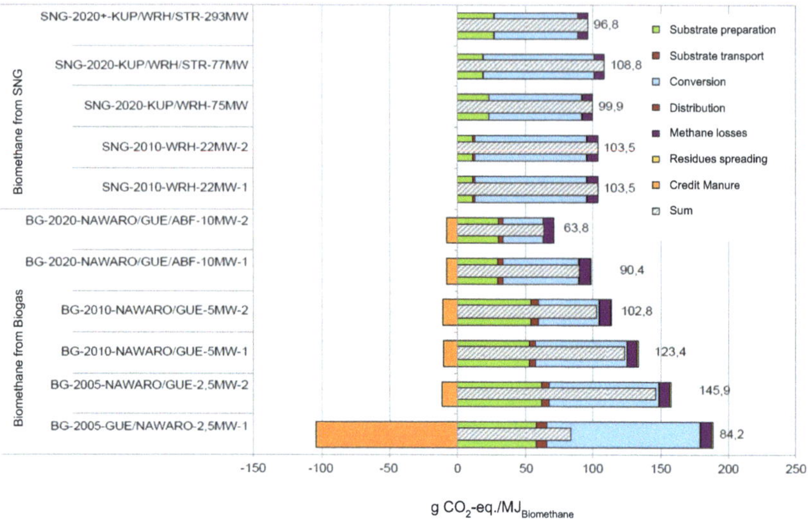

Fig. 8.14 GHG emissions of the biomethane prepared in the various installations. *SNG* synthetic natural gas, *BG* biogas, *NAWARO* renewable resources, *KUP* short rotation coppice, *WRH* forest residues, *GUE* manure, *ABF* waste material. Adopted from Müller-Langer et al. (2009)

synthetic natural gas. In the latter case, the cellulosic substrates were used (such as wood, straw, etc.).

The results of the LCA with respect to GHG emissions are presented in Fig. 8.14. Generally, greenhouse gas emissions generated in the future Bio-SNG plants are lower than in current plants, using agricultural substrates. However, the use of manure allows minimising its GHG emissions which would occur if this manure had been used as a fertilizer without previous anaerobic digestion in the biogas plant. Therefore, avoided emissions (recognized in the calculations as the manure credit) minimize the overall GHG emissions from the classical biogas plants.

Figure 8.15 compares the two possible utilization pathways for the produced biomethane. In the first case, biomethane substitutes the heating with oil and natural gas. In the second case, it is burned in the CHP plant to produce electricity and heat. As one can see, the GHG reduction is enormous, usually reaching around 70 % in comparison to the fossil energies (heating purposes only); for the combined heat and power generation, the GHG savings compared to the German electricity mix reach around 80–90 %.

The information provided by the DBFZ studies as those presented above may become a sound basis for political decision-makers with respect to promoting the most environmentally friendly energy sources and thus contributing to the fulfilment of political targets to mitigate climate change. The next section of the chapter will discuss further possibilities offered by the LCA methodology in the framework of the Clean Development Mechanism.

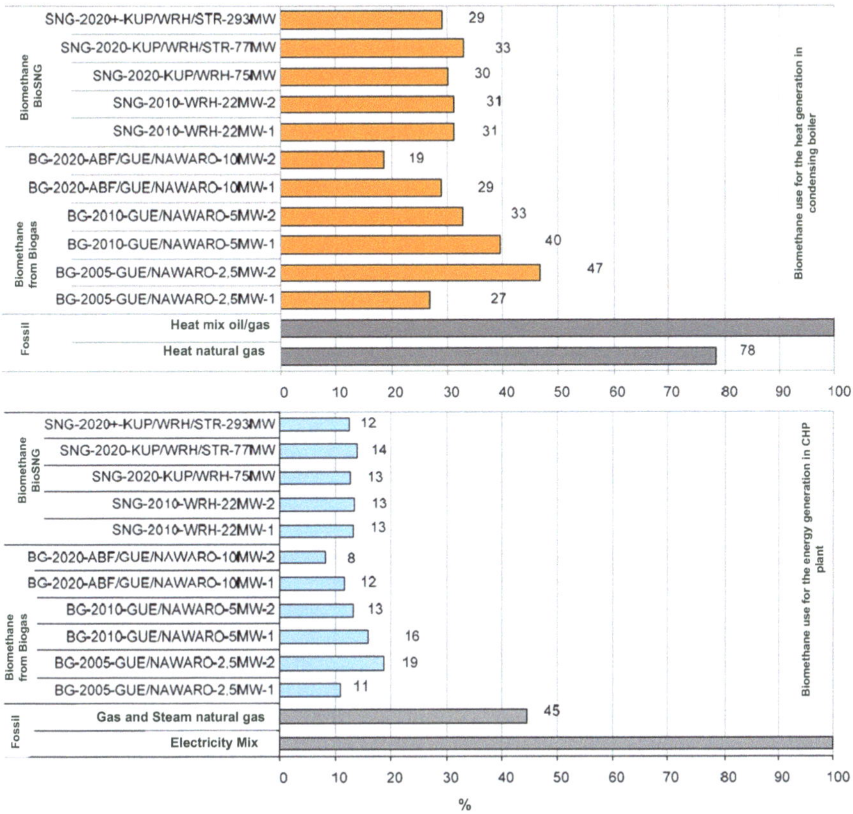

Fig. 8.15 Comparison of the various biomethane pathways with the fossil energy carriers. *SNG* synthetic natural gas, *BG* biogas, *NAWARO* renewable resources, *KUP* short rotation coppice, *WRH* forest residues, *GUE* manure, *ABF* waste material. Adopted from Müller-Langer et al. (2009)

8.4.4 LCA in the CDM Framework

The Clean Development Mechanism (CDM) belongs to one of the flexible mechanisms of the Kyoto protocol, which allows developing countries to meet their GHG gas reduction commitment in a cost-effective way. By investing in developing countries, developed countries may obtain so-called Certified Emission Reductions (CERs) which equal 1 metric ton of carbon dioxide equivalent and may be accounted to the reduction of their domestic GHG emissions (Pew Center on Global Climate Change 2009). Furthermore, these voluntary measures shall facilitate the sustainable development, as well as achieving additional (not being realized without the CDM project), real, measureable and long term benefits of the implemented projects (ibid). The LCA methodology seems to be the most appropriate and easy to apply tool for checking the effectiveness of those actions. This may be done by

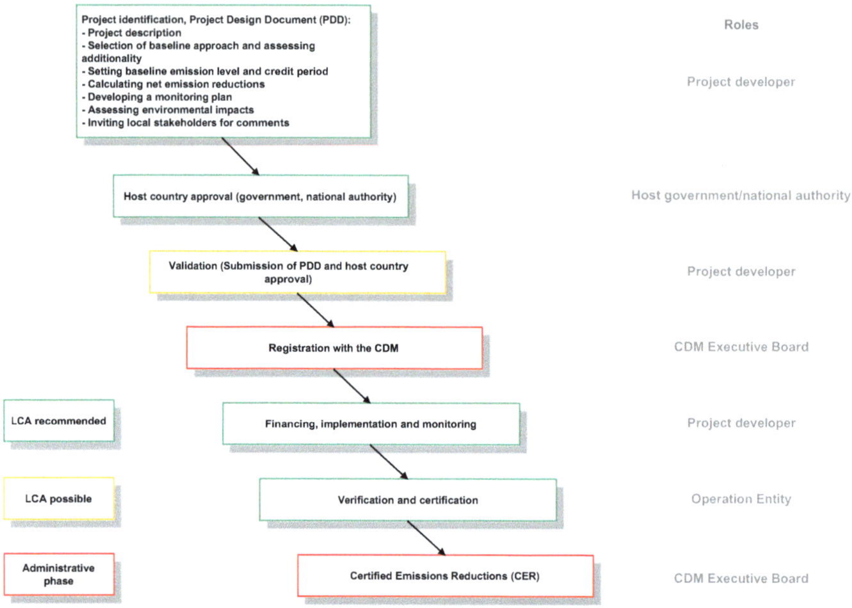

Fig. 8.16 The CDM project flow. Adopted from UNDP (2003)

creating baseline scenarios for the current situation and evaluating the environmental aspects (especially the GHG reductions) before and after the project implementation, being simultaneously helpful as the planning and the monitoring tool.

Figure 8.16 presents activities and responsibilities along the entire CDM project flow. The possibilities of the LCA application occur at almost every stage of the project management. For example, the project developer is responsible for creating the initial documentation in the form of the Project Design Document (PDD) which includes not only the basic description, but also the calculation of the net emission reductions and assessment of the environmental impacts. At a later stage focusing on the project implementation and monitoring, the LCA technique may serve as a supporting tool. The other stakeholders of the process, such as i.e. the host country's government or the national authority may use the LCA during the project approval and validation. Furthermore, the verification and certification may also use the results of the life cycle analysis.

Currently it is estimated that the number of CDM projects will reach more than 5,500, including projects from the beginning of the crediting procedures until the end of the year 2012. The areas of application range from renewable resources (such as hydropower, wind and biomass energy), through the building industry (cement production), to agriculture branches (reforestation and afforestation projects) (UNEP 2010). Certainly, the expertise knowledge necessary for the implementation of the LCA in the CDM framework may not be restricted to only one field (such as renewable energies or the industrial production). Furthermore, the specificity of the CDM and its focus on the specific single projects (which implies high transaction costs and impedes using the advantages of the economy of scale) does not allow applying it in a larger number of studies (Teng and He 2008). It seems that in the coming years the biggest challenge of the CDM will be facilitating the use of its mechanism through developing the uniform methodological base. This should include the life-cycle perspective, thus shifting the focus from the single-projects solutions to the systemic approaches. Such a comprehensive approach is already applied in the case of the renewable energy support i.e. in Germany, thus allowing combating climate change on the global scale with better overall efficiency.

8.5 Future Challenges of the LCA: Methodological Approaches and Uncertainties

8.5.1 Introduction

The high flexibility of the LCA and the possibility to apply the tool to model various products and systems was the basis of its success on the global scale. Though there exist many examples of this successful application, some problem areas remain, which need to be solved or specified in order to create a sound basis for the sake of transparency and comparability of various product systems and policy scenarios, especially in the area of bioenergy modelling and certification. This section will focus on two problem areas in relation to bioenergy systems, namely the question of modelling approaches and setting the LCA boundaries, as well as the uncertainties and assumption-based differences in the results.

8.5.2 LCA Boundaries and Modelling Approaches

Many factors influence the availability of biomass and subsequently determine whether it may be provided in a sustainable manner. It is necessary to take into consideration the environmental aspects (such as climate change or land use restrictions), economical aspects (the market conditions on the global market, the energy prices and the profitability of bioenergy generation) as well as the existing policy framework (Fig. 8.17). These factors are interconnected with each other on the regional and global scale. Moreover, these interactions are very often non-linear and indirect, thus being very difficult to quantify and predict. This complexity poses the major challenge to scientists and modellers, trying to incorporate global effects in the modelling results.

Examples of such complex interactions and the way to quantify them are widely discussed with respect to indirect land-use effects, associated with the increased production and use of biomass worldwide. The discussion was initiated by the

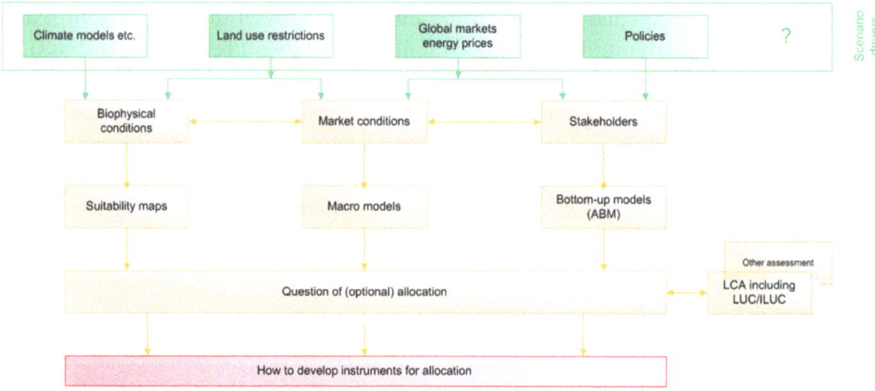

Fig. 8.17 Interrelations of the biomass product systems, their influencing factors and the role of the LCA tool (Thrän and Gawor 2012)

articles of Searchinger et al. (2008) and Fargione et al. (2008), which challenged the scientific hypothesis and at the same time the conviction that substituting gasoline with practically any biofuel will reduce GHG emission due to carbon sequestration through the growth of feedstock. The previous analysis, however, did not take into consideration the global interconnections of the agricultural and trade markets. Searchinger and Fargione identified additional emissions outside the boundaries of the classical LCAs in countries where biofuels are produced and used. These emissions occurred due to rising prices of the biofuel substrates, which consequently made the production of those substrates highly profitable. The rising prices triggered the conversion of rainforests, peatlands, savannas, or grasslands into agricultural fields to produce crop-based biofuels in Brazil, Southeast Asia and the United States. The land converted for these purposes very often could be categorized as having high carbon content. As a consequence, the conversion-related carbon releases created so-called carbon debts, which have to be afterwards balanced by the advantageous GHG balances of biofuels. However, the conversion of carbon-rich soils released huge amounts of CO_2, and subsequently the payback time for biofuels may be counted even in several hundreds of years (Searchinger et al. 2008; Fargione et al. 2008).

Obviously the discussion of these indirect land-use changes currently involves scientists, decision-makers, large biofuels and biomass producers, who often have contradictory interests. Furthermore, the complexity of the topic with respect to the modelling approaches, setting baseline scenarios, the choice of assumptions, etc. does not allow quick decisions about the inclusion of the ILUC effects, which at the same time would gain at least an acceptance of the majority of the stakeholders. Therefore, so far only the US EPA introduced the ILUC effect in its standards for renewable fuels (US EPA 2010). The European Commission has launched the process of public consultation on possible methodologies and proceedings with respect to the ILUC debate (European Commission 2010a). However, the ILUC discussion will probably shape the bioenergy markets at least in the same way as the GHG and climate change discussion have changed the energy generation patterns in the last few years.

8.5.3 Uncertainties and Assumption-Based Differences in the Results

The uncertainties and assumption-based differences may also alter the results of studies made on the basis of (theoretically) well-described product systems. For example, in the modelling of the bioenergy production on the basis of biogas/biomethane, the LCA modeller has to specify the following key issues:

- The amount of methane emissions along the production chain (i.e. for the fermentation process or the biomethane upgrading process);
- The losses of substrate along the production and distribution chain;
- The amount of fertilizer used for the production of the substrate;
- Field emissions of methane from the distribution of the fermentation residues;
- The credit for substitution of manure and its use as a biogas substrate;
- Transportation distances;
- The type of allocation procedures between co-products; etc.

The values for all those plant characteristics may be found in the literature and databases. In most of the cases, however, the differences between the values given are very high. For instance, the field emissions of laughing gas (N_2O) vary according to the local conditions such as soil type, temperature, etc. Therefore, it is very difficult to determine a single set of standard values that would be valid for the comparison between various LCA results. Thus, specifying such standard values will remain one of the main challenges of the biomass LCAs in the nearest future.

8.6 Summary and Conclusions

The ambitious targets for the climate change abatement require many organizational and financial efforts. The political decision-makers in the developed countries implemented several laws regulating especially the use of biomass and other renewable energy sources on the national and international level, as well as promoting ecodesign rules on the single-product level as a mean to reduce emissions of greenhouse gases. The Life Cycle Assessment methodology seems to be an appropriate and easy to use method which may be applied for monitoring and assessing the implemented rules and laws. This chapter has demonstrated by using example of single products (gas turbines) and product groups (biomass used for the generation of the electricity, heat and automotive fuels) that LCA is clearly an excellent tool for monitoring the environmental performance of products on the one hand, and predicting the effects of decision-making (i.e. promoting the specific fuel and energy conversion option) on the other hand. Thus, from the environmental point of view, the best decisions may be determined before their effects will occur in the real world.

The LCA as a tool is characterized in the meanwhile by a mature methodological basis, including the ISO 14040 standards and many computer-based supporting tools, such as modelling software and databases. However, recent debates have also shown that some important aspects could not be taken into consideration within the classical LCA system, since the indirect effects of the product use may strongly diminish the initial advantages from the environmental point of view. Those indirect effects often occur together with the improper land use, agricultural and forest management policies. Therefore, the importance of globally binding sustainability criteria and standards relating to the LCA principles seems to be one of the biggest challenges for scientists, politicians and nations during the next decades.

8.7 Perspectives and Directions of Future Developments

Although the LCA technique has been developed and applied for more than 50 years, there are still some unsolved issues concerning the methodology of use, especially in the context of bioenergy. In the near future, the main focus of public attention will be put on the following issues:

- The consistent methodology for modelling the specific pathways of the bioenergy production, i.e. with respect to the inclusion or exclusion of specific processes within the LCA system;
- Development of the universally applicable allocation methodology for a system producing more than one product (in the simplest case, electricity and heat; in more complicated ones, several comparable products, as in the case of biorefineries);
- Development of legally binding rules for the LCA of industrial products, especially the development of the best-case technologies, which will be then promoted by political decision-makers;
- Creation of the methodologies to include the effects, normally being outside the product system, in the environmental considerations, especially the indirect land-use changes;
- Creation of the sustainability standards for all bioenergy carriers and the political efforts to establish them as prerequisite for biofuels that shall be traded internationally;
- Support of the sustainable policies and actions in the developing countries with respect to the agricultural and forest management practices, since the improper

management policies can easily overweight the positive effects of the climate change efforts in the developed countries.

Biography Dr.-Ing. Marek Gawor, born in 1981, studied Environmental and Resource Management at the Brandenburg University of Technology in Cottbus (BTU) (2000–2005). In 2010 he obtained his Ph.D. at the BTU, with the thesis focusing on the application of Life Cycle Assessment with respect to Environmental Management Systems, EU environmental law, and systemic analysis of the Polish enterprise systems. From 2005 to 2010 he worked at the BTU Cottbus as scientific assistant and lecturer (Production and Operations Management, LCA, Product-Oriented Environmental Protection, Physical Chemistry in Environmental Protection and Science). Since February 2010 he is employed as project manager at the DBFZ, with his main focus on markets and use of biomass and ecological/Life Cycle Assessments of the biomass sources. Since 2012 he has been offering lectures at the BTU Cottbus in the area of Corporate Environmental Protection, Material Management and Recycling. Marek Gawor is currently working as a visiting Professor at the Department of Waste Management BTU Cottbus-Senftenberg.

References

Deutsches Biomasseforschungszentrum (2010) DBFZ: Research program. Available at http://www.dbfz.de/web/Research-Program.132.0.html?&L=1. Accessed 9 Dec 2010

Elcock D (2007) Life-cycle thinking for the oil and gas exploration and production industry. Argonne National Laboratory, Lemont, IL

European Union (2005) Directive 2005/32/EC of the European parliament and of the council of 6 July 2005 establishing a framework for the setting of ecodesign requirements for energy-using products and amending council directive 92/42/EEC and directives 96/57/EC and 2000/55/EC of the European parliament and of the council

European Union (2009) Directive 2009/28/EC of the European parliament and of the council of 23 April 2009 on the promotion of the use of energy from renewable sources and amending and subsequently repealing directives 2001/77/EC and 2003/30/EC

European Commission (2010a) Energy: public consultation—European commission. Available at http://ec.europa.eu/energy/renewables/consultations/2010_10_31_iluc_and_biofuels_en.htm. Accessed 11 Dec 2010

European Commission (2010b) Report from the commission to the council and the European parliament on sustainability requirements for the use of solid and gaseous biomass sources in electricity, heating and cooling

Fargione J, Hill J, Tilman D, Polasky S and Hawthorne P (2008) Land clearing and the biofuel carbon debt. Available at http://www.sciencemag.org/content/319/5867/1235.abstract. Accessed 11 Dec 2010

Federal Ministry for the Environment, Nature Conservation and Nuclear Safety (2010) Act on granting priority to renewable energy sources (Renewable Energy Sources Act, EEG)

Federal Ministry for the Environment, Nature Conservation and Nuclear Safety (2010) Renewable energy—development of renewable energy sources in Germany in 2009—graphics and tables. Available at http://www.bmu.de/english/renewable_energy/doc/39831.php. Accessed 6 Dec 2010

Fritsche UR, Hünecke K, Hermann A, Schulze F, Wiegmann K, Adolphe M (2006) Sustainability standards for bioenergy. World Wildlife Fund (WWF), Switzerland

Gawor M (2005) Comparative life cycle assessment of gas turbines V64.3 and SGT-1000 F. BTU Cottbus

Gawor M (2010) Application of life cycle assessment in the context of classical environmental management system and with respect to the implementation of the EuP directive. BTU Cottbus

Gloria T (2010) Software—LCA links. Available at http://www.life-cycle.org/?page_id=125. Accessed 1 Dec 2010

Goedkoop M and Spriensma R (2001) The eco-indicator 99. A damage oriented method for life cycle impact assessment. Methodology report. PRé Consultants bv PRé North America Inc Stationsplein 121 3818 LE Amersfoort The Netherlands.

Goedkoop M, Oele M, An de Schryver M (2008) SimaPro database manual. Methods Library. PRé Consultants bv PRé North America Inc Stationsplein 121 3818 LE Amersfoort, The Netherlands

Horn RE (2009) Life cycle assessment: origins, principles and context. In life cycle assessment. Principles, practice and prospects. CSIRO Publishing, Australia

International Organization for Standardization (ISO) (1997) ISO 14040: 1997 (E) Environ-mental management—life cycle assessment—principles and framework

International Organization for Standardization (ISO) (2006a) ISO 14040:2006 environmental management—life cycle assessment—principles and framework

International Organization for Standardization (ISO) (2006b) ISO 14044:2006 environmental management—life cycle assessment—requirements and guidelines

International Organization for Standardization (ISO) (2007) ISO 14040 model

International Organization for Standardization (ISO) (2010a) ISO's customer focus. Annual report

International Organization for Standardization (ISO) (2010b) ISO—ISO Standards—ICS 13.020.10: environmental management. Available at http://www.iso.org/iso/iso_catalogue/catalogue_ics/catalogue_ics_browse.htm?ICS1=13&ICS2=20&ICS3=10&published=on. Accessed 29 Nov 2010

Lippiat BC (2007) BEES 4.0: building for environmental and economic sustainability. Technical manual and user guide. National Institute of Standards and Technology, Gaithersburg, MD

Ministry of Housing, Spatial Planning and the Environment (2000) Eco-indicator 99. manual for designers. A damage oriented method for life cycle impact assessment

Müller-Langer F, Rönsch S, Weithäuser M, Oehmichen K, Scholwin F, Höra S, Scheftelowitz M and Seiffert M (2009) Economic and ecological assessment of natural gas substitutes from renewable raw materials. Deutsches Biomasse Forschungs Zentrum (DBFZ)

Pew Center on Global Climate Change (2009) Clean development mechanism backgrounder. April 2009 Status Report

Schnau K (2010) Requirements on Certification Systems in Germany (German approach). Federal Office for Agriculture and Food (BLE), Germany

Searchinger T, Heimlich R, Houghton RA, Dong F, Elobeid A, Fabiosa J, Togkoz S, Hayes D and Yu T (2008) Use of U.S. croplands for biofuels increases greenhouse gases through emissions from land-use change | Science/AAAS. Available at http://www.sciencemag.org/content/319/5867/1238.abstract. Accessed 11 Dec 2010

Siemens AG (2007) Press photos—Siemens global website. Available at http://www.siemens.com/press/de/pressebilder/?press=/de/pressebilder/bilder-photonews/2007/_pn200701.php. Accessed 16 Dec 2010

Siemens (2010) Siemens energy sector. Available at http://www.energy.siemens.com/entry/energy/hq/en/. Accessed 9 Dec 2010

Teng F, He J (2008) Possible development of a technology clean development mechanism in a post-2012 regime. Harvard Kennedy School, Cambridge, MA

Thrän D, Gawor M (2012) Biomass provision and use—sustainability aspects. In encyclopaedia of sustainability science and technology. Springer, Heidelberg

United Nations (UN) (1992) United nations framework convention on climate change

United Nations Development Program (UNDP) (2003) The clean development mechanism: a user's guide. One United Nations Plaza, New York
United Nations Environmental Program (UNEP) (2010) UNEP Risoe CDM/JI pipeline analysis and database. Available at http://cdmpipeline.org/cdm-projects-type.htm. Accessed 10 Dec 2010
United States Environmental Protection Agency (US EPA) (2010) Glossary of climate change terms. Available at http://www.epa.gov/climatechange/glossary.html#C. Accessed 9 Nov 2010
United States Environmental Protection Agency (US EPA) (2012) Recent climate change. Available at http://www.epa.gov/climatechange/science/recentcc.html. Accessed 10 Nov 2010
US Environmental Protection Agency (US EPA) (2010) Regulation of fuels and fuel additives: changes to renewable fuel standard program; final rule. Available at http://www.regulations.gov/search/Regs/home.html#documentDetail?R=0900006480ac93f2. Accessed 26 Nov 2010
van Dam J (2010) Overview of relevant sustainability certification systems worldwide. Brussels
Verify Technologies Limited (2010) Verify sustainability—sustaining ethical profit growth. Available at http://www.verifysustainability.com/. Accessed 10 Nov 2010
Vis M, Vos J, van den Berg D (2008) Sustainability criteria and certification systems for biomass production. Biomass Technology Group, The Netherlands

Chapter 9
Framework for Analysing Institutional Capacity for Wetland Management: The Case of the Gemenc Floodplain

Hendrike Clouting, Wim Douven, Elena Ostrovskaya, Beata Pataki, and Klaas Schwartz

9.1 Introduction

Water is one of the most important natural resources in the context of adaptation to global climate change (GWP 2007). The changing global climate is leading to significant effects on the availability and allocation of water in most river basins. Water collection, filtration and storage in wetlands will play an increasingly important role to secure drinking water and fish supply, availability of water for irrigation and wastewater management or even flood protection. Wetlands provide important services for local communities (e.g. food, drinking water, wild products and raw materials for industry) and play an important role in water purification, flow regulation, and spreading of waterborne diseases. At the same time many wetlands are vulnerable to climate change, e.g. to changes in water allocation or nutrient loading. Junk (2002) and Zsuffa et al. (2008) predict that the increased droughts, water consumption and waste water production will result in further degradation of wetland ecosystems worldwide.

The EU-funded WETwin project studies the position of wetlands in river basins and their interrelation from a physical and institutional perspective. One of the tasks of the project includes an analysis of institutional capacity for the management of

H. Clouting (✉) • W. Douven • K. Schwartz
Department of Integrated Water Systems and Management, UNESCO-IHE Institute for Water Education, PO Box 3015, 2601 DA Delft, The Netherlands
e-mail: h.clouting@unesco-ihe.org; w.douven@unesco-ihe.org; k.schwartz@unesco-ihe.org

E. Ostrovskaya
Astrakhan Center for Hydrometeorology and Environmental Monitoring, 37 Solnechnaya Street, 414028 Astrakhan, Russia
e-mail: eostrovskaya@yahoo.com

B. Pataki
Department of Civil and Structural Engineering, University of Debrecen, Ótemető str. 2-4., Debrecen 4028, Hungary
e-mail: pataki.bea.unideb@gmail.com

E. Albrecht et al. (eds.), *Implementing Adaptation Strategies by Legal, Economic and Planning Instruments on Climate Change*, Environmental Protection in the European Union 4, DOI 10.1007/978-3-540-77614-7_9, © Springer-Verlag Berlin Heidelberg 2014

several wetlands in Europe, Africa and South-America. For this purpose, an analytical framework was developed and implemented in six wetlands in Europe, Africa and Latin America. In this chapter, the framework is presented and its application for the case of the Gemenc floodplain in Hungary is briefly discussed.

9.2 Objectives, Definitions and Methodology

Due to the aforementioned problems, primary responsible, competent and advisory institutions in wetland management will have to proactively adapt their management objectives and practices to (often uncertain) changing environmental conditions. Adapting to a changing situation requires considerable institutional capacity (Box 9.1) of organisations involved in wetland management. The overall objective of the research in the 'management practices and institutional settings' work package of the WETwin project was to develop a framework for adaptive institutional capacity as response to the changing situation in wetland management.

Box 9.1 Definitions of institutions and capacity (Baser and Morgan 2008)

Institutions
Structures, norms and rules of behaviour, both formal and informal, that shape social order of a human community.

Capacity
That emergent combination of individual competencies, collective capabilities, assets and relationships that enables a human system to create value.

Capacity development
The process of enhancing, improving and unleashing capacity; it is a form of change which focuses on improvements.

While applied in the climate change literature the concept of capacity reflects the increased recognition of people's ability to face climate-related and other natural hazards, which was not captured in the mainly negative concept of vulnerability (Gaillard 2010). Gupta et al. (2010: 459) stressed that 'institutions, traditionally conservative and reactive, will now have to support social actors to proactively respond through planned processes and deliberate steps, but also through cherishing and encouraging spontaneous and autonomous change, as well as allowing for institutional redesign'. Institutional capacity refers to the resources and assets communities possess to resist, cope with and recover from disaster shocks they experience, however it is not the opposite end of vulnerability on a single spectrum, because highly vulnerable communities may display a large range of capacities (Davis et al. 2004). The concept of capacity also encompasses the ability to either use and access needed resources and thus goes beyond the sole availability of these resources (Kuban and MacKenzie-Carey 2001). Capacities to adapt to climate and environmental change are seen as rooted in resources which are indigenous to the community and which rely on traditional knowledge, indigenous skills and technologies and solidarity networks (Gaillard 2010). This is also valid for institutions for wetland management.

After a review of the concept of institutional capacity in water management (Sect. 9.3), a framework to analyze institutional capacity in river basins and wetlands was developed in a second step (Sect. 9.4). In a next step, we describe the institutional capacity using this framework for the Gemenc floodplain (Sect. 9.5). Finally recommendations for the case-specific management and institutional requirements for the implementation of Ramsar guidelines in the Gemenc floodplain will be presented (Sect. 9.6).

9.3 Institutional Capacity in Water Management

Institutional capacity for wetland management will become increasingly important in a changing environment. Some authors operationalise the term 'capacity' at different levels—the 'individual', the 'organisation' and the 'enabling environment' that includes institutional setting (Hilderbrand and Grindle 1997; Olsen 2003; van Loon et al. 2010; Alaerts and Kaspersma 2009), while others address the interrelation between capacity and performance (Morgan 2006; Hooper 2005; OECD-DAC Network on Governance 2006). Baser and Morgan (2008) argue that these interrelations are complex and must be considered in relationship to the socio-political dynamics of the context within which they take place (external context, stakeholders, external interventions and internal features and resources).

In the water sector, the concept of capacity development was defined during the 1991 UNDP symposium on 'Water Sector Capacity Building" by its three elements of:

- the creation of an enabling environment with appropriate policy and legal frameworks;
- institutional development, including development of community participation or institutional strengthening of communities; and
- human resources development and the strengthening of managerial systems.

These three elements mainly address Olsen's (2003) first order, and to some extent the second order. In Dinar and Saleth (2004), decomposing water institutions as a method used in institutional analysis is performed at two levels. At the first level, the water institution is decomposed in terms of three broad institutional components: 1. water law, 2. water policy and 3. water administration or organization. At the second level, each of these institutional components is decomposed further to identify its constituent institutional aspects. Based on Olsen's (2003) orders of outcomes, Huntjens et al. (2009) developed a methodology with a score card and indicators to assess what order outcomes are achieved for comparing water

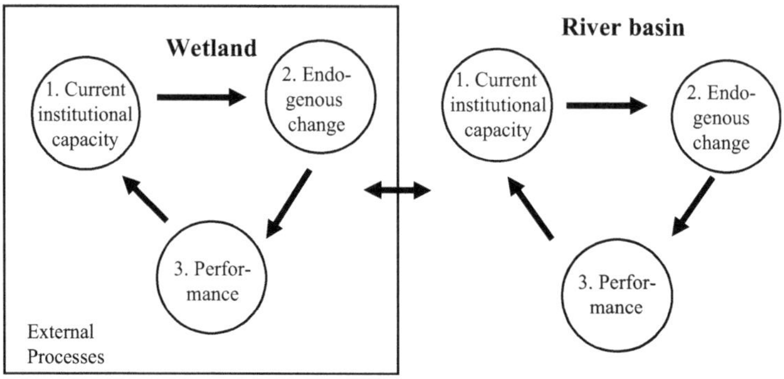

Fig. 9.1 Conceptual framework for analysis of institutional capacity (Douven and Schwartz 2009, modified after Baser and Morgan 2008)

management and governance regimes in river basins. This was an inspiration to select indicators in WETwin.

The order models of Olsen (2003) and Baser and Morgan (2008) give an important perspective on capacity for use in WETwin, for a number of reasons: they link capacity, change and performance; incorporate adaptive capacity; and allow for a development perspective over time (gradual development of capacity, moving over time from one order to the other), which is realistic in a project with European and southern wetland sites. According to Baser and Morgan (2008), capacity is considered in terms of the overall ability of a system to create public value and apply this understanding to the specific area of wetland management. Consequently, the framework focuses on the overall capacity of institutions[1]:

- to use and manage wetlands wisely;
- to integrate wetlands into river basin planning and management;
- to better adapt to changes including climate and environmental changes.

The WETwin framework for assessment of wetland management structure and practice addresses primarily the first two orders of outcomes of Olsen (2003) (institutional capacity and endogenous change) and only secondarily the third order (performance) (Fig. 9.1).

9.4 Adaptive Institutional Capacity for Wetland Management

This section presents the conceptual framework of institutional capacity proposed for WETwin. First, the link to the theoretical Driving Forces-Pressure-State-Impact-Response (DPSIR) system (OECD 2001; EEA 2004) will be addressed

[1] Not considered are organizational or individual capacities.

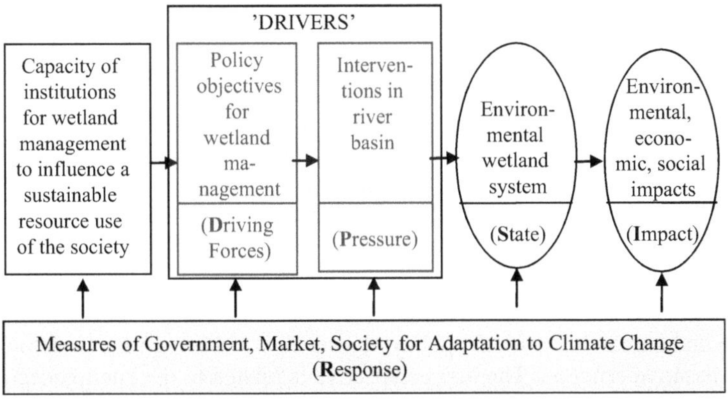

Fig. 9.2 Interrelation of the DPSIR approach and institutional capacity (modified from Helbron 2008)

(Fig. 9.2). Based on the reviewer's opinion which will have to be accepted by the Consortium in the WETwin project, the distinction between 'Driving Forces' and 'Pressures' was eliminated and they will be called uniformly 'Drivers', so the DPSIR chain is modified to 'DSIR'.

Secondly, the 'R' (Response) of the DSIR framework and its link to institutional capacity in WETwin will be outlined. The DSIR framework is quite similar to the Hydrology-Ecology-Livelihood-Policy (HELP) framework that is applied to basins and watersheds (UNESCO) (http://www.unesco.org/new/en/natural-sciences/environment/water/ihp/ihp-programmes/help/). These integrated approaches in analysing wetland systems are of importance in order to assess systems and evaluate impacts of responses considering all relevant aspects (system components, cross scale/sector interactions).

It is envisaged that the enhanced (institutional) capacity of the wetland sites and river basins will lead to change and better performance (Fig. 9.1). Baser and Morgan (2008) advise, however, to be modest and acknowledge that these interactions between capacity, change and performance are complex and highly influenced by external factors. There are two options for the operationalisation of the term institutional capacity: the four levels indicated by Alaerts and Kaspersma (2009) (civil society, enabling environment, organisation and individual) or the two levels of 'enabling environment' and 'organisation' [in line with van Loon et al. (2010) and Morgan (2006)].

The 'R' of DSIR can be a government, market or social response. It is a response to undesired situations, or future undesired situations. All three types of responses will be driven by planning processes taking different shapes. In general, different perspectives (or theories exist) which explain how (policy) decisions come into being. Leknes (2001) describes three theoretical perspectives which aim to explain how decisions are made:

- a limited and bounded rational perspective, which assumes that public decisions are based on evaluating the goal achievement of different solutions;
- a new institutional perspective where laws and regulations, standards, established procedures and norms and values in organisations can explain the decisions; and a negotiation perspective where decisions are the result of the participant's resources, interests, antagonisms and alliances.

Figure 9.3 presents the WETwin conceptual framework linking DPSI to Response (based on Ramsar's Critical path method, see http://www.ramsar.org/). For a good response (institutional) capacity is needed. Capacity, change and performance are interrelated. Performance will give an indication as to what extent the P, S and I are in line with the policy objectives set (this requires feed-back loops and an iterative process). The first perspective is related to the intelligence-design-choice structure of (bounded) rational decision-making (Simon 1960), which is the basis for policy analysis approaches (Miser and Quade 1988), and depicts a number of decision-making steps in policy making assuming that planning takes place in a more or less procedural manner. This is a useful policy analysis approach. However, it should be kept in mind that this approach also has its limitations.[2]

Pressure (P), which as direct driver is part of the 'Drivers' in the WETwin approach originates from interventions affecting the environmental wetland system (S). These can be developments of the river or wetland (upstream) impacting on functions/services of the wetland or river (downstream), such as a dam construction, development of industry, drainage, and wastewater discharge. Both the developments and their effects on wetland functions and services have economic, social and health impacts (I) on different levels: the spatial, temporal, jurisdictional and institutional scale.

Table 9.1 indicates the relation of the framework for the assessment of the wetland management structure and practice to the four-order model of Olsen (2003) and Baser and Morgan's model (2008). The framework addresses the first order (enabling environment), but not as detailed as Dinar and Saleth (2004). The framework addresses the second order (change) mainly through choosing for a more in-depth analysis of cases to better understand practice and the separate section on adaptive capacity. The framework addresses the third order (Performance) partly.

Beside formal regulations and procedures, the framework reviews relevant informal mechanisms for wetland management.

[2] A variety of authors have argued that the reality of the policy process is very different and much more irrational and incremental than the rational approach would suggest (Lindblom 1959; Cohen et al. 1972; Kingdon 1984; March and Olson 1984) or as Turner and Hulme (1997, p. 58) argue 'any lingering idea that policy is some highly rational process in which expert technicians are firmly in control using finely tuned instruments to achieve easily predicted outcomes' must 'be banished'.

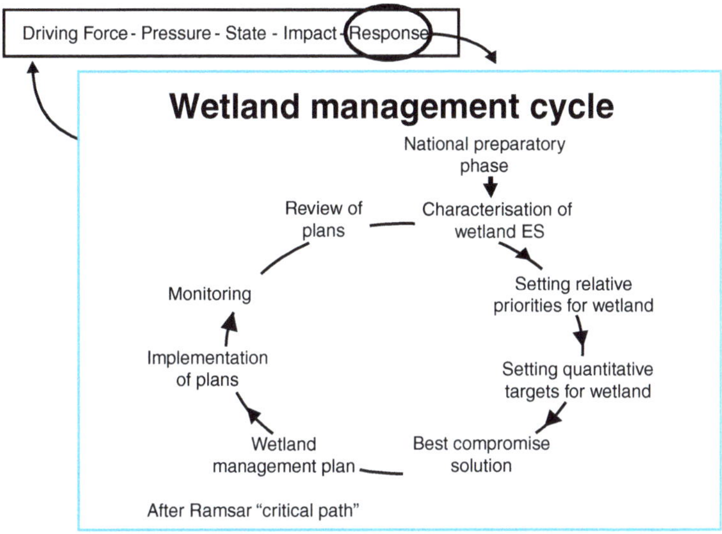

Fig. 9.3 Relation DPSIR and Ramsar planning process

9.5 Case of the Gemenc Floodplain in Hungary

In this subchapter, first findings from the application of the above framework and selected indicators in one of the case study wetlands, the Gemenc floodplain in the Danube River Basin in Hungary, will be presented. This wetland is of international importance and is partly located in the Danube-Dráva National Park (DDNP) and designated as a Ramsar site due to its high value for biodiversity (e.g. its water bodies and alluvial forests).

The ecological importance of the Gemenc is very high as it is part of the ecological biotope network and a habitat for several endangered species such as the Black Stork or the White-tailed Eagle. Important ecological links to African wetlands exist, where a significant percentage of the Gemenc Black Storks overwinters.

Human overexploitation of land and resources, the change of the flood regime of the river with more frequent and more intensive floods, and desiccation (the combined consequence of sedimentation on the floodplain and incision in the Danube river bed definition) (Zsuffa et al. 2008) will gradually lead to the disappearance of characteristic aquatic and semi-aquatic habitats (oxbow lakes, side arms, swamps) towards poor and dry systems within a few decades. Sustainable forest management is generally implemented, but clear-cutting and planting of alien tree species is still occurring in the floodplain, which conflicts with the national park regulation.

The institutional structure, capacity and subsidiarity in wetland management at regional and local decision-making levels in Hungary is in transition. The country's change from the former system to an EU member state was with priority initiated at

Table 9.1 Framework for the analysis of institutional capacity for wetland management

WETwin framework for the assessment of the wetland management structure and practice (28 May 2009)			
Components	Elements/questions	Order model of Olsen (2003)	Baser and Morgan (2008)
General description of the river basin	• Location and area • Main economic activity and land use • Main stakeholders—water use • Main problems—the causes of those problems—the main current threats	Pressure-state-impact	–
General description of the wetland	• Location and area—physical description • Main economic activity and land use • Population around wetland—ownership—uses and users • Main problems—the causes of those problems—the main current threats	Pressure-state-impact	–
Management structure of river basin	• Description of current legislation and policies related to water management at national, provincial and local level. • Responsibilities organizations have in managing this basin	First order—enabling conditions • Governmental commitment; authority agreement, funding • Legal/institutional capacity to implement; • Clear policy and goals; ▪ Constituencies present at local and national levels.	Institutional capacity
	• Have reforms changed the tasks and responsibilities of different organizations in the basin	Second order—changes in behaviour • Changes in behaviour of institutions and stakeholder groups • Changes in behaviour directly affecting resources of concern • Changes in investment strategies	Change
Management structure of wetland	• Management structure • Tasks and responsibilities • Legislation • Yearly budget and funding	First order (see above)	Institutional capacity

	• Formal relation between the wetland management structures and other management structures (horizontal, vertical) • Main stakeholders in the wetland? Involvement in management. • Formal structures for stakeholders to participate in the decision making process • Management structure appropriate to tackle current wetland problems		
Actual planning and management process (based on two cases) Here example planning is given	• Regular wetland management plan development plus goals and objectives • How is the management plan developed? Who participates? Influence of stakeholder participation on the outcome of the planning process? • The main components of the plan: • Performance indicators (economic) • Value of the wetland for ecology and livelihood? mechanisms in the plan to protect the wetlands from developments that can degrade the ecosystem • The economic social or environmental considerations • Is the management plan developed in relation with river basin management • Status in planning process in the case the wetland is a Ramsar site • Integration plan with other plans at local, regional or national level • To what extent is the plan actually implemented? • Is there any monitoring system in place • Who controls compliance with the plan?	Second order (see above) Third order—attainment of IWRM objectives • Desired social and/or environmental qualities maintained, restored or improved	Institutional capacity Performance
Use of guidelines and indicators in planning and management	• Available guidelines at national, provincial/basin municipal levels • Do the guidelines used address the major problems of the wetland • Practicality/use of guidelines—awareness stake holders	First order (see above)	Institutional capacity

(continued)

Table 9.1 (continued)

WETwin framework for the assessment of the wetland management structure and practice (28 May 2009)			
Components	Elements/questions	Order model of Olsen (2003)	Baser and Morgan (2008)
	of (inter)national guidelines • Need for additional guidance, and type of guidance • Sufficient human capacity to implement the existing guidelines. • Existing financial capacity to implement existing guidelines. • Organizational capacity to implement the existing guidelines		
	• Are guidelines used by the stakeholders/actors who are engaged in (informal) management of the wetland	Second order (see above)	Change
Adaptive capacity	• Existing management structure appropriate to face perceived future threats • Capacity organizations to obtain sufficient relevant information regarding possible changes to the wetland • Information available concerning the functioning of the wetland • Do the organizations have a strong external network through which it can access information? • Can changes in plans and in operational management be implemented quickly? • Are the human resources in the organizations responsible for wetland management such that they can quickly address new challenges facing the wetland? • Are the financial resources in the organizations responsible for wetland management such that they can quickly address new challenges facing the wetland?	First and second order (see above)	Institutional capacity Change

national policy level (top-down), and will take time to be fully implemented at local level. Although this process started around 20 years ago, it is assumed by the authors that it still takes some more years until the governmental system has reorganised itself. Currently, institutions responsible for wetland management in the Gemenc floodplain operate according to a 'learning-by-doing' approach. They are continuously adapting to legal and policy requirements from the EU with time, first at policy level, then at lower regional and local levels ('trickling down').

This process is going in parallel with changes in the perception/awareness and behaviour of water managers. From June 2010 on, the structure of the managing organisations has been changing and it is not yet clear how this will affect the lower institutional levels. At the highest national level, the harmonisation with the Ramsar and EU framework for wetland management is more or less achieved, but the implementation gap of higher legislations/policies at regional/local level is obvious.

The management structure is very complicated with often overlapping responsibilities. The Danube-Drava National Park is responsible for the nature conservation management of the whole floodplain. The Gemenc Zrt. is in charge for the management of the forests and games in the floodplain (with the permission of DDNP). The local Inspectorate for environmental protection, nature conservation and water management is authorized for planning, while the local Environment and Water Directorate is the operational body that is responsible for river basin management planning. However, the higher policy framework provides limited operational methods and tools for the lower scales, Sustainability objectives for the water management of the Gemenc wetland should therefore be implemented in the river basin management plan at sub-basin unit level. The fact that river basin plans prepared in 2010 only include a few sentences about problems and management targets for wetlands, but instead focus on the river body, seriously hampers the plan implementation at the wetland level.

Additionally, conflicts arise between competing sectors and land uses. In the case of the Gemenc floodplain, conflicts exist between the national park authority, being in charge of the Ramsar site and National Park, and the economic development of forestry. The competent wetland management institutions lack capacity in human resources, financial means, technical tools etc. to avoid or mitigate further harm of the environment and degradation of the wetland. For instance, rehabilitation measures initiated by DDNP are being financed by the World Bank under the Reduction of Nutrient Resources Project.

Beside new formal legislations and procedures being in place, instruments and planning culture from the past are still used by wetland managers. In these conditions the need to implement policy for adaptation to climate change from international (EU) and national frameworks places an additional challenge for the wetland managers. The main goal of the proposed and on-going interventions is to restore the seriously degraded hydrological regime of the floodplain, since it determines the state of ecosystems and the quality of their services. The plans of measures were designed to adapt the system to the present hydrological regime. Considering current conditions and even historical data series does not help to predict long-term impacts on the floodplain. Our model-based analysis allows us to conclude that the restoration project can bring back the desired aquatic habitats only in the short run, even if water retention is pushed to its limit. In the long term, the conditions will quickly deteriorate, especially with regard to aquatic and semi-aquatic ecosystems.

Defined outcome targets in WETwin (Table 9.2) cover both formal regulations and procedures, as well as relevant informal mechanisms for wetland management.

9.6 Conclusions and Recommendations

Drawing on a review of relevant literature, the Ramsar convention guidelines, the framework of four elements of capacity, indicators and scoring of indicators, combined with field experience and expert judgments, this chapter analyzed capacity to manage wetlands for the case study of the Gemenc floodplain. The framework (for sorting out different influences within the capacity concept) turned out to be useful for identifying gaps in the institutional setup for wetland management and deficiencies in the actual performance of the management system. The method has certain limitations such as subjectivity in scoring indicators. It provides a general picture of the capacity, a kind of warning light that there is a need for deeper analysis or further investigation. However, it allows for gaining insights on the problems facing wetland managers and contributes to improved understanding of drivers and barriers to sustainable use of wetland resources.

Application of the framework to the analysis of the case study of the Gemenc floodplain allows for making the following conclusions:

- Overall capacity for wetland management is insufficient and particularly exists at the strategic level of the 'enabling environment', but actual implementation of the policies is lagging behind.
- The overall 'enabling environment' is rather challenging as the mix of new formal legislation and procedures, having been put in place recently, and informal instruments and planning/management practices from the past, which are still used by wetland managers and stakeholders, hinders their performance. Practices are sometimes deviating from what current policies and laws prescribe. There are also contradictions between legislations and overlaps among mandates of involved authorities. Vertical and horizontal coordination among the sectors and stakeholders involved in wetland management is hampered by bureaucratic barriers and conflicts of interests. Participation of local stakeholders in decision-making is limited.
- The fact that the Gemenc floodplain is still degrading indicates that measures undertaken for its protection and preservation were not effective enough. The floodplain sees more threats than opportunities: the wetland dries up caused by human-induced pressures, and climate change is projected to exacerbate the process. This means that international conventions (such as Ramsar), national

Table 9.2 Outcome targets and elements of institutional capacity with results for the Gemenc floodplain

Outcome targets	Wetland's distance to target	Ramsar handbooks 1–16 (2007)
(1) Enabling environment		
(a) Formal instruments: water and environmental policy, institutional and legal framework, planning regulations, financing		
Policy and legislation for IWRM are developed and strengthened	☺	Hb 7, p. 12
Wetland management issues are incorporated into water or river basin management policies	☺	Hb 7, p. 13
Legal framework for spatial planning and environmental assessment is established	☺	Hb 3, p. 83 f., Hb 13 on EIA
Formal regulations and processes for the involvement of stakeholders, community participation and public awareness are in place	☺	Hb 7, p. 14
Adequate financial resources to ensure effective operation of organizations charged with planning and management of wetlands conservation are provided	☺	Hb 7, p. 14; 20; 28
(b) Informal instruments: planning culture and traditions, participation		
Stakeholders are involved, community participation and public awareness are an important goal	↓	Hb 7, p. 14
(2) Organizational capacity and availability of means for operation		
(a) Assessment and enhancement of wetland functions		
Studies to identify the wetlands functions and benefits to water management are undertaken	↓	Hb 7, p. 20
The wetland functions are enhanced or restored	↓	Hb 7, p. 20
Status of wetlands and their biodiversity is assessed and actions needed to provide better protection measures are undertaken	☺	Hb 7, p. 28
(b) Integration of data on current and future water supply and demand (water allocation)	?	
Current and future supply and demand for water are identified	?	Hb 7, p. 20
Studies to determine the minimum and ideal flows and flow regimes required to maintain natural riverine wetland ecosystems are undertaken	☺	Hb 7, p. 23
Sustainable water allocation plans for the users including allocating water to maintain wetlands are developed	?	Hb 7, p. 27
(c) Mitigation of impacts of land use and water development projects on wetlands		
Impacts of land use and development projects are minimized	↓	Hb 7, p. 22
Impacts of water development projects are minimized	☺	Hb 7, p. 23
(d) Vertical and horizontal coordination and cooperation		
Coordination and cooperative governance between the water and wetlands sectors exists	↓	Hb 7, p. 37
Communication between water and wetlands sectors is improved	?	Hb 7, p. 36

(continued)

Table 9.2 (continued)

Outcome targets	Wetland's distance to target	Ramsar handbooks 1–16 (2007)
Knowledge and expertise exchange between science and policy is improved	?	Hb 1, p. 18–19
(e) Partnerships with relevant conventions, organizations and initiatives (knowledge and expertise sharing)		
Partnership with relevant conventions, organizations and initiatives is established	☺	Hb 7, p. 33
(3) Adaptive capacity	↓	–
The management system allows for learning from past experiences and continual improvements	↓	Hb 16, p. 17–18, 48
Cooperation among stakeholders is established	☺	Hb 16, p. 18–19
(4) Effectiveness (trend in a wetland status)	↓	Hb 16, p. 48–49

Key
? Not enough information is available to make an assessment
☺ The measures in place are close or similar to the Ramsar recommendations
↓ The measures in place are insufficient

policies and local regulatory experience have not resulted in the sustainable management of the floodplain. Capacity of the wetland management institutions in the case to deal with environmental changes such as climate change appeared to be limited. Organizations responsible for managing wetlands have inflexible planning and operational arrangements. Given that the major challenges facing the studied wetland in the near future are climate-induced, the lack of adaptive capacity is reason for concern. Conflicts between nature conservation and economic activities in the floodplain undermine the capacity of the managers to adequately respond to existing and future threats to the wetland.

It can be summarised that the Gemenc floodplain requires an improved capacity as an integrated part of wetland management with the overall objective to bring the human-induced and climate-increased degradation to a halt. The capacity for wetland management should be further strengthened with the special emphasis on wider stakeholder/public engagement, knowledge development and operational management. Adaptive capacity of the organizations involved in the wetland management should be given special attention. These have to be combined with legal restrictions to resource use in the floodplain for the benefit of valuable ecosystems and habitats.

However, it should be clear for the wetland managers that the lack of the capacity is inexcusable for inaction. Their ambition should be to continuously change, to improve the institutional capacity, and to accept that such a change takes time. With each step, capacity will improve, so as to bring about a progressive strengthening of actions over time.

Biography Dr. Hendrike Clouting is a Lecturer in Environmental Planning and Management at UNESCO-IHE.
Dr. Wim Douven is Associate Professor in Integrated River Basin Management (IRBM) at UNESCO-IHE.
Dr. Elena Ostrovskaya is now Director at Astrakhan Center for Hydrometeorology and Environmental Monitoring (Russia).
Ms. Beata Pataki is an Assistant Lecturer at University of Debrecen and has also worked as Research Associate at the VITUKI Environmental Protection and Water Management Research Institute (Hungary) while having contributed to this paper.
Dr. Klaas Schwartz (M.A. Public Administration, University of Twente; PhD. Faculty of Social Sciences, Erasmus University Rotterdam) is Senior Lecturer in Urban Water Governance in the Department of Integrated Water Systems and Governance at the UNESCO-IHE Institute for Water Education.

References

Alaerts GJ, Kaspersma JM (2009) Progress and challenges in knowledge and capacity development. In: Blokland MW, Alaerts GJ, Kaspersma JM (eds) Capacity development for improved water management. CRC Press, Netherlands, pp 3–30

Baser H, Morgan P (2008) Capacity, change and performance study report, Discussion Paper No 59B, European centre for development policy management

Cohen M, March J, Olsen J (1972) A garbage can model of organizational choice. Admin Sci Quart 17(1):1–25

Davis I, Haghebeart B, Peppiatt D (2004) Social vulnerability and capacity analysis. discussion paper and workshop report, Provention consortium: Geneva

Dinar A, Saleth RM (2004) The institutional economics of water: a cross-country analysis of institutions and performance. Elgar, Cheltenham

Douven W and Schwartz K (lead authors) (2009) Development of a research approach for researching institutional capacity and requirements for guideline and DSS development: a concept note. Deliverable 4.2/4.3; Draft version 2; Date 11 November 2009

EEA (2004) Impacts of Europe's changing climate: an indicator-based assessment. EEA Report No 2/2004, Copenhagen. URL: http://www.eea.europa.eu/publications/climate_report_2_2004. Last Accessed 13 July 2012

Gaillard JC (2010) Vulnerability capacity and resilience: perspectives for climate and development policy. J Int Dev 22:218–232

GWP – Global Water Partnership (2007) Climate change adaptation and integrated water resources management – an overview. Global Water Partnership, Technical Committee Policy Brief 5

Gupta J, Termeer C, Klostermann J, Meijerink S, van den Brink M, Jong P, Nooteboom S, Bergsma E (2010) The adaptive capacity wheel: a method to assess the inherent characteristics of institutions to enable the adaptive capacity of society. Environ Sci Policy 13:459–471

Helbron H (2008) Strategic environmental assessment in regional land use planning: indicator system for the assessment of degradation of natural resources and land uses with environmental potential for adaptation to global climate change (LUCCA). Doctoral Thesis at the Faculty of Environmental Sciences and Process Engineering at BTU Cottbus. http://opus.kobv.de/btu/volltexte/2008/584/. Last Accessed 13 July 2012

Hilderbrand ME, Grindle MS (1997) Building sustainable capacity in the public sector: what can be done? In: Grindle MS (ed) Getting good government: capacity building in the public sectors of developing countries. Harvard University Press, Harvard, pp 30–61
Hooper BP (2005) Integrated river basin governance—learning from international experience. IWA Publishing, London
Huntjens P, Aekeraj S, Henocque Y (2009) State-of-the- art of (adaptive) IWRM programs/ projects. Scorecard for ASEM Waternet Product of WP5 Water Governance
Junk WJ (2002) Long-term environmental trends and the future of tropical wetlands. Environ Conservat 29(4):414–435
Kingdon J (1984) Agendas, alternatives, and public policies, http://www.jstor.org/action/ showPublicationInfo?journalCode=admisciequar HarperCollins Publishers, New York, NY
Kuban R, MacKenzie-Carey H (2001) Community-wide vulnerability and capacity assessment (CVCA). Office of Critical Infrastructure Protection and Emergency, Preparedness, Ottawa
Leknes E (2001) The roles of EIA in the decision-making process. Environ Impact Assess Rev 21: 309–334
Lindblom C (1959) The science of 'muddling through. Public Admin Rev 19(2):79–88
March J, Olson J (1984) The new institutionalism: organizational factors in political life. Am Political Sci Rev 78:734–49
Miser H, Quade ES (ed) (1988) Handbook of systemsanalysis; craft issues and procedural choices. Wiley, ChichesterMizrahi Ye (2004) Capacity enhancement indicators. Review of the literature. WBI Working Papers. World Bank Morgan P (2003) Draft background paper on methodology DAC study on capacities, change and performance, ECDPM, June 2003
Morgan P (2006) The concept of capacity: capacity, change and performance. European Centre for Development Policy Management, Maastricht
OECD (2001) Towards sustainable development. Environmental indicators. Paris http://www. oecd.org/dataoecd/37/1/33703867.pdf. Last Accessed 13 July 2012
OECD-DAC Network on Governance (2006) The challenge of capacity development: working towards good practice. http://www.oecd.org/dataoecd/4/36/36326495.pdf. Last Accessed 13 July 2012
Olsen SB (2003) Frameworks and indicators for assessing progress in integrated coastal management initiatives. Ocean Coastal Manag 46:347–361
Ramsar Handbooks 1-16 (2007) Wise use of wetland. Ramsar convention secretariat. http://www. ramsar.org/cda/en/ramsar-pubs-handbooks/main/ramsar/1-30-33_4000_0__. Last Accessed 13 July 2012
Simon HA (1960) Management by machines: how much and how Soon? Manag Rev 49(12–19): 68–80
Turner M, Hulme D (1997) Governance administration and development. Kumarian Press, West Hartford
van Loon L, Driessen PPJ, Kolhoff AJ, Runhaar HAC (2010) An analytical framework for capacity development EIA—the case of Yemen. Environ Impact Assess Rev 30(2):100–107
Zsuffa I, Cools J, Vlieghe P, Debels P, van Griensven A, van Dam A, Hein T, Hattermann F, Masiyandima M, Cornejo MP, de Grunauer R, Kaggwa R and Baker C (2008) The WETwin project: enhancing the role of wetlands in integrated water resources management for Twinned River Basins in EU, Africa and South-America in support of EU water initiatives

Chapter 10
Adaptation to Climate Change in Developing Countries: A Need in the Niger Delta Region of Nigeria

Chika Ubaldus Ogbonna

10.1 Introduction

Climate change is a worldwide environmental phenomenon that needs urgent attention by minimizing the impacts through different adaptation processes particularly in developing countries (DCs). The global climate is subject to increasing change, and this has become more evident over recent years (Humpreys 1998; IPCC 2001a; Masika 2002; McMichael and Lindgren 2011; McMichael et al. 2012). In particular, the atmospheric concentrations of greenhouse gases have increased, augmenting global warming. According to the International Panel on Climate Change (IPCC) these concentrations of carbon dioxide, methane, and nitrous oxide are higher now than at any time during the last thousand years, and the weight of scientific evidence suggests that observed changes in the earth's climate are at least in part due to human activities (IPCC 2001b; Remme and Blesl 2008; Bidwai 2012). According to the IPCC third Assessment report the global average atmospheric temperature rose 0.6 ± 0.2 °C (1.1 ± 0.4 °F) in the twentieth century. The report further projected that global temperature will increase between 1.4 and 5.8 °C (2.5–10.5 °F) between 1990 and 2100. This projection was based on the Special Report on Emissions Scenarios. A recent report of the United Nation's Framework Convention on Climate Change (UNFCCC) states that the best estimates indicate that the Earth could warm by 3 °C by 2100 (UNFCCC 2011). Estimates on how fast global warming will occur are necessary to assess the future climate change and take measures against any adverse effect (Nishioka and Harasawa 1998).

O'Brien and Leichenko (2000) noted that scientists and policy makers have become embroiled in extensive debates about potential changes brought about by

C.U. Ogbonna (✉)
Department for Civil Law and Public Law with References to the Law of Europe and the Environment, Faculty for Environmental Sciences and Process Engineering, Brandenburg University of Technology Cottbus-Senftenberg, Erich-Weinert-Str. 1, Cottbus, Germany
e-mail: ogbonchi@tu-cottbus.de

E. Albrecht et al. (eds.), *Implementing Adaptation Strategies by Legal, Economic and Planning Instruments on Climate Change*, Environmental Protection in the European Union 4, DOI 10.1007/978-3-540-77614-7_10, © Springer-Verlag Berlin Heidelberg 2014

an increase in anthropogenic greenhouse gas emissions along with strategies for mitigation and adaptation. However, the UNFCCC has already identified two ways to address climate change: first through mitigation of climate change by reducing greenhouse gas emissions and enhancing sinks and secondly through adaptation to the impacts of climate change. Mitigation comprises all human activities aimed at reducing the emissions or enhancing the sinks of greenhouse gases such as carbon dioxide, methane and nitrous oxide (Klein et al. 2005; UNFCCC 2009). Adaptation in the context of climate change refers to any adjustments that takes place in natural or human systems in response to actual or expected climatic stimuli or their effects or impacts, aimed at moderating harm or exploiting beneficial opportunities (IPCC 2001c; Klein et al. 2005). In a nutshell, adaptation means managing the unavoidable that may occur due to the changes. These adjustments encompass changes in processes, practices and structures to reduce potential damages that take advantage of opportunities associated with climate change (IPCC 2007). Assessment of the global and regional impacts of climate change has formed the basis for climate policy debates. It is essential to note that these debates have acknowledged the fact that some areas are more vulnerable to climate change than others. However, Adger (2001) suggests that adaptation is a dynamic process. The ability of societies to adapt is more or less determined by the ability to act collectively.

10.2 The Climate Challenge for Developing Countries

Developing Countries (DCs) are entering a new era of vulnerability as a result of climate change. The effects of climate change pose threats to livelihood, human and socio-economic development of urban and rural population in the most DCs of the world particularly in Africa. It has become the new reality in the poorest and most vulnerable regions of the world where the risk-coping options and resources are limited. Moreover, the inhabitants of most DCs depend on climate-sensitive natural resources for their livelihood. For example, the most damaging human disaster of the closing decades of the twentieth century was the result of the draught in East Africa in 1984. It was concentrated in three particularly impoverished countries: Ethiopia, Sudan and Chad. About 500,000 people died while far more suffered from malnutrition or lost their homes (Giddens 2009). Similarly, DCs in the Horn of Africa are presently feeling the impact of climate change as a result of prolonged draught, which has caused severe food crisis already affecting millions of people in Kenya, Ethiopia, Djibouti and Somalia. In the northern part of Nigeria farmers experience frequent drought-related crop failure (Ekpoh 2010). This poses a threat to food security and people's livelihood. In 2000, a major disaster occurred in Mozambique when rains, floods and cyclones affected 800,000 people, and caused 700 deaths, made 250,000 homeless, increased the incidence of malaria and other tropical diseases and impacted negatively on food production (Masika 2002). During the year 2000, a large number of people were affected by drought in India; about 7,500 villages spread over 145 blocks in 15 districts were severely affected (UNCAPSA 2007). Climate change will perhaps further accelerate such crises. As increased precipitation may lead to excessive flooding, populations will need to abandon long-inhabited floodplain or construct expensive dam or move from the environment entirely. For instance, according to Eckert and Waibel (2009) a significant rise in weather extremes such as tropical typhoons and flooding events as consequence of climate change has been observed in Vietnam. The flooding events in the recent past have already placed severe strains on Vietnam metropolis (Eckert and Waibel 2009).

Climate change, according to most scenarios, will place added demands on urban infrastructure, particularly in DCs. According to the IPCC, it could accelerate urbanization, as people migrate from low-lying coastal areas to interior areas, from drought-stricken farms to cities (IPCC 1990). For instance, Han (1989) pointed out that unabated, sea-level rise will have devastating consequences for densely populated river delta areas in Egypt, India, Bangladesh and elsewhere. The author noted that inhabitants will need to migrate to mainland interior areas to escape flooding. For example, a 1-m sea-level rise would seriously affect nearly a hundred million people along the coast of China alone. Hundreds of millions of urban dwellers in low- and middle-income nations are at risk from the direct and indirect impacts of climate change (IIED 2007).

The increase in frequency and intensity of climatic changes in parts of Asia and Africa in recent decades has led to several environmental impacts. These environmental impacts are summarized in Table 10.1.

10.3 Adaptation Constraints and Opportunities

Adapting to climate change has become a necessity in any comprehensive climate PPPs, particularly in DCs. Beg et al. (2002) noted that climate change does not yet feature prominently within the economic or environmental policy agendas of most DCs. They are said to have fewer resources to adapt: socially, technologically and financially. The new situation has provided challenges in achieving the Millennium Development Goals (MDGs) and the international community efforts to reduce poverty and achieve sustainable development in DCs. According to Luterbacher and Sprinz (2001), substantial population growth has occurred everywhere, especially in the DCs of Asia and Africa and has accentuated pressures on human environment. The author further states that increase in the use of fossil fuels and fluorocarbons and widespread deforestation, along with increase in agricultural production, urban development and industrial production have led to a great rise in the quantity of green house gases. At the same time environmental degradation and the increased use of natural resources have been associated with social problems, such as mass migration.

As noted by Lwasa (2010), climate change has added a new layer to the existing challenges in Africa that adaptation is high on the development agenda. Adaptation and vulnerability to the adverse impacts of climate change are the most crucial

Table 10.1 Climate change impacts in the DCs

Environmental impacts	Socio-economic resources and sectors affected
• Changes in rainfall patterns	• Water resources
• Increased frequency and severity of:	• Agriculture and forestry
Floods	• Food security
Droughts	• Human health
Storms	• Infrastructure (e.g. transport)
Heat waves	• Settlements: displacement of inhabitants and loss of livelihood
• Changes in growing seasons and regions	• Coastal management
• Changes in water quality and quantity	• Industry and energy
• Sea level rise	• Disaster response and recovery plans
• Glacial melt	

Adapted from UK Parliament Publications (2006)

environmental concerns of many DCs in recent times. Consequently, some DCs have begun to develop alternative policy frameworks, for example through national adaptation programs. These have focused on climate-proofing infrastructure projects, such as transport and irrigation systems, improved disaster monitoring and management and better land-use planning (UN-DESA 2009). For instance the Bangladesh Climate Change Strategy and Action, adopted by the government of Bangladesh in 2009, seeks to guide activities and programs related to climate change in Bangladesh. The strategy contains 44 programs formulated around six themes which include: food security/social protection/health, comprehensive disaster management, mitigation/low carbon development, and capacity/institutional strengthening. Thirty-four programs listed under five themes are wholly or partially focused on adaptation (World Bank 2010). Adaptation measures are also incorporated into disaster preparedness in Bangladesh. Non-governmental organizations working with the Nigerian government made a recent effort by introducing a program called the Nigeria National Adaptation Strategy and Plan of Action (NASPA) which will be designed to respond to an urgent need for a coordinated and integrated approach to climate change response in Nigeria. According to the 2010 NASPA report, the program will enable federal, state and local governments, civil society, the private sector and various agencies and institutions to effectively integrate climate change adaptation concern into their development policies and programs such as water and other natural resource, agriculture, health and infrastructure. NASPA will give priority to community-level input as an important source of information, recognizing that grassroots communities are key stakeholders and providing a voice to the most vulnerable (including women and youth), ensuring that everyone is represented in the Plan (NASPA 2010; BNRCC 2010). Furthermore, Orindi and Murray (2005), acknowledged the progress being made in East Africa on integrating adaptation into the most vulnerable sectors. In their National communication to the UNFCCC, Tanzania, Uganda and Sudan identified a number of adaptation measures, including:

- Increase irrigation to boost crop production
- Introduce low-water-use crops and adopt sustainable water resource management policies (seasonal rainfall harvest; water quality control)
- Increase capital investment in reservoirs and infrastructure
- Reduce water loss through water conservation technologies
- Make water resource management an attractive career and field of investment
- Institute policy mechanism to control unsustainable forest clearing and forest consumption
- Promote techniques for tackling emergency food shortage
- Promote the use of liquefied petroleum gas (LPG) for cooking and solar cookers, instead of inefficient woodstoves and charcoal stoves
- Adjust farming areas and reduce animal population
- Conduct a comprehensive study of malaria

However, according to the United Nations report there are difficulties in scaling up adaptation projects and strategies in DCs because of funding and institutional shortcomings, as well as the failure to adopt a more broadly developmental approach. On the other hand, Perlmutter and Rothstein (2011) noted that DCs are not yet major polluters, the authors pointed out that a decade will probably pass before DCs become important contributors to global warming. The authors therefore argued that what is needed at the moment is technical assistance and enough foreign aid or private investment to begin implementing adaptation policies. Reid et al. (2012) affirms that climate change is happening now and is leading to several impacts, including changing rainfall patterns, increases in the number of floods, droughts and storms, and slower onset changes such as a rise in sea levels. Such impacts affect food security and water resources, resulting in more disasters, especially amongst the world's poorest and most vulnerable communities (Reid et al. 2012). Certainly, DCs need international assistance to support adaptation in the context of national planning for sustainable development; there is need for more capacity building and transfer of technology, funds and alternative policy frameworks. According to the 2010 UNFCCC report, systematic planning and capacity building are needed to reduce the risk of disasters and raise the resilience of communities to increasing extreme events such as droughts, floods and tropical cyclones.

10.4 Climate Change Vulnerability in Nigeria

Climate change poses a great challenge to the sustainable development of Nigeria. The potential impact of climate change at the national level could be significant. This will inevitably add to more challenges in the development efforts in the country. The Fourth Assessment Report of the IPCC states that Sub-Sahara Africa, where Nigeria is located, is particularly vulnerable to climate change in comparison to developed nations. Both physical and socio-economic factors predispose Nigeria to the adverse effects of climate change (BNRCC 2010). According to the Nigerian Environmental Study Action Team Report (2010) some of these factors include: (1) Nigeria's long (800 km) coastline, prone to impacts from rise in sea-level; (2) Nigeria's North prone to drought and desertification; (3) threatened water and energy resources; (4) more than 60 % of the population depends on threatened agricultural and fishing resources; (5) high population of 150+ million people; (6) weak policies and programs, especially among vulnerable communities and in vulnerable regions. Nigeria is vulnerable to the impacts of climate change largely because approximately 70 % of Nigerians are engaged in smallholder rain-fed agricultural production, population growth is very high and infrastructure is being strained beyond capacity. Nwajiuba (2008) envisage that important coastal cities in Nigeria such as Lagos, Port Harcourt, Calabar, Warri, and others are threatened. A study conducted by the United Kingdom Department for International Development (DFID) used an integrated analytical assessment model to show projected economic impact of climate change in Nigeria. The study predicts a potential sea level rise from 1990 level to 0.3 m by 2020 and 1 m by 2050, and rise in temperature of up to 3.2 °C by 2050 under a high climate change scenario. The prediction is based on IPCC climate change assumptions, latest research findings and results of a consultation exercise in Nigeria (FMEAN 2010). The low estimate predictions are for sea level rise of 0.1 m and 0.2 m by 2020 and 2050 respectively, and a temperature increase of 0.4–1 °C over the same time periods. The study states that climate change could result in a loss in GDP of between 6 and 30 % by 2050, worth an estimated US$100–460 billion dollars. By 2020, if no adaptation is implemented, between 2 and 11 % of Nigeria's GDP could potentially be lost (DFID 2009; FMEAN 2010).

10.5 Overview of Climate Change Vulnerability in the Niger Delta

The Niger Delta region is the richest part of Nigeria in terms of natural resources, for example large oil and gas deposits. The ecosystem of the area is highly diverse and supportive of numerous species of terrestrial and aquatic flora and fauna and human life. The region is divided into four ecological zones, namely coastal inland zone, mangrove swamp zone, fresh water zone and lowland rain forest zone (Uyigue and Agho 2007). In the Niger Delta, the southern part of Nigeria, human activities are probably what drive the change in human environmental system. Such activities include improper land use, vegetation burning, industrial activities, oil spillage, transportation, agricultural practices and gas flaring. According to the 2008 World Bank report, Nigeria accounts for roughly one sixth of worldwide gas flaring. Nigeria flares about 75 % of her gas and all takes place in the Niger Delta region (World Bank 2008). In a research study conducted by Awosika et al. (1992), it was estimated that a sea level rise of 0.2 m could lead to a land

Table 10.2 Total loss (km^2) due to coastal erosion and inundation estimated from different scenarios of sea level rise

	Low estimate				High estimate			
Sea level rise (m)	0.2	0.5	1.0	2.0	0.2	0.5	1.0	2.0
Niger Delta	2,846	7,453	15,125	18,398	2,865	7,500	15,332	18,803

Source: Awosika et al. (1992) and Uyigue and Agho (2007)

loss of 2,846 (km^2) in low estimate scenario while a 2.0 m sea level rise is expected to affect 18,803 (km^2) of land in high estimate scenario (see Table 10.2). This might lead to the displacement of between 1,000 and over half a million people in the region (Awosika et al. 1992).

Some parts of the Niger Delta region are usually subjected to seasonal flooding when rivers overflow their banks. Given the scientific prediction of rise in sea level that would be occasioned by climate change, it means that the lowland of the Niger Delta shall be exposed to higher risks with increasing changes in climate. The Niger Delta frequently experiences flood problems: For example an extreme 10-h rainfall in July 2006 drove 10,000 residents out of their home and caused widespread traffic chaos in Port Harcourt city (Gupta 2004). According to Oku (2003), soil type, vegetation depletion and climate factors like rainfall are some of the properties that render the Niger Delta prone to flooding. Annual rainfall in the Niger Delta area varied 2,000–3,000 mm (Yakubu et al. 1998). The delta has such high levels of rainfall because it is so close to the Atlantic Ocean. Coastal erosion and flooding in general may impact negatively on the livelihood of many communities in the region. Flooding and erosion remove topsoil, destroy roads, affect fresh water resources and thereby threaten lives and properties.

However, the changing climate has created uncertainty in the rainfall pattern (timing and amount of rainfall) in every part of Nigeria. The problem is more severe in the rain forest zone of the Niger Delta where rain-fed agriculture is mainly practiced (Uyigue and Agho 2007). The situation has led to delays in the planting of food crops in the region. Farmers now wait until there is enough rainfall before planting, hence irrigation is seldom practiced. Considering the recent study by the DFID and other studies by different scholars, several sectors may be affected in the region as a result of the climate change and variability. The key sectors may include: agriculture and forestry, water resources, human health, human settlement, urban environment, terrestrial ecosystem and biodiversity and coastal zones. Therefore, any effective national development planning process and effort must take climate change into account and, more particularly, must facilitate adaptation to the effects of climate change (Oladipo 2010).

10.5.1 Agriculture and Forestry

Agriculture is the mainstay of the local economy in the Niger Delta region. The inhabitants of the region depend mostly on climate-sensitive sectors such as agriculture and fishery. Climate change would have an effect on agriculture, livestock and fish production. Agriculture production could be affected in a number of ways. Extreme weather events such as thunderstorms, heavy winds, and floods are devastating farm lands leading to crop failure. Pest and crop diseases migrate in response to climate variations and will potentially pose a threat to livestock. Food security is usually vulnerable to extreme weather events such as floods and change in rainfall patterns. Climate change threatens livelihood, food security and programs aimed at elimination of poverty in the region (Onyenechere and Igbozurike 2008). There have been some observable changes in agricultural practices in the region in recent years. For example, Onyenechere (2010) noticed that in the southern parts of Nigeria where the Niger Delta is located agricultural (particularly farming) activities normally start from March; this is because by this period of the year rainfall, an active driving factor of agriculture should have arrived. But today there is abnormality in the commencement of agricultural activity for instance, within the second and third week of April in 2010, there is inadequate rainfall to plan and start agricultural practices in consonance with the arrival of rains. Furthermore, Munonye et al. (2008) found that pattern of rainfall has affected crop yields particularly yam and cassava over the years. A variety of food crops are produced in the Niger Delta region, all dependent on rainfall. Food production on the whole, however, has not kept pace with Nigeria's population increase.

There could be changes in fishery resources which are of particular importance to the inhabitants as they provide a considerable amount of dietary protein and the sector also serves as a major source of employment and labor for many. Slight changes in key environmental variables such as change in temperature, salinity and precipitation could affect ecological processes of aquatic resources such as productivity and species interaction (Kennedy et al. 2002). For instance both inland and ocean fisheries are very sensitive to varying degrees of climate fluctuations (Urama and Ozor 2010). Many researchers have identified the importance of forest in climate front. Forest resources are essential in the global carbon cycle because they store huge amounts of carbon in the biomass and soil. Poor logging practices, log poaching, fuel wood collection and uncontrolled forest fires are some of the features of Nigeria's forestry practices that contribute to forest degradation (Ezegbule 2008). The changing climate pattern may further exacerbate such stresses on forest resources in the region.

10.5.2 Ecosystem and Biodiversity

Biological diversity is essential to maintaining ecosystem processes which provide various services to maintain life on earth. Climate change poses a significant influence on biological diversity and ecosystem (Parmesan and Yohe 2003; Wiegleb and Bröring 2005; Manley 2008). Current and expected shifts in climate are threatening global biodiversity and are forcing managers and planners to re-evaluate how they plan for the protection of species and ecosystems (Rose and Burton 2009). Changes in the natural ecosystem are easily identified in a changing environment; as a natural ecosystem is usually adapted to specific climatic conditions in a particular locality. Ecosystems are a vital part of the climate system as they help regulate the climate, including through sequestering greenhouse gases (mitigation through emission removals), and regulating water flow, which can aid adaptation to flooding and drought; and therefore ecosystems should be an important part of strategies relating to climate change (Doswald and Osti 2011). Studies show clearly that changes in distribution and behavior of a large number of species are the consequences of shifts in local or regional climate, weather patterns and resulting changes of vegetation and habitat quality (UNEP/CMS 2006). Over 70 Protected Areas (PAs) in the Niger Delta region have lost substantial portions of their area, which translates to loss of biodiversity (Phil-Eze and Okoro 2009). Many species of plants and animals are rapidly becoming extinct in the region which may also lead to changes in bird and animal migratory patterns (due to the need for new habitats or new food sources). Extinction of rare and endangered species of plants and animals would also increase as a result of the predicted climatic changes. Adverse environmental change could lead to the destruction of forest crops, economic trees and animal habitat. This may affect people's livelihood as many communities in the region depend on forest resources for their sustenance (Dadiowei 2009). Global analyses show that recent climate change is already affecting species and ecosystem and will continue to do so (Yates et al. 2009).

10.5.3 Urban Environment

Niger Delta urban areas could be vulnerable to the current and projected impact of climatic changes. The impacts may be felt by people and urban infrastructure. The United Nations estimates that more than half of the world's populations already live in urban areas; and it is expected that the proportion of the city dwellers in the world population will have risen to three quarters by 2050, with almost all of the growth occurring in the developing world (UN-DESA 2009). This prediction could be associated to the urbanization trend in the Niger Delta region as a result of rural-urban migration. Most environmental changes, including those induced by climate change, are intensified by population growth. In Nigeria about 20 million people (14.3 % of the national population) live along the coastal zone where most of the economic activities that form the backbone of the national economy are located (UNEP/GRID-Arendal 2011). According to Ogba and Utang (2007) the Niger Delta coastal settlements, which are already under stress of demographic pressure and unsustainable oil exploration, are equally under the threat of sea level rise. Urban and rural settlements along the coast could also be affected by increased coastal erosion and flooding. Several incidences of flooding have already been reported in the region. Sea level rise will threaten the coastal zones and the low-lying islands, which are already constantly plagued with floods and erosions. Some settlements are known to have already relocated further inland from their original sites (Etuonovbe 2007). The predicted impact of climate change will exacerbate such a situation. Ogba and Utang (2007) pointed out that global projections of sea level rise put the area under future inundation of 100 km in land, which could however, lead to loss of land, property, economic activities and livelihoods.

However, it is important to know that cities, urban and rural residents in the region may not just be victims of climate change but also a part of the problem. It is obvious that dealing with the expected climate change threats in the region will also depend on changes in consumption patterns of the urban inhabitants. Therefore, rural and urban resident must be part of the solutions to climate and environmental challenges. At the same time, there is rising consensus that cities must take immediate adaptation measures to reduce the impact of climate vulnerability and risks. There is a great need to integrate climate change policies in urban settlement development plans to prevent future loss and destruction of lives and properties in the Niger Delta region.

10.5.4 Human Health

The impact of changing climate on health and sanitation may be significant in the region. The World Health Organization report confirmed that the health impacts of climate change are potentially huge. Many of the most important global killers such as malaria, diarrhea and protein-energy malnutrition are highly sensitive to climatic conditions (WHO 2007). However, the actual health impacts of climate change will be strongly influenced by local environmental conditions and socio-economic systems (IPCC 2001d). The changing climate will no doubt compound the serious problems of sustainability of the environment and management of resources, so as to human health in the region. Extreme climatic events will lead to an increase in mortality rate and disruption of livelihood. This could also lead to an increase in some diseases caused by insects, such as dengue, malaria (mosquitoes) and sleeping sickness (tsetse fly), yellow fever, resulting from a shift in pattern of rainfall and temperature.

A recent study conducted by Duru and Ume (2008) in Ahiazu Mbaise, a Local Government in the Niger Delta region has shown that pathogen which causes malaria is very sensitive to changing climatic variables. The study attributed high incidence of malaria to climate change especially global warming, high temperature and the presence of stagnant water as a result of excessive flooding and high humidity. Some of the direct impacts of climate change would include deaths, various illnesses and injury due to increased exposure to heat waves. Indirectly climate change and sea level rise could increase the transmission of contagious diseases including cholera and influenza. There may be exposure to increased incidences of waterborne diseases such as hepatitis A and typhoid fever. Increase in incidence of asthma, allergic disorders and cardio-respiratory diseases would also probably occur due to climate induced changes in atmospheric pollen and spore percentage and due to temperature increases that enhance the formation, persistence and respiratory impact of certain air pollutants (Ravindranath and Sathaye 2002). Studies have shown that asthma increases in the wet season in many parts of Nigeria. Thus increase in wetness anticipated for the southern part of Nigeria where Niger Delta is situated could trigger increased rate of asthmatic attack. Human health will be vulnerable when food production, livestock, agriculture, forestry; are all affected as a result.

10.6 Adaptation Policy Options: Needs and Issues

Just as climate change patterns vary across regions and countries, so do risk and adaptation patterns. However, to tackle the challenges of climate change in the region adaptation deserves due consideration in Nigeria environmental policy. In general, the effectiveness of environmental policies in Nigeria, as well as their potentials to support adaptation and mitigation measures, is yet to be fully realized, while climate change is mentioned in some key government policies, there are yet to be working policies or strategies for climate change adaptation and mitigation sector activities (FMEAN 2010). According to Oladipo (2010) Nigeria faces a number of policy-related, institutional and informational challenges, however, policy coordination, in particular, remains a major challenge for Nigeria's climate policy. The non-effectiveness of this major issue under the Nigerian national environmental policy in recent years is a negation of the global commitment signed by the country in the United Nations Conference on Environment and Development (UNCED) conference in Rio de Janeiro in 1992 to contribute to programs that could facilitate sustainable development. Nigeria has, however, made important progress in advancing critical initiative to building a national adaptation strategy, with both houses of the National Assembly passing bills to create a National Climate Change Commission.

10.6.1 Sustainable Urban Environment Adaptation

There is a need for sustainable urban adaptation planning in the region. Climate change presents a unique challenge for Nigerian built environment. Effective urban and regional planning is believed to have a key role for a city's adaptation to climate change (UN-Habitat 2009a). Many researchers are of the opinion that sustainable urban planning strategies can be used in a context of extreme vulnerability (physical, social, capacity and institutional) to mitigate and adapt to climate change. Climate change is likely to pose significant threats to the physical infrastructure and social fabric of towns and cities in DCs and in Nigeria. Settlements and infrastructures will be affected by changes in climate conditions including extreme weather events. The predicted climate change impact in urban areas in DCs could result to damage of buildings, energy services, telecommunications, transport structures (e.g., roads, railways, ports, bridges) and water services (Thom et al. 2007). For the security of human settlements in the region it is absolutely essential to integrate climate change concerns in the process of urban development planning. Adapting to the impact of climate change in Nigeria requires effort from researchers and policy makers (Nwajiuba 2008).

Adaptation planning efforts to manage high temperature at the neighborhood scale should focus on providing cool and attractive outdoor areas such as gardens and open spaces. If well designed, adaptation at this scale can also benefit internal spaces, for example streets provide evapo-transpirative cooling outdoors, as well as shading buildings (Shaw et al. 2007). Trees are essential for planning of settlements because they can help reduce flooding, reduce soil erosion, shelter buildings, streets and parks, provide habitat for wildlife, provide scenery and reduce heat on human environment; however it takes a long time for trees to grow in order to help in the adaptation process. Increase of 'green 'and 'blue' spaces and urban tree planting will be beneficial to biodiversity (McEvoy et al. 2006; Berry et al. 2008). Blakely (2007) points out that cities have three essential roles in the climate change arena. The first is to reduce the risk of climate change; second, developing risk profiles for the range of risk they face based on geography and geology of their location, and finally, developing strategies for adapting to climate change on a micro or macro-scale. According to the International Institute for Environment and Development (IIED) in many cities in DCs existing infrastructure may not be able to withstand the adverse impacts of climate change, particularly where it involves changes in the intensity and frequency of extreme events such as cyclone, storms surges, floods, heat stress and so on. This is because climate change related extreme events and gradual changes are rarely considered in city planning or in designing infrastructure in DCs (IIED 2011).

Urban planning has become very important as cities grow rapidly in developing countries. This is because urban planning plays a major role in protecting critical infrastructure and services such as electricity, water and sanitation, telecommunication, transportation system and health service (UN-Habitat 2009b). Urban planning is very essential for adapting to the effects of climate change; apparently well-planned cities provide a better foundation for sustainable development than unplanned cities (UN-Habitat 2007). Building climate change considerations into planning process and systems allow early action, which should be more cost-effective than responding to changes as they happen or retrospectively (Wilson 2006). Urban planning can contribute immensely in implementing climate change adaptation policies. Recently, there is a gradual shift of focus on mitigation actions to a concern with long term adaptation measure of, for example, preventing flooding and landslides, protecting or relocating vulnerable settlements, improving drainage and preventing new developments in areas likely to be affected by sea level rise or floods, among others (Kong and Yuen 2009).

10.6.2 Adaptation Through Sustainable Land Use Planning

The IPCC (2007) report has already identified one way of increasing adaptive capacity through introducing the consideration of climate change development planning, by including adaptation measures in land-use planning and infrastructure design and measures to reduce vulnerability in existing disaster zones. Thus adaptation is necessary to build resilience if cities, communities and sub-urban areas respond to climatic changes. Therefore there is need for effective locally driven adaptation planning to reduce the future consequences of these changes in the region. According to Cote (2011) adaptation plan refers to any land-use planning and policy actions by a regional or municipal agency that directly attempts to address the physical impacts of climate change. Adaptation planning at the municipal level involves acquiring and regulating land, incorporating new institutional management techniques, and folding and interpreting complicated science into land-use planning strategies (Smit et al. 2000; Cote 2011). Land use planning can play a key role in reducing current and future community risks associated with climate change, notably by enhancing prevention and preparedness and/or facilitating response and recovery in a community (Bajracharya et al. 2011).

Most developing countries are still in the early stages of identifying appropriate responses to climate change risks, limiting practical experience of mainstreaming climate change adaptation into national development planning (Tearfund 2006). However there are barriers to the development and implementation of climate change adaptation considerations in planning. The system of planning, control and urban development in the Niger Delta region of Nigeria is not yet prepared for the unavoidable challenges. According to Onyenechere (2010), Nigeria lacks a centralized institution that can champion and coordinate climate change and spatial plan. There is need for efficient planning before spaces are allotted to different uses as a result of the projected impacts of climate change in Nigeria. Addressing global environmental issues, like climate change, requires proactive, sustainable urban management and structure (Irvine 2008). According to Agwu et al. (2000) Nigeria lacks a comprehensive planning system. While most urban centers have development plans, others have none. Rural areas are generally unplanned and development uncontrolled. Therefore, there is need to start developing national and regional adaptation policies implemented by viable development plans. A sustainable urban development and management practices is one of the effective ways in which the consequences of climate change in the region could be reduced. Urban management, planning and land use controls can prevent people from building in zones at risk of flooding, erosion and landslides, for example restrictions on building within 50 year floodplains in South Africa (El Sioufi 2010). However, the question lies on the readiness and capacity of the rapid growing cities in the Niger Delta region in mainstreaming existing policies, strategies and plans to address climate change response. There is also a need for appropriate weather and climate forecast that can help inform decisions for land-use planning in the face of predicted climate variability and change.

10.7 Environmental Assessment as a Tool for Adaptation Planning

Environmental assessment (EA) is a procedure that predicts the environmental effects of proposed projects before they are carried out. Climate change is a major challenge which needs to be considered at all sectors particularly in urban and rural planning in the Niger Delta region. Many researchers have reiterated the importance of integrating climate change adaptation into planning process. This is because onward planning can minimize threats and maximize opportunities (Collingwood 2011). According to Albrecht (2005) environmental consequences should be considered at the planning stage and not at the time at which projects are already implemented, which often turns out to be too late to take environmental matters into account. There is need to integrate Environmental Impact Assessment (EIA) and Strategic Environmental Assessment (SEA) in adaptation programs, plans and strategies. EIA is well established in Nigeria, it came into existence almost 23 years after the United States of America had passed the National Environmental Policy Act (NEPA) of 1969, the EIA Decree No. 86 of 1992 and the Urban and Regional Planning Decree No. 88 of 1992 provided principles guiding development of projects that may likely affect the environment significantly (Federal Republic of Nigeria 1992a, b; Agwu et al. 2000). However, some pitfalls have been observed in Nigerian EIA practices, for example according to Ugochukwu (2008), the public is not given the opportunity to comment on the screening report of any of the projects subjected to full EIA. Rather public involvement takes place after the submission of the final draft of EIA report. During this stage, members of the public are allowed to peruse and comment on the reports within 21 working days on the displayed reports. On the other hand, a good-quality SEA process informs planners, decision makers and affected public on the sustainability of strategic decisions, facilitates the search for the best alternative and ensures a democratic decision making process (Gauthier and Simard 2009). Such evaluation allows for transparency and accountability.

The difference between EIA and SEA is that EIA has only limited application possibilities as it can only be applied to specific development projects while SEA can be applied to different regions or sectors of development (DEAT 2004; Schmidt et al. 2010; UNEP 2004). Policies, plans and programs are usually assessed by SEA.

SEA can be used to avoid, minimize or mitigate any negative influence on the driving forces of the generation and maintenance of biodiversity (Wiegleb and Bröring 2005). The range of SEA is far wider than that of EIA, SEA could assist in the integration of the concept of sustainability into strategic decision-making through, and for example, the determination of limits of acceptable change and the identification of sustainability targets and indicators, ensuring that development is within sustainable limits (DEAT 2004; UNEP 2004). SEA processes or elements are in place already in Eastern and Central Europe and other countries like Brazil, Chile, South Africa and China have a new EIA law that includes provision for SEA of plans and programs (UNEP 2004; Albrecht 2005). Laudable as government economic transformation agenda and programs for achieving sustainable development in the Niger Delta are, none of these PPPs have been subjected to any form of SEA either at the national, regional or sectorial/industrial level (Adebanji 2010). However, SEA could be a useful tool in identifying and assessing the environmental consequences of climate change adaptation PPPs especially during the preparation stage and before implementation.

10.8 Conclusion and Recommendations

The potential challenges of climate change in Nigeria should concern every Nigerian. A major barrier to planning for climate change in Nigeria is the lack of functional national and regional climate change policy. Nigeria equally lacks a centralized institution that can champion and coordinate climate change and spatial planning (Onyenechere 2010). Urban plans and policies for cities should be prepared considering the potential impacts of climate change. The government, policy makers, stakeholders and urban planners in the Niger Delta region are challenged to integrate and develop sustainable development policies based on assessing vulnerabilities and identifying workable adaptation measures. The urgency of climate change adaptation promises to revitalize practice in urban development around which innovations in planning, institutional readiness and community resilience are crucial to successful adaptation (Lwasa 2010). The need for implementing National Climate Change Adaptation Strategy for Nigeria has become very important. Appropriate legal frameworks are necessary in coping with the challenging situation. For example Strategic Environmental Assessment (SEA) could supplement Environmental Impact Assessment (EIA) in Nigeria in improving decision making related policies, plans and programs, thereby improving sustainable urban and regional development. The major advantage of SEA over EIA is that it allows the consideration of a much wider range of mitigation and enhancement measures, particularly measures to avoid impacts at an earlier stage of decision-making (Albrecht 2005). For better results there is need to adjust the EIA legislation in Nigeria. SEA can be taken as an extension of EIA beyond single impacts and projects (Secretariat of the CBD 2002; Wiegleb and Bröring 2005). SEA in harmonization with EIA will help in making decisions based on the implementation of coastal defense programs and protection of key infrastructure or any proposed plan that may be required for adaptation. The application of SEA into the evolving National Adaptation Strategy and Plan of Action (NASPA) could help in evaluating the impact of policies, plans and programs on the environment. The introduction of the National Adaptation Strategy and Plan of Action is in the right direction; however strong political will is required for the successful establishment and implementation. Therefore, SEA can be an effective tool for developing and refining PPPs for climate change adaptation (OECD 2010), particularly in Nigeria.

Perhaps the biggest obstacle is lack of awareness and knowledge. The government and NGOs should effectively use the various communication channels to spread the awareness of climate change to the masses. The public needs to be educated and informed about climate change and how it can change adversely their lives. The effects of climate change could undermine the conservation and further developmental process in the region. There could be a lot to lose if proper adaptation strategies are not implemented and mainstreamed into current plans, strategies and policies so as to include climate change adaptation and mitigation measures. By so doing, the so called win-win adaptation targets could be achieved. The government should start supporting climate research and relevant institutions in the region in order to discover the best possible ways of mainstreaming climate change adaptations into existing development PPPs. It is also necessary to discover new methods of agricultural practices, for example improved variety of crops that could adapt to a potential change. In order for agricultural production to be sufficient to meet the demands of the ever growing human population in Nigeria; the impact of the climate must be understood and integrated in any future planning (Ogbuehi et al. 2008). Actions are needed now to support and deliver innovations such as the National Adaptation Strategy and Plan of Action in response. However, in reducing these potential impacts and challenges there is a need for adaptation measures through sustainable urban development planning and provision of adequate policy and regulatory instruments. Hence, climate change has been widely recognized by different scholars as a priority for future planning of cities, urban and rural areas and the protection of communities from climate risks in DCs.

Biography Chika Ubaldus Ogbonna holds a Bachelor's degree in Urban and Regional Planning (B.URP) from Imo State University Owerri, Nigeria, and an MSc in Environmental and Resource Management from the Brandenburg University of Technology (BTU) Cottbus, Germany. His PhD research deals with regulatory and policy frameworks to climate change adaptation in developing countries with a special focus on Nigeria's Niger Delta Region. Additional research areas include environmental policy, urban planning, climate change, sustainable development issues, renewable energy, green economy.

References

Adebanji A (2010) Environmental and resource conflicts in the Niger delta: an impediment to Nigeria's transition to the green economy. In: 'IAIA10 conference proceedings' the role of impact assessment in transitioning to the green economy, International Conference Centre Geneva, Switzerland, pp 1–5

Adger WN (2001) Social capital and climate change. Tyndall Centre Working Paper 8, Tyndall Centre for Climate Change Research and CSWEGE School of Environmental Sciences University of East Anglia, p 19

Agwu EIC, Ezedike CE, Egbu AU (eds) (2000) The concept and procedure of environmental impact assessment. Crystal, Owerri, pp 93–97

Albrecht E (2005) Legal context of the SEA directive-links with other legislation and key procedures. In: Schmidt M et al (eds) Implementing strategic environmental assessment, vol 2, Environmental protection in the European Union. Springer, New York, pp 31–53

Awosika LF, French GT, Nicholla RJ, Ibe CE (1992) The impact of sea level rise on the coastline of Nigeria. In: Proceedings of IPCC symposium on the rising challenges of the sea, Venezuela, pp 14–19

Bajracharya B, Childs I, Hastings P (2011) Climate change adaptation through land use planning and disaster management: local government perspectives from Queensland. In: 17th Pacific Rim Real Estate Society conference climate change and property: its impact now and later, Gold Coast, 16–19 Jan 2011, pp 4–5

Beg N, Morlot JC, Davidson O, Afrane-Okesse Y, Tyani L, Denton F, Sokona Y, Thomas JP, La Rovere EL, Parikh KJ, Parikh K, Rahman AA (2002) Climate policy: linkages between climate and sustainable development. In: Climate policy 2, OECD/IEA, Paris, pp 129–144

Berry PM, Paterson J, Cabeza M, Dubuis A, Guisan A, Jaattela L, Kuhm I, Musche M, Piper J, Wilson E (2008) Adaptation and mitigation measures and their impacts on biodiversity. In: MACIS. Minimization of and adaptation to climate change impacts on biodiversity

Bidwai P (2012) The politics of climate change and the global crisis: mortgaging our future. Orient Blackswan, New Delhi

Blakely EJ (2007) Urban planning for climate change. Working paper, Lincoln Institute of Land Policy. https://www.lincolninst.edu/pubs/dl/1310_Blakely%20Final.pdf. Last accessed 27 Nov 2011

Building Nigeria Response to Climate Change – BNRCC (2010) Local action on climate change: BNRCC's Quarterly News Letter October 2010. BNRCC, Ibadan, Nigeria, pp 1–12

Collingwood (2011) Climate change adaptation capability statement. Collingwood Environmental Planning, London, http://www.cep.co.uk/files/Download/CEP%20Climate%20Change%20Adaptation%20Capability%20Statement%2023_02_11.pdf. Last accessed 15 Nov 2011

Cote M (2011) Climate adaptation planning: anticipating the legal issues for urban planners. Int J Clim Change Impacts Responses 2(3):80–102

Dadiowei T (2009) Environmental impact assessment and sustainable development in the Niger Delta: the Gbarain oil field experience. Niger Delta Economies of Violence Working paper no. 24, pp 1–50

DEAT (2004) Strategic environmental assessment, integrated environmental management, Information series 10, Department of Environmental Affairs and Tourism (DEAT), Pretoria, pp 4–7

Doswald N, Osti M (2011) Ecosystem-based approaches to adaptation and mitigation – good practice examples and lessons learned in Europe. Federal Ministry of Environment, Nature Conservation and Nuclear Safety, Bonn, Germany, pp 3–4

Duru P, Ume N (2008) Climate change and the incidence of malaria in Ahiazu Mbaise, Imo State. In: Nwajiuba C (ed) Climate change and adaptation in Nigeria. Margraf, Weikersheim, pp 95–100

Eckert R, Waibel M (2009) Climate change and challenges for urban development of Ho Chi Minh City/Vietnam. Pacific News January/February, No 31, pp 18–20

Ekpoh IJ (2010) Adaptation to the impact of climate variations on agriculture by rural farmers in North-Western Nigeria. J Sustain Dev 3(4):194–202
El Sioufi M (2010) Climate and sustainable cities: major challenges facing cities and urban settlements in the coming decades. International Federation of Surveyors, United Nations Habitat, Kenya, pp 1–2
Etuonovbe AK (2007) Coastal settlement and climate changes: effects of climate change/sea level rise on the People of Awoye in Ondo State Nigeria. Economic benefits of hydrography. Strategic integration of surveying services FIG working week 2007, Hong Kong SAR, China, pp 1–14
Ezegbule E (2008) Forestry development and the challenge of climate change in Nigeria, Nigeria Environmental Study/Team Action Team (NEST) Ibadan, Nigeria. In: Nwajiuba C (ed) Climate change and adaptation in Nigeria. Margraf, Weikersheim, pp 7–14
Federal Republic of Nigeria (1992a) Environmental Impact Assessment Decree (No. 86), Gazette 79 (73)
Federal Republic of Nigeria (1992b) Nigerian Urban and Regional Planning Decree (No. 88), Gazette 79 (75)
FMEAN (2010) National environmental, economic and development study (NEEDs) for climate change in Nigeria, Federal Ministry of Environment Abuja, Nigeria (Special Climate Change Unit), pp 1–20
Gauthier M, Simard L (2009) Public participation in strategic environmental assessment (SEA): critical Quebec approach. In: Environmental Evaluators Network Forum, Conferences Centre, Ottawa, pp 1–13
Giddens A (2009) The politics of climate change. Polity, Cambridge, pp 176–181
Gupta KR (2004) Urban development debates in the new millennium studies in revisited theories and redefined praxes, vol 1. Atlantic, New Delhi, pp 19–21
Han M (1989) Global warming induced sea level rise in China: response and strategies. Presentation to world conference on preparing for climate change, Cairo, Egypt, December 19
Humpreys S (1998) 'Equity in the CDM' Dakar. In: Masika R (ed) Gender, development and climate change. Oxfarm, Oxford, pp 1–9
International Institute for Environment and Development (IIED) (2007) Climate change and cities: adapting to climate change in urban areas: the possibilities and constraints in low-and middle-income nations. Human settlement discussion paper series, pp 1–10
International Institute for Environment and Development (IIED) (2011) Capacity strengthening in least developed countries for adaptation to climate change (CLACC) cities studies. International Institute for Environment and Development. http://www.iied.org/climate-change/key-issues/adapting-cities/capacity-strengthening-least-developed-countries-for-adapt. Last accessed 14 Nov 2011
IPCC (1990) Potential impacts of climate change: human settlement; energy, transport and industrial sectors; human health, air quality and changes in ultraviolet B-radiation. Intergovernmental Panel on Climate Change, Working Group II (A. Izrael Chair) Geneva WMO/unep, p 34
IPCC (2001a) Climate change 2001: the scientific basis: contribution of Working Group I to the Third Assessment Report of the Intergovernmental Panel on Climate Change. Cambridge University Press, New York, pp 1–99
IPCC (2001b) IPCC Third Assessment Report: climate changes 2001 summary for policy makers. http://www.ipcc.ch. Last accessed 28 Aug 2011
IPCC (2001c) Climate change 2001: Working Group II: impacts, adaptation and vulnerability. http://www.grida.no/climate/ipcctar/wg2/index.htm. Last accessed 10 Nov 2011
IPCC (2001d) Special report on emission scenarios. A special report of Working Group 111. WMO, UNEP, USA, p 26
IPCC (2007) Climate change 2007 – impacts, adaptation and vulnerability: contribution of Working Group II to the Fourth Assessment Report of the IPCC
Irvine J (2008) A changing climate for urban design: an examination of the New Zealand regulatory approach. New Zeal J Environ Law 12:277–313

Kennedy KS, Twilley RR, Kleypas JA, Cowam JH Jr, Hare SR (2002) Coastal and marine ecosystems and global climate change: potential effects on U.S. resource. Pew Center, Arlington, VA, pp 1–41
Klein JTR, Schipper ELF, Suraje D (2005) Mitigation and adaptation into climate and development policy: three research questions. Environ Sci Pol 8:579–588
Kong L, Yuen B (2009) Climate change and urban planning in Southeast Asia. SAPIENS 2(3): 1–12, http://sapiens.revues.org/881. Last accessed 14 Nov 2011
Luterbacher U, Sprinz D (eds) (2001) International relations and global climate change. MIT Press, Cambridge, MA, pp 5–6
Lwasa S (2010) Adapting urban areas in Africa to climate change: the case of Kampala. Curr Opin Environ Sustain 2:166–171
Manley P (2008) Biodiversity and climate change (May 20, 2008). U.S. Department of Agriculture, Forest Service, Climate Change Resource Center. http://www.fs.fed.us/ccrc/topics/biodiversity.shtml. Last accessed 24 May 2012
Masika R (ed) (2002) Gender, development and climate change. Oxfam, Oxford, pp 14–30
McEvoy D, Lindley S, Handley J (2006) Adaptation and mitigation in urban areas: synergies and conflicts. Proc Inst of Civ Eng Municip Eng 159:185–191
McMichael AJ, Lindgren E (2011) Climate change: present and future risk to health, and necessary response. Intern Med 270(5):401–413
McMichael C, Boarnett J, McMichael AJ (2012) An III wind? Climate change, migration, and health. Environ Health Perspect 120(5):647–653
Munonye J, Okoli V, Nwajiuba CU (2008) Rainfall variability and arable crop production in the humid South of Nigeria. In: Nwajiuba C (ed) Climate change and adaptation in Nigeria. Margraf, Weikersheim, pp 45–52
NASPA (2010) Developing Nigeria's Climate Change Adaptation Response: Nigeria's 2010 National Adaptation Strategy and Plan of Action (NASPA). http://www.naspanigeria.org/docs/NASPA_BkgrndDoc.pdf, pp 1–9. Last accessed 11 Nov 2011
Nigeria Environmental Study Action Team (NEST) (2010) Local action on climate change: BNRCC's Quarterly Newsletter, pp 1–15
Nishioka S, Harasawa H (eds) (1998) Global warming: the potential impact on Japan. Springer, Tokyo, pp 3–8
Nwajiuba C (2008) Adapting to climate change: challenges and opportunities. In: Nwajiuba C (ed) Climate change and adaptation in Nigeria. Margraf, Weikersheim, pp 1–6
O'Brien KL, Leichenko RM (2000) Double exposure: assessing the impacts of climate change within the context of economic globalization. Global Environ Change 10:221–232
OECD (2010) Strategic environmental assessment and adaptation to climate change. DAC Network on Environment and Development Co-Operation (ENVIRONET) Organization for Economic Co-operation and Development, Paris, pp 9–10
Ogba CO, Utang PB (2007) Vulnerability and adaptation of Nigeria's Niger Delta coast settlements to sea level rise. Strategic integration of surveying services FIG working week 2007, Hong Kong SAR, China, 13–17 May 2007, pp 1–15
Ogbuehi H, Onuh M, Ohazurike N (2008) Consequences of climate change on physiological processes in agricultural production. In: Nwajiuba C (ed) Climate change and adaptation in Nigeria. Margraf, Weikersheim, pp 17–27
Oku HB (2003) The Niger Delta environment, its local geography. Prelyn, Port Harcourt, pp 153–166
Oladipo E (2010) Towards enhancing the adaptive capacity of Nigeria: a review of the countries state of preparedness for climate change adaptation. A report submitted to Heinrich Böll Foundation Nigeria, Ilorin, Kwara State, pp 1–43
Onyenechere E (2010) Climate change and spatial planning concerns in Nigeria: remedial measure for more effective response. J Hum Ecol 32(3):137–148

Onyenechere E, Igbozurike U (2008) Women crop farmers' adaptation to rainfall variability and climate change in Amaraku Community, Imo State, Nigeria. In: Nwajiuba C (ed) Climate change and adaptation in Nigeria. Margraf, Weikersheim, pp 65–72
Orindi V, Murray L (2005) Adapting to climate change in East Africa: a strategic approach, Gatekeeper series 117. IIED, London, pp 1–16
Parmesan C, Yohe G (2003) A globally coherent fingerprint of climate change impacts across natural system. Nature 421:37–42
Perlmutter DD, Rothstein RL (2011) The challenge of climate change: which way now? Wiley, Chichester, UK, pp 189–218
Phil-Eze PO, Okoro IC (2009) Sustainable biodiversity conservation in the Niger Delta: a practical approach to conservation site selection. Biodivers Conserv 18:1247–1257
Ravindranath NH, Sathaye JA (2002) Climate change and developing countries: advances in global change research. Kluwer, Dordrecht, pp 80–82
Reid H, Ampomah G, Olazábal Prera MI, Rabbani G, Zvigadza S (2012) Southern voices on climate policy choices: analysis of and lessons learned from Civil Society Advocacy on Climate Change. International Institute for Environment and Development, London, pp 20, 38
Remme U, Blesl M (2008) A global perspective to achieve a low-carbon society (LCS): scenario analysis with the ETSAP-TIAM model. Clim Change Policy 8:60–75
Rose NA, Burton PJ (2009) Using bioclimatic envelopes to identify temporal corridors in support of conservation planning in a changing climate. Forest Ecol Manag 258:64–74
Schmidt M, Palekhov D, Atkinson R (2010) Environmental impact assessment and strategic environmental assessment. Learning materials. BTU Eigenverlag, Cottbus, pp 93–113
Secretariat of the CBD (2002) Vorläufige Leitlinie für die Einbeziehung von Biodiversitätsaspekten in die Gesetzgebung und/oder das Verfahren von Umweltverträglichkeitsprüfung und Strategischer Umweltprüfung. In: Beschluss VI/7A) der Vertragparteien des Übereinkommens über die biologische Vielfalt auf ihrem sechsten Treffen, Den Haag, 7–19 Apr 2002
Shaw R, Colley M, Connell R (2007) Climate change adaptation by design: a guide for sustainable communities. TCPA, London, pp 17–41
Smit B, Burto I, Klein RJT, Wandel J (2000) The anatomy of adaptation to climate change and variability. Clim Change 45:223–251
Tearfund (2006) Overcoming barriers: mainstreaming climate change adaptation in developing countries. Climate change briefing paper 1, Tearfund, Teddington, pp 3–24
Thom B, Crane J, Cox R, Farrell C, Hayes P, Kay R, Kearns A, Low Choy D (2007) Planning for a Sustainable Future: UK Department of Communities and Local Government. United Kingdom Government white paper, London
Ugochukwu CNC (2008) Sustainable environmental management in the Niger Delta region of Nigeria: effects of hydrocarbon pollution on local economy. PhD dissertation, Brandenburg University of Technology, Cottbus, vol 7, pp 136–153
UK Department for International Development (DFID) (2009) Impact of climate change on Nigeria's economy, pp 1–3
UK Parliament Publications (2006) Adapting to climate change in developing countries. Parliamentary Office of Science and Technology: Postnote October 2006, no. 269, pp 1–4. http://www.parliament.uk/post. Last accessed 24 May 2012
UNCAPSA (2007) United Nations Economic and Social Commission for Asia and the Pacific: Background paper on Drought and Assessment of Asian and Pacific progress. ESD/RIM/2007/INF.4, pp 5–6. http://www.un.org/esa/sustdev/csd/csd16/rim/escap_drought.pdf. Last accessed 7 Nov 2011
UN-DESA (2009) World economic and social survey 2009: promoting development, saving the planet. United Nations, New York, pp 1–56
UNEP (2004) Environmental impact assessment and strategic environmental assessment: towards an integrated approach, Geneva, vol 5, pp 85–109. http://www.unep.ch/etu/publications/textONUbr.pdf. Last accessed 28 Oct 2011

UNEP/CMS (2006) Migratory species and climate change: impacts of changing environment on wild animals. United Nations Environmental Program (UNEP) and the Secretariat of the Convention on the Conservation of Migratory Species (CMS), pp 8–60

UNEP/GRID-Arendal (2011) The regional impacts of climate change: African Coastal Zones – IPCC special reports on climate change. http://www.grida.no/publications/other/ipcc_sr/?src=/climate/ipcc/regional/030.htm. Last accessed 27 Nov 2011

UNFCCC (2009) Fact sheet: need for mitigation, pp 1–7. http://unfccc.int/files/press/backgrounders/application/pdf/press_factsh_mitigation.pdf. Last accessed 7 Nov 2011

UNFCCC (2011) Climate change: impacts vulnerabilities and adaptation in developing countries, pp 8–81. http://unfccc.int/resource/docs/publications/impacts.pdf. Last accessed 5 Nov 2011

UN-Habitat (2007) UN HABITAT pitches for urban planning, Bonn. http://www.unhabitat.org/content.asp?cid=4836&catid=550&typeid=6&subMenuId=0. Last accessed 14 Nov 2011

UN-Habitat (2009a) Planning sustainable cities: global report on human settlements. United Nation Human Settlements Program. Earthscans, London, pp 46–47

UN-Habitat (2009b) Cities and Climate Change Initiative. Launch Conference Report, Oslo. http://www.unchs.org/downloads/docs/6520_35892_OSLO%20Report.pdf. Last accessed 27 Aug 2011

Urama KC, Ozor N (2010) Impacts of climate change on water resources in Africa. The role of adaptation. African Technology Policy Studies Network (ATPS), pp 1–22

Uyigue E, Agho M (2007) Coping with climate change and environmental degradation in the Niger Delta of Southern Nigeria. Community Research Development Center (CREDC), Edo State, Nigeria, pp 1–31

WHO (2007) Global climate change: implications for international public health policy. Bull World Health Organ 85(3):161–244

Wiegleb G, Bröring U (2005) Biodiversity programmes on global, European and national levels related to SEA. In: Schmidt M et al (eds) Implementing strategic environment assessment, vol 2, Environmental protection in the European Union. Springer, New York, pp 539–553

Wilson E (2006) Adapting to climate change at the local level: the spatial planning response. Local Environ 11(6):609–625

World Bank (2010) Economics of adaptation to climate change: Bangladesh, pp 1–86. http://climatechange.worldbank.org/sites/default/files/documents/EACC_Bangladesh.pdf. Last accessed 9 Nov 2011

World Bank (2008) World Bank development report: agriculture for development. World Bank, Washington, DC

Yakubu AF, Sikoki FD, Horsefall M Jr (1998) An investigation into the physicochemical conditions and planktonic organisms of the lower reaches of the Nun River, Nigeria. JASEM 1(1): 36–42

Yates CJ, Elith J, Latimer AN, Le Maiter D, Midgley GF, Schurr FM, West AG (2009) Projecting climate change impacts on species distributions in mega-diverse South African regions: opportunities and challenges. Ecol Soc Austr 35:374–391

Chapter 11
Climate Change Adaptation and Biodiversity Conservation: An Economic Perspective

Frank Wätzold

11.1 Introduction

Even in the unlikely case that the world soon adopts an ambitious policy to mitigate emissions from gases that cause climate change, it is highly likely that there will still be a substantial increase in global temperature (IPCC 2007). This increase will lead to significant changes in climatic conditions which are expected to become a major threat for biodiversity (Sala et al. 2000). According to a study by Thomas et al. (2004) who estimated extinction risks for sample regions that cover some 20 % of the Earth's terrestrial surface, 15–37 % of species will be committed to extinction by 2050.

There is a large debate in the ecology literature about how exactly climate change will influence biodiversity and what might be adequate response strategies, i.e. how should we modify existing conservation strategies and policies in a way that the impact of climate change on biodiversity is mitigated as much as possible (see Heller and Zavaleta 2009 for a review). However, until today, this debate is not mirrored in the economic literature in which very little can be found about appropriate strategies and policies in order to mitigate climate change impacts on biodiversity.

The aim of this article is to contribute to filling this gap by analyzing two major conservation policies—the designation of reserves and payments to compensate land owners for carrying out measures to conserve biodiversity—on a conceptual level from an economic point of view. As to my knowledge this is the first attempt in this direction, I do not aim for comprehensiveness but rather for presenting initial ideas that hopefully motivate more research in this field.

The geographical background of this analysis is Europe and the presented analysis should be seen against this background. This implies that strategies and

F. Wätzold (✉)
Department of Environmental Economics, Brandenburg University of Technology Cottbus-Senftenberg, P.O. Box 101344, 03013 Cottbus, Germany
e-mail: waetzold@b-tu.de

E. Albrecht et al. (eds.), *Implementing Adaptation Strategies by Legal, Economic and Planning Instruments on Climate Change*, Environmental Protection in the European Union 4, DOI 10.1007/978-3-540-77614-7_11, © Springer-Verlag Berlin Heidelberg 2014

policies are discussed which are primarily designed to conserve biodiversity in a European context. To what extent insights from this analysis can be transferred to other regions in the world is a matter of future research.

The article is structured as follows: The next chapter briefly reviews the ecological literature on why climate change causes the extinction of species and what might be adequate response strategies to mitigate the impact of climate change on species. Chapter 3 starts by briefly describing the two major conservation policy instruments in Europe, namely reserves and compensation payments for conservation measures which in Europe mostly exist in the context of agri-environment schemes. Then, economic criteria to evaluate conservation policies under climate change are presented and, finally, these criteria are applied to analyzing the two conservation policy instruments. Chapter 4 concludes.

11.2 Ecological Background

11.2.1 Climate Change and Range Shift of Species

According to IPCC (2007) global temperatures have risen on average by 0.13 °C per decade over the last 50 years. This increase has already caused a significant shift poleward and upward in elevation for many species ranges (cf. Parmesan 2006 for a review of related studies). Consider as an example the study by Hickling et al. (2005) who analyzed 37 species of British dragonflies and damselflies and found that the range of 34 species had shifted northwards with a mean shift of 74 km over the last 40 years. As a consequence, local communities are disaggregating and changing toward more warm-adapted species (Parmesan 2006).

Species are likely to have problems to adapt to further shifts of their habitat for several reasons: Climate change in the future is likely to be faster than in the last decades and migration may not be sufficiently fast for many species threatening their persistence (Davis and Shaw 2001). For some species, migration may also not be possible. This concerns particularly those species whose habitats are at high latitudes and high elevations, on low-elevation islands and the coast. For other species, migration may be possible in principle, but habitat fragmentation severely restricts the movement of these species in the landscape (Opdam and Wascher 2004). To what extent species can adapt also depends on the synchrony of effects on species' food and habitat resources (Parmesan 2006). Coordination in timing between the life cycles of predators and prey, parasitoids and their host insects, herbivorous insects and their host plants, insect pollinators and flowering plants is important for the abilities of species to adapt to climate change (Harrington et al. 1999). In summary, the ability of species to adapt to climate change depends on both species characteristics like physiological tolerances, life-history strategies, probabilities of population extinctions and colonizations, and dispersal abilities

(Parmesan and Yohe 2003; Best et al. 2007; Münkemüller et al. 2009) as well as landscape characteristics (Opdam and Wascher 2004).

11.2.2 Proposed Responses to Cope with Climate Change

In principle, nature conservation policies and strategies can be designed in a way that the impact of climate change on species is mitigated. This section follows Vos et al. (2008) who have suggested three major strategies to mitigate the impact of climate change on species in European landscapes depending on what type of bottlenecks exist. According to Vos et al. (2008) a bottleneck occurs if low connectivity or too little habitat area prevents the appropriate spatial response of species to climate change.

(I) The first strategy is based on the idea that losses in the distribution range of species due to unsuitable climate may be compensated by new regions where climate conditions are becoming suitable (so-called 'new climate space'; Pearson et al. 2002). A bottleneck arises here if a landscape is fragmented and a suitable habitat in a new climate space is too distant from an existing habitat network to be colonized by a species. This bottleneck calls for conservation policies that link existing habitat networks with unconnected habitats in new climate space to enable species to colonize new habitats.

(II) The second strategy is based on the same idea—to enable species to move to new climate space—, but the bottleneck is different. A bottleneck emerges here if the size of the population in the 'overlap zone' is too small. Vos et al. (2008, p. 1726) define the overlap zone as the area which contains suitable habitat for the species in successive time frames (for example, an area whose habitat is suitable in 2012 as well as 20 years later). If such a bottleneck exists, appropriate conservation policies should focus on increasing the amount of suitable habitat in the overlap zone. This promotes a greater colonization capacity.

(III) Rather than on the aspect of migration, the third strategy focuses on improving the species' persistence in the landscape. Here, the bottleneck is a lack of protection in those parts of a species' range where suitable habitat will continue to exist after successive time frames. Such areas form climate refugia where the species is protected in a changing climate—at least for some time. Appropriate conservation policies focus on enlarging habitats in such areas or improving the spatial conditions of habitats in this area.

11.3 Adopting an Economic Perspective

In order to adopt an economic perspective on assessing strategies and policies for biodiversity conservation a 'typical' economic approach to the assessment of environmental and resource policy instruments is adopted (cf. Endres 2011).

First, the main policy instruments for conservation in Europe are briefly described, then the criteria to analyse these instruments are sketched, and finally the criterion to evaluate the policy instruments in situations of climatic change are applied in a conceptual manner.

11.3.1 Conservation Policy Instruments in Europe

Two policy instruments are of particular relevance for biodiversity conservation in Europe; reserves and compensation payments for land use measures to conserve biodiversity. Summarising Wätzold and Schwerdtner (2005), these instruments are briefly described in the following.

(I) Reserves, particularly in the context of the EU-wide Natura 2000 network: Many endangered habitats and species in Europe require reserves for their protection where human influence is limited to the extent necessary (consider as an example for a species the black stork, *Ciconia niger*, that requires undisturbed forest patches as breeding habitat). Regarding reserves, the two most important European legal instruments are the Habitats Directive (Council Directive 92/43/EEC) and the Birds Directive (Council Directive 79/409/EEC). Together, the two directives are designed to create a network of protected areas, the Natura 2000 network.

(II) Compensation payments for land use measures to conserve biodiversity which exist in Europe mostly in the context of agri-environment schemes: For the persistence of many other endangered species and habitats, a certain type of human land use is required (consider as an example meadow birds which require certain mowing or grazing regimes). This is particularly relevant in agricultural landscapes where in former times a great diversity of agricultural land use led to a corresponding variety of species and habitats. Today, due to economic and technological developments, many types of land use are no more economically viable. Agricultural intensification resulting in uniform production areas and farmland abandonment in marginal areas are now seen as major threats to the conservation of European biodiversity. In order to reserve the trends of intensification and land abandonment the instrument of agri-environment schemes has been developed. Within such schemes, farmers are compensated for maintaining (unprofitable) agricultural land use which, however, is beneficial to biodiversity, the environment or the landscape. Agri-environment schemes are now based mainly on Council Regulation (EC) 1257/99, with the design details of agri-environment schemes being left to decide by the Member States.

11.3.2 Evaluation Criteria

The two major standard evaluation criteria for environmental policy instruments are ecological effectiveness and cost-effectiveness. These are explained below and adapted to an evaluation of policy instruments for biodiversity conservation in a situation of climate change.

Ecological effectiveness: In general, ecological effectiveness refers to the ability of a policy instrument to ensure that an ecological goal is actually achieved. Regarding the conservation of species in a changing climate, I consider that the ecological goal is to ensure the persistence of a viable species population under changing climatic conditions. Hence, ecological effectiveness here refers to the ability of a conservation instrument to ensure the persistence of a species population in a changing climate in a landscape.

Cost-effectiveness: In general and depending on the policy context, the criterion of cost-effectiveness refers either to the ability of a conservation policy instrument to achieve an environmental goal at least cost or to maximize conservation outputs for given costs (Wätzold and Schwerdtner 2005). Adapting this criterion to the assessment of conservation policy instruments in a changing climate implies the following modifications of the definition: In this context, cost-effectiveness is understood as the ability of a conservation instrument to ensure the persistence of a viable species population under changing climate in a landscape at least costs. Or, alternatively, cost-effectiveness means the ability of a conservation policy instrument to maximize the persistence of a species in a landscape under changing climatic conditions for given costs.

11.3.3 Evaluation of Policy Instruments

Ecological effectiveness: The criterion of ecological effectiveness may require that in order to ensure the persistence of a species in a landscape under a changing climate, a modification of the location of the area(s) for conservation is necessary. In this context, a key problem of reserves is that their spatial design tends to be static. It is usually a long and politically controversial process to designate an area as a reserve or even several areas as a reserve network. Depending on which of the three strategies proposed by Vos et al. (2008) is most suitable, the location differs of where additional reserves have to be designated.

Strategy I (enabling colonization of new habitats) might be primarily appropriate if reserves with suitable habitats exist in areas where the species currently can be found as well as in 'new climate space' and species are unable to migrate between these areas due to habitat fragmentation. For this strategy, new areas need to be designated as reserves that enable migration between current habitat and 'new climate space'. If suitable habitats in new climate space are not protected, an additional challenge is to designate such areas as reserves.

Strategy II (increasing habitat in the overlap zone) requires that new areas are designated as reserves in the overlap zone between current habitat and new climate space, and strategy III (improving the species' persistence in the landscape) demands that new reserves are created in areas which improve the species' persistence in the landscape and which will continue to be suitable habitat under climate change.

While creating new reserves is always difficult, it is an empirical matter for which of the three strategies it is most difficult to designate areas as reserves in a given socio-economic and ecological context. However, we can reasonably assume that it is likely that the difficulty is higher with fewer alternatives available. The number of (realistic) alternatives is again an empirical matter and case specific. However, a careful generalization suggests that for strategy I more specific areas may be required as additional reserves than for strategies II and III. This implies that for the latter two strategies the ecological effectiveness is likely to be higher than for strategy I.

In general, the ecological effectiveness of compensation payments for conservation measures respectively agri-environment schemes suffers from the voluntary nature of (most of) such schemes. If farmers who own or use land which is potentially valuable as habitat and requires conservation measures do not want to participate in an agri-environment scheme (for whatever reasons) they cannot be forced to do so. Furthermore, contracts with farmers made on the basis of agri-environment schemes are usually signed only for a certain period (often for 5 years). This implies that even if farmers have once decided to participate in the scheme, there is again uncertainty about participation after that period.

This lack of permanence, however, turns into an advantage if the impact of climate change is taken into account. All three proposed strategies suggest a spatial reconfiguration of areas to be conserved which, in principle, is possible after each contract period. Agri-environment schemes targeted at a particular species need not be continued in areas which are no longer suitable for this species. Instead, they can be shifted to areas which are favorable to mitigate the impact of climatic changes on this species.

For all three strategies this requires a spatial targeting of measures, which has to be included in the design of the schemes (cf. e.g. Drechsler et al. 2010). Note that most current agri-environment schemes are spatially homogeneous which means that every farmer in a region is allowed to participate in the scheme and receives the same amount of money for the conservation measure to be supported regardless of his or her costs and where the measure is carried out.

For strategy I (enabling colonization of new habitats) the locations of measures need to be very specific and it is difficult to design a scheme which supports measures on such specific locations. In principle, there are two possibilities: the first option is to design a corridor which includes potentially valuable areas for conservation and pay only those farmers who carry out conservation measures in this corridor. The second option is to individually negotiate with farmers about carrying out conservation measures on their land (programs in which contracts are negotiated individually already exist in some countries, cf. Whitby and Saunders

1996). The ecological effectiveness of both options may suffer if suitable land is limited to the land managed by a few farmers and these farmers are unwilling to participate in agri-environment schemes.

For the other two strategies (increasing habitat in the overlap zone and improving a species' persistence in the landscape), the location of measures is likely to be less restricted. Here, the scheme just needs to be changed in a way that additional areas are added to those areas on which farmers are eligible to participate in the scheme. If the size of these additional areas is sufficiently big, enough farmers are likely to participate and the ecological effectiveness may be reasonably good.

Cost-effectiveness: Applying the criterion of cost-effectiveness to the selection of reserves implies that those reserves are chosen which achieve a certain conservation target at least costs or for which for given costs of reserve designation a conservation target is maximized (cf. Ando et al. 1998; Wätzold and Schwerdtner 2005). For assessing the cost-effective modification of the reserve network under changing climatic conditions, the cost-effectiveness of the three strategies proposed by Vos et al. (2008) has to be compared. To focus on the key mechanisms of cost-effectiveness the potential problem of political resistance against designating reserves (which was mentioned in the part on ecological effectiveness) is ignored here.

Evaluating the cost-effectiveness of the three strategies in a setting of given costs for conservation requires an analysis to what extent which strategy can be applied for a certain amount of costs. For each strategy, it is assessed which (most ideal) areas can be newly designated as reserves. Then—for each of the three strategies—the contribution of the newly designated areas to the persistence of a species population in a changing climate on a landscape level is assessed, and the strategy which contributes best to this goal is the most cost-effective one.

If the setting is to ensure the persistence of a viable species population under a changing climate in a landscape, the necessary changes in the reserve network have to be determined for the three strategies and the corresponding costs have to be estimated. The strategy which achieves the goal of species persistence at least cost is the cost-effective one.

The approach to assess the cost-effectiveness of agri-environment schemes under climate change is similar to estimating the cost-effectiveness of reserves. In a setting of given costs, an appropriate compensation payment scheme needs to be designed for each of the three strategies and it has to be estimated for each payment scheme which areas will be enrolled by farmers for the given amount of costs (cf. Drechsler et al. 2007). Then the effects of these areas on the persistence of the species under climate change need to be assessed and the payment scheme which generates the highest impact on species persistence is the most cost-effective one.

If the policy goal is to ensure the persistence of a viable species population under a changing climate in a landscape, for each strategy a payment scheme has to be designed which is able to reach this goal (if feasible). Then the costs arising from these payment schemes have to be estimated for each scheme and the payment scheme with the lowest costs is the cost-effective scheme.

11.4 Summary and Conclusion

The discussion of strategies and policies which aim to mitigate the negative impacts of climate change on biodiversity conservation has been dominated by ecologists. The purpose of this paper was to contribute to the discussion from an economic point of view. To this aim three strategies that have been suggested in the ecology literature to mitigate the impact of climate change on species populations have been identified. Then typical economic criteria to analyse environmental and conservation policy instruments have been modified so that they can be used to assess conservation policy instruments in a changing climate. Finally, these criteria have been applied to assess the conservation instruments of reserves and compensation payments for conservation measures (respectively agri-environment schemes) on a conceptual level.

The results of this analysis should be understood as a first step from economics to gain a better understanding of how conservation policy instruments need to be modified in a changing climate. More comprehensive research is needed both on a conceptual, theoretical level but also with regard to applied research like case studies for particular species in real landscape (cf. Drechsler et al. 2007) to be able to address the threat that many species will become extinct due to anthropogenic climate change.

Biography Frank Wätzold is Professor of Environmental Economics at Brandenburg University of Technology Cottbus-Senftenberg. His research fields include the economic analysis of environmental policy instruments and policy processes. A particular focus is on the development of ecological-economic models to evaluate strategies and policies to conserve biodiversity.

References

Ando A, Camm J, Polasky S, Solow A (1998) Species distribution, land values and efficient conservation. Science 279:2126–2128

Best AS, Johst K, Münkemüller T, Travis JMJ (2007) Which species will successfully track climate change? The influence of intraspecific competition and density dependent dispersal on range shifting dynamics. Oikos 116:1531–1539

Davis MB, Shaw RG (2001) Range shifts and adaptive responses to quaternary climate change. Science 292:673–679

Drechsler M, Johst K, Ohl C, Wätzold F (2007) Designing cost-effective payments for conservation measures to generate spatiotemporal habitat heterogeneity. Conserv Biol 21(6):1475–1486

Drechsler M, Johst K, Wätzold F, Shogren JF (2010) An agglomeration payment for cost-effective biodiversity conservation in spatially structured landscapes. Resour Energy Econ 32:261–275, Special Issue on Spatial Natural Resource and Environmental Economics, edited by Albers H, Ando A, Shogren JF

Endres A (2011) Environmental economics – theory and policy. Cambridge University Press, Cambridge

Harrington R, Woiwod I, Sparks T (1999) Climate change and trophic interactions. Trends Ecol Evol 14:146–150
Heller NE, Zavaleta ES (2009) Biodiversity management in the face of climate change: a review of 22 years of recommendations. Biol Conserv 142:14–32
Hickling R, Roy DB, Hill JK, Thomas CD (2005) A northward shift of range margins in British Odonata. Glob Change Biol 11:502–506
IPCC (2007) Working group I report “the physical science basis”. http://www.ipcc.ch/publications_and_data/ar4/wg1/en/contents.html. Accessed 20 Apr 2011
Münkemüller T, Reineking B, Travis J, Bugmann H, Johst K (2009) Disappearing refuge in time and space: how environmental change threatens species coexistence. Theor Ecol 2:217–227
Opdam W, Wascher D (2004) Climate change meets habitat fragmentation: linking landscape and biogeographical scale levels in research and conservation. Biol Conserv 117:285–297
Parmesan C (2006) Ecological and evolutionary responses to recent climate change. Annu Rev Ecol Evol Syst 37:637–669
Parmesan C, Yohe G (2003) A globally coherent fingerprint of climate change impacts across natural systems. Nature 421:37–42
Pearson RG, Dawson TP, Berry PM, Harrison PA (2002) SPECIES: a spatial valuation of climate impact on the envelope of species. Ecol Model 154:289–300
Sala OE, Chapin FS, Armesto JJ, Berlow E, Bloomfield J, Dirzo R, Huber-Sanwald E, Huenneke LF, Jackson RB, Kinzig A, Leemans R, Lodge DM, Mooney HA, Oesterheld M, Poff NL, Sykes MT, Walker BH, Walker M, Wall DH (2000) Biodiversity – global biodiversity scenarios for the year 2100. Science 287:1770–1774
Thomas CD, Cameron A, Green RE, Bakkenes M, Beaumont LJ, Collingham YC, Erasmus BF, de Siqueira MF, Grainger A, Hannah L, Hughes L, Huntley B, van Jaarsveld AS, Midgley GF, Miles L, Ortega-Huerta MA, Peterson AT, Phillips OL, Williams SE (2004) Extinction risk from climate change. Nature 427:145–148
Vos CC, Berry P, Opdam P, Baveco H, Nijhof B, O'Hanley J, Bell C, Kuipers H (2008) Adapting landscapes to climate change: examples of climate-proof ecosystem networks and priority adaptation zones. J Appl Ecol 45:1722–1731
Wätzold F, Schwerdtner K (2005) Why be wasteful when preserving a valuable resource? – a review article on the cost-effectiveness of European biodiversity conservation policy. Biol Conserv 123:327–338
Whitby M, Saunders C (1996) Estimating the supply of conservation goods in Britain. Land Econ 72:313–325

Chapter 12
Economic Instruments for Integrating Climate Change Considerations into Development Strategies of Industrial Regions in Ukraine: Experience and Issues

Ludmila Palekhova

12.1 Introduction

Ukraine is an active participant in global action addressing climate change for the benefit of present and future generations. Ukraine signed and ratified the United Nations Framework Convention on Climate Change (UNFCCC), in which it was included to the list of Annex I countries as an industrial country with economy in transition. Furthermore, Ukraine's participation in the Kyoto Protocol to the UNFCCC demonstrates its support of international commitments on climate protection, as well as asserts its aspiration for EU membership in a long-term perspective.

Kyoto Protocol was ratified by Ukrainian Parliament in February 2004 and since then became an integral part of Ukrainian legislation (MEP 2006). However, Ukraine accepted commitments on the UNFCCC and Kyoto Protocol before establishing the full-fledged framework for their implementation. In particular, activities aimed at the implementation are carried out almost exclusively at the national level, while implementation dynamics at regional level remains largely insufficient.

One of the distinctive features of the Ukrainian economy is the dominance of mining and metallurgical sectors, as well as high degree of their concentration within industrial regions in an aggregate fashion. In this context, the challenges of reducing emissions of pollutants in Ukraine, including greenhouse gas emissions, are directly linked to the solution of economic problems of its industrial centres. And there is no doubt that climate protection activities require a broad discussion involving wide range of stakeholders, which would help to integrate climate change considerations into development strategies of industrial regions.

L. Palekhova (✉)
Department of Marketing, Institute of Economics, National Mining University (NMU), Karl Marx Av., 19, 49005 Dnepropetrovsk, Ukraine
e-mail: pall@hotmail.ru

E. Albrecht et al. (eds.), *Implementing Adaptation Strategies by Legal, Economic and Planning Instruments on Climate Change*, Environmental Protection in the European Union 4, DOI 10.1007/978-3-540-77614-7_12,

Analysis in this chapter explores the relationship between issues related to the transformation of the economy in Ukraine and the problems of greenhouse gas emissions. Aim of this analysis was to highlight which climate-relevant aspects should be considered during development planning in industrial regions of Ukraine. The chapter also provides an overview of Ukrainian experience in applying flexible mechanisms of the Kyoto Protocol and explores reasons for their insufficient application in industrial regions. The discussion is, therefore, focused on reviewing the appropriate methods and instruments for strengthening climate change considerations in development strategies of industrial regions in Ukraine.

12.2 The Ukrainian Economy and Greenhouse Gas Emissions

Ukraine joined the global action on climate change during difficult times in its history, under conditions of fundamental socio-economic transformations. It should be noted that Ukrainian republic, as part of the former USSR, was a leader in Europe in extraction of iron ore, production of steel and cast iron, mineral fertilizers, grain, etc. After gaining independence in 1991, the country experienced a deep economic recession, which lasted for more than a decade. By the year 2000, the real GDP of Ukraine decreased to 41 % of the 1990 level (Goskomstat 2012).

The drastic decline in industrial production led to the decrease of anthropogenic pressure on the environment. The National Inventory Report of Greenhouse Gas Emissions and Sinks for 1990–2009[1] revealed that the actual emissions of greenhouse gases in terms of CO_2 equivalents (CO_2eq) declined in 2009 by 60 % compared to the base year.[2] Nevertheless, Ukraine is still one of the leading countries in greenhouse gas emissions: according to data published by the United Nations Climate Change Secretariat, the total emissions of greenhouse gases excluding emissions/removals from land use, land-use change and forestry in 2009 equalled to 374,120 Gg CO_2eq (UNFCCC Secretariat 2011). According to this index, Ukraine ranked 11th among 41 countries included in Annex I.

It is worth noting that the spatial distribution of greenhouse gas emissions in Ukraine has certain peculiarities. According to the National Inventory Report, 68 % of greenhouse gas emissions are generated in four industrial regions—Donetsk, Dnipropetrovsk, Luhansk and Zaporizhia *oblasts*,[3] which comprise only ca. 18 %

[1] National Inventory Report of Anthropogenic Emissions by Sources and Removals by Sinks of Greenhouse Gases in Ukraine for 1990-2009, published by the Ministry for Environmental Protection of Ukraine (MEP 2011).

[2] The base year according to the UNFCCC is 1990 except for Bulgaria (1988), Hungary (average of 1985–1987), Poland (1988), Romania (1989) and Slovenia (1986), as defined by the Conference of the Parties decisions 9/CP.2 and 11/CP.4.

[3] Oblast is a territorial and administrative unit in Ukraine, which is subdivided into 24 *oblasts*. The term 'oblast' is a loanword in English, but it is nevertheless often translated as 'region', 'province', 'area', or 'zone'.

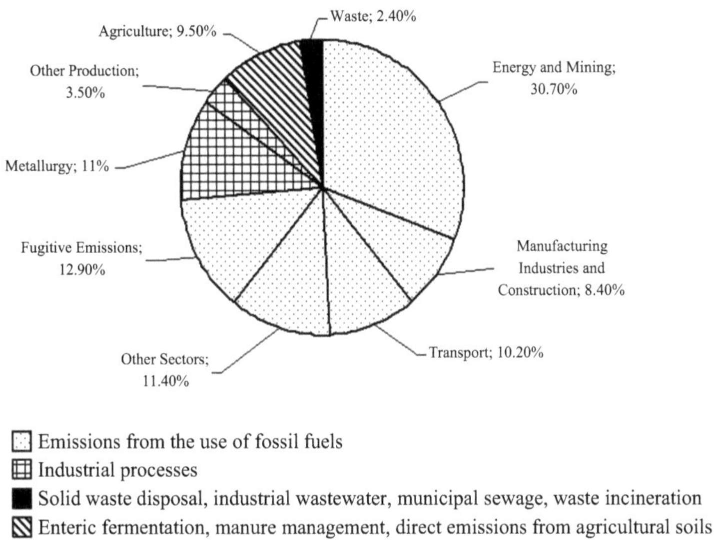

Fig. 12.1 Contribution of various sectors towards greenhouse gas emissions in Ukraine, 1990–2009, in per cent (source: MEP 2011)

of the total area of the country. The largest in Ukraine industrial parks and enterprises are concentrated in these regions. A considerable majority of these industrial facilities were built in the 1950s–1970s and now do not comply with the current environmental requirements (Palekhov et al. 2008), thus providing significant potential for decreasing anthropogenic pressure on climate in case of their modernisation.

It should be mentioned that the crisis period of 1990–2000 has strengthened the structural deformations and distortions of the economy towards the raw materials sector. The mining industry has become one of the basic forces of the Ukrainian economy in recent years. For instance, Dnipropetrovsk Oblast is accounted for ca. 20 % of national GDP and ranks the second in Ukraine according to this index (Goskomstat 2012). The traditional industrial sector represents approximately 80 % of the regional economy: metallurgy—35 %, mining industry—30 %, energy—15 % (Dnipropetrovsk Oblast 2012). Figure 12.1 shows how these sectors contribute towards greenhouse gas emissions in Ukraine.

The major issue of Ukrainian industry is its energy intensity, which is two to three times higher than in Western Europe. This situation is caused by a constellation of factors, including: unsustainable patterns of energy supply, high energy intensity of technological processes, inefficient energy usage and power consumption. As the result, 73.6 % of greenhouse gas emissions come from the use of fossil fuels, including 12.9 % as fugitive emissions[4] due to leaks and energy losses (see Fig. 12.1).

[4] Fugitive emissions are intentional or unintentional releases of greenhouse gases mostly during industrial activities, e.g. they may occur during the extraction, processing and delivery of fossil fuels to the point of final use (IPCC 2006).

Table 12.1 Current and projected structure of the primary energy consumption in Ukraine, in per cent

Year	Gas	Coal	Oil	Others[a]	Total
2005	43.7	21.7	12.9	21.7	100
2010	42.6	28.0	10.2	19.2	100
2030	18.8	33.3	11.3	36.6	100

Sources: Goskomstat (2012); Cabinet of Ministers Resolution No. 145-r of 15.03.2006

[a]Included are: nuclear energy, hydro energy, alternative energy

In recognition of these problems, the Energy Strategy of Ukraine for the Period up to 2030[5] focuses on improvement of energy efficiency, development of the domestic energy production and transition to alternative energy sources (see Table 12.1). These measures should also provide for the reduction of greenhouse gas emissions by at least 50 % below 1990 levels by 2020 (EBRD 2011).

The economic crisis in Ukraine was accompanied by the drastic reduction in capital investments. This led to a gradual technical degradation of the production assets. For example, in Dnipropetrovsk Oblast the level of wear of fixed assets has reached 79 % (Dnipropetrovsk Oblast 2012).

Therefore, as can be seen from the discussion above, the success of actions to mitigate anthropogenic climate change in Ukraine is largely dependent on the success of economic reforms in the main industrial regions. Moreover, industrial regions should play a more constructive role in tackling the climatic challenges, and for that actively use opportunities arising from international environmental cooperation in technological modernisation.

12.3 Application of the Kyoto Investment Mechanisms for Development of Industrial Regions

In Ukraine, traditionally, regional economic policy is poorly linked to the climatic issues. Its environmental component was generally limited to setting emission limits and fees for emission of pollutants, which had little to no effect. Only after ratification of the Kyoto Protocol an opportunity arose to provide some real and practical solutions for this issue. As the result, Ukraine made a number of steps towards institutional development, including establishment of the National Environmental Investment Agency as a central executive body in 2007[6] (later renamed to State Environmental Investment Agency—SEIA). Also the national inventory system for accounting greenhouse gas emissions and sinks was created.

[5]The Energy Strategy of Ukraine for the Period up to 2030 was approved by the Cabinet of Ministers Resolution No. 145-r of 15.03.2006.

[6]Cabinet of Ministers Decree No. 612 of 04.04.2007 "On establishment of the National Environmental Investment Agency of Ukraine". The main responsibility of the SEIA is to execute the UNFCCC provisions, implement the Kyoto Protocol mechanisms, as well as to attract investments to the environmental protection.

Table 12.2 The participation of Ukrainian industrial regions in the green investment projects

Ukrainian regions	Number of projects	% of the total number	Budget of projects, million UAH	% of the total budget
All regions	987	100	3,700.0	100
Main industrial regions	94	9.5	279.3	7.6
Donetsk Oblast	30	3.0	43.2	1.2
Dnipropetrovsk Oblast	–	–	–	–
Zaporizhia Oblast	28	2.8	60.3	1.6
Luhansk Oblast	36	3.7	175.8	4.8

Source: SEIA (2011)

In doing so, Ukraine has gained access to 'flexible mechanisms', which created opportunities for attraction of foreign investments in projects on modernisation of the production and increasing energy efficiency. As is well known, the flexible mechanisms of the Kyoto Protocol include: 'Joint Implementation Projects' (JI), defined in Article 6 of the Protocol; 'Clean Development Mechanism' (CDM), Article 12; and 'Emissions Trading' (ET), Article 17. Ukraine participates only in the ET and JI projects.

Ukraine has potential to become one of the largest suppliers of emission credits and allowances at the international GHG reduction market and derive substantial economic benefits from implementation of environmental projects. During the first Kyoto Protocol period (2008–2012) the country was planning to sell more than 1 billion of assigned amount units (AAUs), however only 47 million AAUs were actually sold.

As of 20.12.2011 Ukraine attracted ca. 470 million EUR in targeted investments in Green Investment Scheme (GIS) projects. SEIA approved 987 GIS projects with a total budget of 3.7 billion UAH,[7] among which only 30 % of projects were fully coordinated with other relevant authorities, and just 37 GIS projects with a budget of ca. 113 million UAH were completed (SEIA 2011). Therefore, Ukraine was able to fulfil its plan for development of green projects by less than 5 % and used less than 3 % of the attracted funds.

As Table 12.2 indicates, the distribution of GIS projects is unbalanced, with major industrial regions of Ukraine mostly neglected. Currently, these investments are allocated mainly for energy saving projects, e.g. on insulation of public facilities, such as schools and hospitals in regions facing budget deficit. Only one project proposed through the green investment scheme is focused on modernisation of technical facilities—Pilot project on construction of installation for treatment of the mine drainage water at the Voikov Mine, Luhansk Oblast (SEIA 2011). This situation is determined by the fact that Ukraine does not have effective mechanisms for distribution of external environmental investments, such as setting targeted

[7] Ukrainian hryvnia (UAH) is the national currency of Ukraine. Official exchange rate of Euro to Ukrainian hryvnia by the National Bank of Ukraine for 20.12.2011 was: 100 EUR = 1,041.80 UAH (NBU 2012).

Table 12.3 The participation of Ukrainian industrial regions in JI projects, which are currently carried out

Ukrainian regions	Number of projects	% of the total number	Generated ERUs	% of the total ERUs quantity
All regions	72	100	62,857,037	100
Main industrial regions	43	59.7	43,497,069	69.2
Donetsk Oblast	19	26.4	20,609,408	32.8
Dnipropetrovsk Oblast	6	8.3	2,451,656	3.9
Zaporizhia Oblast	7	9.7	2,533,066	4.0
Luhansk Oblast	11	15.3	17,902,939	28.5

Source: SEIA (2012)

incentives. Moreover, local authorities often have insufficient understanding of opportunities created by the surplus allowances trading and do not take any measures for involving industrial enterprises in the carbon market.

At the same time, as of 01.01.2012 Ukraine is a world leader in the number of JI projects (89 out of 349), which generated 62.9 million Emission Reduction Units (ERUs) or 53.1 % of their global quantity (SEIA 2012). Ukrainian enterprises received ca. 600 million EUR from foreign investors for carrying out JI projects. Furthermore, 60 % of the projects are implemented in industrial regions (see Table 12.3).

The majority of JI projects deal with landfill gas, mine methane, energy distribution and transportation. In general, these projects produce a strong multiplier effect with positive economic results.

The project 'Coal Mine Methane Capture and Utilization at Holodnaya Balka Mine in Donetsk Oblast' is a good example of the successful JI project. The project activities included improvement of coal mine methane drainage system and its utilisation through installation of a combined heat and power plant and a flare system. According to the project design document submitted by its developers to the Joint Implementation Supervisory Committee, the project is environmentally sound and resource-saving, and it would provide reduction of greenhouse gas emissions (JISC 2007). It was planned that the generated electricity would be mainly used for mine's own needs, replacing grid electricity. Total estimated emission reductions over the crediting period (01.01.2008–31.12.2012) equals to 201,584 tonnes of CO^2eq. Moreover, the project was supposed to provide additional benefits such as mine's economic efficiency, labour protection and safety, as well as stimulus for initiation of such projects at other similar coal mine sites.

However, despite certain progress towards implementation of the Kyoto Protocol, the investments mechanisms working through the JI scheme are used insufficiently by the industrial regions. Also, analysis of plans for economic development of Ukrainian regions revealed that they do not actively take advantage of opportunities provided by the JI mechanism.

To address these circumstances, regional development programmes should include measures promoting investments through JI mechanism; and JI project proposals, especially for small-scale projects, should be consolidated as packages for better integration and synergy of their results. This will allow attracting additional foreign investments for solution of the variety of environmental and economic problems, what will encourage consistent reorganisation of the regional economy. Furthermore, this approach will contribute to minimisation of risks in JI projects and will improve their implementation through various instruments of regional governance.

12.4 Instruments for the Integration of Climate Change Issues into Development Strategies of Industrial Regions

Analysis of the regional development strategies for industrial regions of Ukraine showed that they support climate protection to a certain extent. However, in general, climate protection measures are scattered among individual projects and do not address causes of the problems. Development strategies do not contain environmental performance indicators or special economic regulators for this process.

For example, the Strategy for development of Dnepropetrovsk region for the period till 2015 is typical for Ukrainian industrial regions. The main economic target stipulated by the Strategy is a sustainable growth of the region's economy which should be achieved through a number of objectives: modernisation of the key sectors of the economy, including mining and metallurgical sector, diversification of the economy, etc. Also, for each of the objectives, the Strategy defined means of achieving them. In particular, development of agriculture and civil construction sector was defined as a basis for diversification of the economy. The civil construction sector is expected to increase its share in the region's economy by 1.5 times by 2020; the production of construction materials should grow 2.5 times (Dnipropetrovsk Oblast 2011).

However, regional development strategies in Ukraine generally are not backed up by appropriate environmental assessments. The strategies are usually missing target and planning indictors for total greenhouse gas emissions in a region and adaptation to climate change. For instance, in the Strategy for development of Dnepropetrovsk region quantitative environmental targets were defined only for a set of projects to modernise six industrial enterprises in the framework of the Kyoto Protocol (Dnipropetrovsk Oblast 2011).

At the same time, in the context of climate protection commitments, there is a need for a wider interpretation of the role of regional planning. The planning should provide for transition to the low-carbon economy through the right balance of targets and indicators guiding the structure and dynamics of the economic development, which should be based on a new model of energy consumption with a gradual decrease in greenhouse gas emissions.

Discussions in scientific literature deal with various instruments for integrating climate change issues into regional development strategies. The following instruments can be considered as particularly helpful for Ukrainian practice: regional climate plan and strategic environmental assessment for development programmes.

Climate plan should: identify major climate change-related issues for their subsequent integration into regional development strategies; prioritise adaptation and mitigation initiatives based on an assessment of physical and economic climate change impacts; highlight additional socio-economic benefits of addressing climate change in the short- and medium-term (UNDP 2009). Since a national inventory system for greenhouse gas emissions and sinks has been established in Ukraine, development of regional climate plans could be provided with a strong data support.

Strategic environmental assessment, as is well known, is used for incorporating environmental concerns into decisions of public authorities about all different spheres of society. Using this instrument can help to systematically integrate climate change issues into regional development programmes, in particular, to verify compliance of development targets with climate protection priorities.

Furthermore, there is a need for instruments aiding balanced regional development strategies. In Ukraine, instruments that are part of state permitting, compensatory and regulatory systems are also used for the solution of environmental problems, including air protection issues. Examples of these instruments include, in particular, licences for the use of natural resources, fees for polluting the environment, environmental taxes (e.g. land tax), etc. And new challenges require development and application of a variety of incentive methods at both national and regional levels. For instance, utilising economic incentives can contribute to sustainable climate protection while strengthening the economy.

12.5 Conclusions and Recommendations

UNFCCC and the Kyoto Protocol formed new strategic perspectives and created additional opportunities for stimulating sustainable growth of the economy in Ukraine.

In this context, industrial regions should play a more constructive role in tackling the climatic challenges. And for that purpose local authorities should use opportunities arising from international environmental cooperation in technological modernisation, i.e. adopt measures to increase participation of industrial enterprises in green investment scheme projects, attract foreign investments through JI projects, what will contribute to the solution of the variety of environmental and economic problems and encourage consistent reorganisation of the regional economy.

At the same time, the role of regional planning should be extended in the context of the clime protection challenges. The planning should provide for transition to the low-carbon economy through the right balance in economic targets and indicators. In Ukrainian practice, regional climate plan and strategic environmental assessment for development programmes can be considered as particularly helpful for achieving such balance. Implementation of the new approach requires also re-examination and improvement of the current compensatory and regulatory instruments governing relations between public authorities and private sector, and introduction of a state system of incentives.

Biography Ludmila Palekhova is associate professor at the Department of Marketing, National Mining University (NMU), Dnepropetrovsk, Ukraine.

At the same time, for the last 15 years she has been a director of the Consulting Centre which prepares and supports the implementation of development programmes for Ukrainian industrial enterprises under conditions of the post-Soviet transition economy. Her research scope includes the issues of strategic management and marketing based on the principles of sustainable development, as well as transition to green economy. In particular, in her latest publications she discussed problems related to: voluntary sustainability standards; potential of environmental certification for marketing of industrial products; marketing and investment mechanisms in the context of the Kyoto Protocol; green innovations in the development of industrialised regions.

References

Cabinet of Ministers Decree No. 612 of 04.04.2007 "On establishment of the National Environmental Investment Agency of Ukraine". Official gazette of Ukraine (Ofitsiyny visnyk Ukrainy) No. 26 of 20.04.2007, p 34

Cabinet of Ministers Resolution No. 145-r of 15.03.2006 "On approval of the Energy Strategy of Ukraine for the Period up to 2030". Official gazette of Ukraine (Ofitsiyny visnyk Ukrainy) No. 14 of 19.04.2006, p 78

Dnipropetrovsk Oblast – Dnipropetrovsk Oblast State Administration (2011) Strategy for development of Dnepropetrovsk region for the period till 2015, approved by the decision of the Dnepropetrovsk regional council No. 132-7/VI of 24th June 2011. http://www.adm.dp.gov.ua/OBLADM/Obldp.nsf/%28DOCWEB%29/997A3C4197D4380EC225788E0047C2D3?OpenDocument. Last accessed 29 April 2014

Dnipropetrovsk Oblast – Dnipropetrovsk Oblast State Administration (2012) Report on carrying out the Programme on socio-economic and cultural development of the Oblast in 2011

EBRD – European Bank for Reconstruction and Development (2011) 2050: Greenhouse gas emissions projections for Ukraine. Executive Summary. Report prepared by factor CO_2 Madrid as part of the Ukraine Carbon Market Facilitation Programme. http://www.ebrd.com/downloads/sector/eecc/GHG_emissions_projections_for_Ukraine_20111021.pdf. Last accessed 29 April 2014

Goskomstat – State Statistics Service of Ukraine (2012) Statistical information. http://www.ukrstat.gov.ua. Last accessed 29 April 2014

IPCC – Intergovernmental Panel on Climate Change (2006) 2006 IPCC Guidelines for National Greenhouse Gas Inventories, Prepared by the National Greenhouse Gas Inventories Programme. Vol. 2: Energy. Institute for Global Environmental Strategies, Kanagawa, Japan

JISC – Joint Implementation Supervisory Committee (2007) Joint Implementation Projects Database. Project 0051: Coal mine methane capture and utilization at Holodnaya Balka Mine in Donetsk Oblast. http://ji.unfccc.int/JI_Projects/DB/S9AHNSKYDUK7GPJ9L07TUWZP7XM45N/PublicPDD/GO3NL7W0TZJCBJBR1VJ5R1UX5QEZU1/view.html. Last accessed 29 April 2014

MEP – Ministry for Environmental Protection of Ukraine (2006) Ukraine's report on demonstrable progress under the Kyoto Protocol. http://unfccc.int/resource/docs/dpr/ukr1.pdf. Last accessed 29 April 2014

MEP – Ministry for Environmental Protection of Ukraine (2011) National Inventory Report of anthropogenic emissions by sources and removals by sinks of greenhouse gases in Ukraine for 1990-2009. State Environmental Investment Agency of Ukraine, Kiev. http://www.seia.gov.ua/seia/doccatalog/document?id=129550. Last accessed 29 April 2014

NBU – National Bank of Ukraine (2012) Foreign exchange market indicators. http://www.bank.gov.ua/control/en/allinfo. Last accessed 29 April 2014

Palekhov D, Schmidt M, Pivnyak G (2008) Standards and thresholds for EA in highly polluted areas – the approach of Ukraine. In: Schmidt M, Glasson J, Emmelin L, Helbron H (eds) Standards and thresholds for impact assessment. Springer, Heidelberg, pp 33–48

SEIA – State Environmental Investment Agency of Ukraine (2011) Projects of targeted environmental (green) investments that are under consideration or approved by the State Environmental Investment Agency, as of 20.12.2011 (in Ukrainian). http://www.neia.gov.ua/nature/control/uk/publish/article?art_id=133668&cat_id=124591. Last accessed 17 Feb 2012

SEIA – State Environmental Investment Agency of Ukraine (2012) Report about the work of the State Environmental Investment Agency of Ukraine in 2011 (in Ukrainian). http://www.neia.gov.ua/nature/control/uk/publish/article?art_id=133753&cat_id=110541. Last accessed 17 Mar 2012

UNDP – United Nations Development Programme (2009) Charting a new low-carbon route to development: a primer on integrated climate change planning for regional governments. http://www.undpcc.org/undpcc/files/docs/publications/Charting_carbon_route_web_final%20%282%29.pdf. Last accessed 29 April 2014

UNFCCC Secretariat (2011) National Greenhouse Gas Inventory data for the period 1990-2009. U.N. Doc. FCCC/SBI/2011/9. United Nations, Geneva. http://unfccc.int/resource/docs/2011/sbi/eng/09.pdf. Last accessed 29 April 2014

Chapter 13
Risk Management and Climate Change: A Question of Insurability

Lars Krause and Terence Onang Egute

13.1 Introduction

In the recent past, the world has experienced an increasing trend of devastating environmental disasters. The scientific community has agreed that the frequent occurrence of these disasters is closely related to anthropogenic climate change. The victims of environmental disasters—whether from the private or public sector—often face enormous damages. The Western world often tries to cover potential damages by means of insurance. But this only works in very few cases. Existing insurance policies may cover only an insignificant part of the enormous damage. As a result, concern has arisen with respect to the insurability of climate change environmental disasters. This paper aims at clarifying the question of whether and under what circumstances it is possible to secure the risks caused by climate change environmental disasters through risk management. It also highlights the role of different actors within risk management.

13.2 The Concept of Risk Management

Risk management is the systematic treatment of risks or, in other words, risk management are all human activities which integrate recognition of risk, risk assessment and developing strategies to manage it (Krause and Borens, 2009a, p. 181; Kirchner 2002, p. 18). According to Renn et al., risk management helps to reduce an intolerable high risk of an event to an acceptable amount

L. Krause (✉) • T.O. Egute
Department of Civil Law and Public Law with References to the Law of Europe and the Environment, Brandenburg University of Technology (BTU) Cottbus-Senftenberg, Erich-Weinert-Str. 1, 03046 Cottbus, Germany
e-mail: lars.krause@tu-cottbus.de; eguteter@tu-cottbus.de

E. Albrecht et al. (eds.), *Implementing Adaptation Strategies by Legal, Economic and Planning Instruments on Climate Change*, Environmental Protection in the European Union 4, DOI 10.1007/978-3-540-77614-7_13,

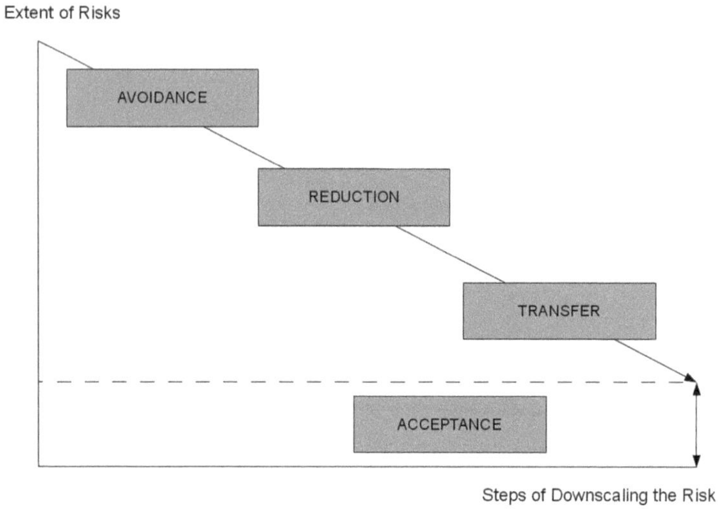

Fig. 13.1 Treatment of risk in a risk management system according to ISO 31000

(Renn et al. 2007, p. 97). Risk management can also help to reduce the probability or gravity of a negative outcome of that event.

Risk management includes all kinds of risks, e.g. physical, legal and financial risks. If you try to find an interdisciplinary definition of risk, it will be the product of likelihood and impact of an event relating to a variation of drawn objectives and it will be counted by the unit of the objectives (Krause and Borens 2009a, pp. 180–181). In short, a risk describes the probability and impact of an event. An event with a low probability of occurrence and a high loss is the same as a high probability and low loss risk. At zero risk there are no differences from the objectives.

Risk management is a discipline which was practiced during the ancient period. In modern times, risk management is used mostly by the financial sector, i.e. to find risks and chances of investments. Since the first risk management codex US-MIL-STD 882 was established by the United States Army in the 1970s, many other institutions tried to fix a standard for risk management. Governments tried to establish risk management acts. One of the most important acts is the BASEL II Codex which came into force in 2009 (Krause and Borens 2009a, pp. 181–182). Towards the end of 2009, the International Organization for Standardization (ISO) published the ISO 31000. The ISO 31000 standardizes the risk management process (Krause and Borens 2009a, p. 184). According to the ISO 31000, risk management is all 'coordinated activities to direct and control an organization with regard to risk'. In ideal risk management, a prioritization process is followed, whereby the risks with the greatest loss and the greatest probability of occurrence are handled first, and risks with lower probability of occurrence and lower loss are handled in descending order (Kirchner 2002, p. 18).

The strategies to solve risk include four kinds of different treatments: transferring the risk to another party, avoiding the risk, reducing the negative effect of the risk, and accepting some or all of the consequences of a particular risk (Fig. 13.1) (Kirchner 2002, pp. 43–46).

Risk transfer is the type of risk treatment that gives answers to the leading question. Risk transfer means causing another party to accept the risk, typically by contract or by hedging (Ehrmann 2005, pp. 89–90). Insurance is one type of risk transfer that uses contracts. The leading question is as follows: Can we trade climate risk by insurance?

13.3 Risk Potential of Climate Change

Before answering the leading question, we should take a look at the risk potential of climate change-related events. The annual reports of the World Economic Forum (WEF) contain several kinds of environmental and other risks. According to their 2008 report, which is in line with the above definition of risk, climate change risk is taken as shown in Fig. 13.2. The reports from 2009 to 2011 show the likelihood and impacts of events caused by climate change in a different way and provide less specific numbers.

The figures show the risk of the listed events to be taken in the years 2007 and 2008 and the outlook for 2009. The numbers between 1 and 5 for likelihood and severity are taken in order to ensure easy understanding, clarity and comparability. The numbers shown are not fixed; the risks are open to mitigation measures which will alter likelihood and severity within 10 years. According to the above definition of risk, risk is more significant if it is calculated by the actual numbers and not clustered; e. g. in 2007, the risk of an extreme weather event in regard to the economic losses was valued at 2.5 billion US$. In 2008, the risk increased up to 3.75 billion US$. The WEF shows the highest risk related to tropical storms: In 2008, it resulted in 10,000 deaths (WEF 2008, pp. 45–51; WEF 2009, pp. 27–31).

According to the World Economic Forum (WEF), until 2050 German private and public households have to expect up to 33 billion US$ costs for climate change (Kemfert 2005, p. 209). In 2006, the Worldwide Fund for Nature (WWF) and the SAM Group published a study about climate-related risks for the industry. According to this study, RWE AG, Europe's biggest emitter of carbon dioxide, has to deal with risks amounting to 17 % of its net equity value if we go ahead without changing our behavior (WWF and SAM 2006, p. 1). In the 2011 Global Risk Report of the WEF, climate change-related risks are presented as the highest risks. Also the risk of climate change is interconnecting with the risk of water security, food security and extreme energy price volatility. North America and Asia are expected to experience most of the impacts (WEF 2011, pp. 1, 44, 46). Since 2006, the NGO Germanwatch embarked on annually releasing the 'Global Climate Risk Index' (CRI). The Index for 2011 affirms that less developed countries are generally more affected by climate change than industrialized countries. The CRI shell serves as a warning signal indicating past vulnerability which may further increase in regions where extreme events will become more frequent or more severe through climate change (Harmeling 2010, p. 5). Table 13.1 shows the top ten countries of the 2011 Long-Term Climate Risk Index beside the CRI score, death

2007						2008						Outlook (O no chang,↑ ↓ in-, decrease)		
Economic Loss			Lives Lost			Economic Loss			Lives Lost					
L	S	R	L	S	R	L	S	R	L	S	R	L	S[$]	S[D]
Extreme weather events														
3	3	9	-	-	-	3,5	3	10,5	3,5	2,5	8,75	0	0	↑
Heat waves														
-	-	-	-	-	-	3,5	3	10,5	3,5	2	7	↓	0	0
Loss of freshwater														
3,5	2,5	8,75	3,5	3	10,5	3	2	6	3	2,5	7,5	0	0	↓
Tropical Storms														
2	3	6	-	-	-	2	3	6	4	3	12	↓	0	↓
Inland Flooding														
2	2,5	5	2	2	4	2	2,5	5	2	2,5	5	0	0	0

Likelihood: 1: < 1%, 2:1–5%, 3: 5–10%, 4: 10–20%, 5: >20%
Severity - economic loss [billion US$]: 1: 2–10, 2: 10–50, 3: 50–250, 4: 250–1,000, 5: >1,000
Severity - lives lost: 1: 1,600–8,000, 2: 8,000–40,000, 3: 40,000–200,000, 4: 200,000–1m, 5: >1m.
Source: Own calculation based on WEF, Risk Report 2008 and 2009

Fig. 13.2 Risk of climate change-related events

Table 13.1 Long-term Climate Risk Index 2011

CRI, 1990–2009	Country	CRI score	Death per 100,000 inhabitants	Total losses in million US$ PPP	Number of events
1	Bangladesh	7.33	5.63	2,068.14	259
2	Myanmar	8.67	14.33	676.35	30
3	Honduras	10.83	5.21	663.57	53
4	Nicaragua	16.17	2.80	263.33	39
5	Vietnam	19.00	0.59	1,861.50	203
6	Haiti	19.67	3.98	164.62	46
7	Philippines	26.83	1.08	684.45	270
8	Dominican Republic	27.67	2.55	185.08	41
9	Mongolia	31.00	0.54	308.65	30
10	Tajikistan	33.50	0.47	311.27	51

Source: Harmeling (2010) Global Climate Risk Index 2011. Germanwatch

per 100,000 inhabitants, total losses in million US$ per purchasing power parties (PPP) and the number of events most affected between 1990–2009. Bangladesh is this year's most affected country as it was in 2009 and 2008.

The annual Global Risk Report is supported by Munich Re with respect to the core data which are among the most reliable and complete data bases on this matter and provide the basis for the report. The CRI score analyses the quantified impact of extreme weather events. It looks at impacts and results in an average ranking of countries in four indicators. The CRI score should be seen as a piece in the puzzle of an analysis of countries' exposure and vulnerability to climate-related risk (Harmeling 2010, pp. 6–7).

Another problem is that more and more peak catastrophes have happened. Compared to the Global Climate Risk Index 2010, not even the year 1991 with the Bangladesh-Cyclone has there been a peak year as remarked by Harmeling in the 2010 report. The 2010 Pakistan flooding event only nine years after the Bangladesh-Cyclone leads to a different evaluation of the occurrence of extreme weather events.

13.4 Can We Trade Climate Risk by Insurance?

The amount of losses from natural hazards has grown and the WEF still expects an increase (WEF 2011). This has been caused by climate change, increased population density and the well-being in the catastrophe-prone areas and additionally other events which have not been realized yet, i.e. the acute flooding risk for all coastal areas sum up the problem. The flooding risk poses potentially a very acute problem for the Netherlands. Scientists expect a flood in the next years that will consume about half of the Netherlands. On the 'Flood Maps' internet page http://flood.firetree.net, the impact on the coastal areas around the world is shown. Countries like Bangladesh are flooded every year in a catastrophic proportion. Flooding which are repeated every 50 or 100 years are as catastrophic as flooding that occurred decades before i.e. the River Elbe flooding in the Dresden area in 2002 or the River Oder flooding in 1997. The last one seems to be an annual event. The first remarkable flooding took place in 1838 and after that, there was another in 1947 which was followed by the one in winter 1981. The interval between two floods is increasing. Since 1997, flooding occurs in the River Odra area almost every year and in 2010 Germany and Poland had to deal with three floodings.

As far as flooding is concerned, insurance companies offer quite limited coverage, forcing governments to assume major flooding risk even though they are not professional risk-bearers. When we examine the case of the 2010 flooding in Pakistan, it will be deduced that the country would have been helpless in taking care of the losses without donations from around the world. A look at the development of costs to the U.S. insurance industry caused by weather-related catastrophes shows that the costs remained constant until 1987 after which an increase of costs can be detected. In 2005, the costs amounted to 50 billion dollars. Insured U.S. weather-related losses are growing ten times faster than premiums and the overall economy, and even faster when compared with the population (Mills et al. 2006, pp. 114–119).

The risk of catastrophes due to climate change has several special features, which must be considered when catastrophe insurance arrangements are being developed. A natural catastrophe is a low-probability event, but when it happens, it generates very large losses. The climate catastrophe has a high potential of resulting losses for an individual in magnitude or it is a high loss for an individual insured. As a result, the insurer should have access to a large pool of liquid capital. Furthermore, a high loss potential develops a danger of insolvency for the insurer. The insurer has to identify these risks and try to develop methods to avoid them. Another special feature of climate-related catastrophes is that nobody knows when and how the loss will happen. Even if the catastrophe is forecasted, only few people believe in it because of the huge losses to be expected: When flooding was to occur at River Odra officials failed to heed to the flooding warning that was given by the Swiss Meteorologist J. Kachelmann. An early warning can save lives and damages.

Concerning the question about the probability of a catastrophic event and the resulting loss compared to traditional insured risk, the result is that the usual policy pricing methods are inapplicable. Another special point is the market incompleteness. Usually, there is no private insurance offer for certain types of catastrophic risks. One reason for this is the lack of capacity of the insurance industry money to cope with some catastrophes due to extremely large loss potential (Mills et al. 2006, pp. 123–138). This is due to the fact that the government, like in the past, has to assume the burden of solving the financial problems caused by climate catastrophe.

13.5 Aspects of Solving the Problems

In relation to stakeholders and their options and possibilities to take care of the problem, four aspects should lead to a solution of the problem. The most important stakeholders of the problem are concerned governments and the insurance industry. They are in every kind of issue related to catastrophes and have the possibilities of taking action with politics and money. That is why the role of the governments should first of all be clear. After this is determined, the impact of climate-related parameter uncertainty on insurance contracts should be determined. Then, a national or international catastrophic insurance fund should be implemented. Last but not least, the insurance consumers should be encouraged to adopt a responsible behavior in order to minimize the loss associated with climate change-related catastrophes.

13.5.1 The Role of the Governments

Especially in cases where there is no private insurance to cope with climate catastrophic events, governments could play three main roles. Firstly, as a co-insurer, hand in hand with the private insurance industry. Secondly, the governments could play the role of a re-insurer directly, or they could lend money for primary insurers to fill up capital. The third possibility is to be a holder of resource pools. For example, a national fund for damages caused by climate change. In Germany, the government takes action in case of a catastrophe but the money has to be taken out of the annual public budget i.e. when river Elbe flooded, the German federation and the hardest-hit states put in place a federal flood fund of over 6 billion €. The amount of money is orientated on the losses but also on the financial capacity of the federal state and the provinces at the time of the disaster. It is unimaginable that in bad times less money would be given.

The heaviest impact, as shown in the Global Risk Report and the Global Climate Risk Index, is faced by developing countries. Most often, they do not share the same luxury and cannot avoid long term economic impacts as easy as Germany e.g. catastrophe caused by climate change often leads to a reduction in aggregate supply and demand with lower level of income for the community which leads to deflation and a higher level of unemployment. That is why economists claim that the necessary instruments of macroeconomic stabilization need to respond to shocks, and contingency plans need to be implemented in order to keep the costs of natural disasters at a minimum (Banuri 2005, p. 11).

13.5.2 Impact of Climate Related Parameter Uncertainty on Insurance Contracts

Climate change brings several kinds of catastrophes and damages with it; the insurance industry may have to cope with a variety of special damages. Flood, windstorm, wildfires, mold and moisture damage, earth movement and coastal erosion and health impact are interacting. Because of the character of a climate catastrophe, it is not easy to set prices for an insurance premium. The insurance industry does not want to extend the insurance coverage without increasing the premium. That is why the insurance industry has to strive to improve loss data collection and enhance the actual analysis. Furthermore, they have to analyze the positive and negative implications of climate change on their business, investments and customers (Mills et al. 2006, pp. 135–136).

Besides, we have to take a look at the impact of climate-related parameter on insurance contracts. This leads to the assumption that the insurer doesn't want to extend its insurance coverage but rather the premium will increase. This can be stopped by a national fund for damages which covers special damages the insurance industry is not willing to insure without a premium increase. The governments should force engineers to construct safer buildings just like earthquake-proof buildings and protect constructions to prevent loss of government money.

Beside the sustainable insurance contracts, an insurance company like the Munich Re is investing more and more into sustainable funds. By October 2009, the Munich Re already invested 100 million € out of their 65 billion € asset. This is for now a very negligible investment, but it is growing every year. The criteria of environmental sustainability have become more and more important (Munich Re 2009).

13.5.3 National or International Insurance Funds

The question is how to organize 'catastrophe insurance' in a sustainable way. Starting in the 1990s, the private insurance industry could not provide coverage against natural hazards in the traditional way without significant insolvency risk because of the very high loss and low probability event (Nell and Richter 2004, p. 3). That is why national or international insurance funds are necessary, useful and practicable. Funds can safeguard private insurers and their customers and they can be an effective tool in case of market incompleteness. It is necessary that the fund has as much money as needed in case of a catastrophe to provide fast direct aid to the catastrophe area (Pollner 1999, p. 3). The federal flooding fund mentioned above was not a sustainable fund from 2002. The fund was only set up for the river Elbe flooding and was closed by the end of 2006. In case of future flooding, Germany has to set up a new fund with money out of the annual public budget.

13.5.4 Encourage Sustainable Behaviour of the Consumer and the Populace

Can the insurance customer be driven to take action to reduce the negative effect of climate change or the climate change by itself to reduce losses? This idea is to compensate responsible behavior of customers when they reduce their loss potential or when they exhaust CO_2 emission, they get a discount on premium. The insurance industry in Germany, for instance, already tried this several years. In 2007, the DEVK insurance company gave owners of the BahnCard a bonus to their car insurance premium of up to 30 %. With a BahnCard, its owner gets a discount of the costs of a train and public traffic charge of 25 %, 50 % or 100 %. On the one hand, a certain aspect of the customer's behavior out of the insurance area drives the insurance premium. But on the other hand, an owner of the BahnCard uses more public traffic to compensate the costs of the BahnCard and drives less and the risk of a car accident decreases. This is an example of how less risky behavior can lead to a reduction of the negative climate effects. More examples like this are found in other insurance contracts. Other insurance companies decouple the insurance contract and the risk behavior of the customer. In 2009, the DA direct insurance company gave savings of up to 30 % to costumers of new environmentally-friendly cars. The company asks for several ecological criteria of the car. Here, the risk behavior of the customer has nothing to do with the calculation of the insurance premium, i.e. the normal price-driving factor.

A big problem is the high density of population in some flooding-prone areas. With a sustainable urban land use planning and relocation, the government can take action to prevent the people not to construct any buildings in dangerous areas. In 2004, several years after the Elbe flooding, the German Federal State of Saxony revised its water protection law (SächsWG). Since then, it is forbidden to build or change any buildings in admitted flood-prone areas (§100a SächsWG). However, the residents were not relocated, and the construction of new buildings is forbidden. For residents of flood-prone areas, this means that they have to expect new damages from flooding and have to bear the burden of resolving the risk themselves.

13.5.5 Using Risk Management for Assistance

It may also be a combination of the solutions discussed above which, when put together in a risk management system according to ISO 31000, could lead to a sustainable solution and help to reduce the loss of climate-related catastrophe. In case of climate-related risks, using a risk management system can help to bring some kind of structure in the chaotic climate catastrophic events, their negative impact, an assessment of likelihoods and the dealing with it. Implementing a risk management system helps to find an understanding of climate-related catastrophes by analyzing it in all aspects. It should be taken in all involved organizations and helps to create a communication and reporting mechanism to learn more about annual or periodic events on a national or international level. It should be noted that the most important aspect of a risk management system is that it has to be established as a cycle with steps reviewing of earlier steps. Also, earlier events and their impact will be reviewed in a cycle. A review should also lead to a continuous improvement of the whole risk management system (Krause and Borens 2009b, p. 231).

As discussed above, using a risk management system according to ISO 31000 in general is helping also to develop strategies to treat the impact of risks by avoiding, reducing, transferring and accepting, as shown in Fig. 13.1. The solutions mentioned above to solve problems caused by climate-related catastrophes can be put in the system of transferring, avoiding, reducing as shown in Fig. 13.3.

The stakeholders of catastrophes are each apart from the overall system of risk treatment. First of all, sustainable behavior should avoid the risk of climate change in the long run. Citizens should not build houses in flood-prone areas. Furthermore, insurance pricing system could drive this necessary sustainable behavior to avoid the risk. Protection law by national governments, such as land planning law, will reduce the risk of a catastrophe by minimizing the impact. Unavoidable and not easily reduced risks must be transferred to insurance companies and or to national and international funds as shown above. After all, the remaining risk accepted intentionally must be borne by the affected persons.

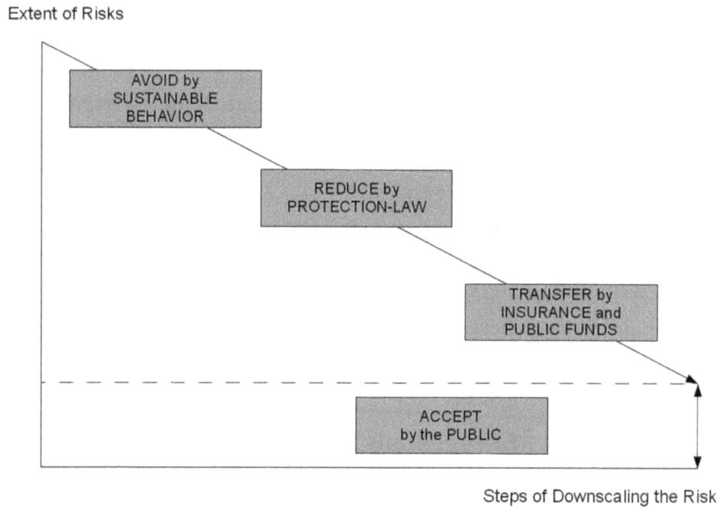

Fig. 13.3 Treatment of climate-related catastrophe risks according to ISO 31000

13.6 Conclusion

To sum up, the insurance industry has to learn more about climate change and its extent of losses to be able to set up more resilient premium. Furthermore, with this new information, they don't longer have to be surprised by catastrophes and face insolvency even if there are more and more peak catastrophes. Also, citizens are charged higher premiums because of frequent occurrence rates and enormous damage (WEF 2007, 2008, 2009). Therefore, they should be advised not to build in risk-prone areas. The way Germany addresses the problem through water protection law is right. It is also a fact that in developing countries, the settlement areas are near big rivers because of the fertile nature of land in these areas and not all citizens can be tracked to live in safer areas. That is why governments should set up funds so that in case of a disaster the current budget should not be over-stressed or overused and the macroeconomic harm remains low. Using a risk management system according to ISO 31000 may put all aspects together in a co-act solution.

Biography Lars Krause has been a research fellow at the Department of Civil and Public Law with References to the Law of Europe and the Environment of the Brandenburg University of Technology since 2010. After his studies with a focus on Environmental Law, he has been doing his doctorate.
Krause's research work focuses on risk in terms of law as well as risk management. In 2008 he translated the strategic chapter of ISO 31000 into German and commented on it. Lately he has been occupied with Chemical and Atomic Energy

Law. Further focus of Krause's research comprise Waste Law, Renewable Energies Law and Civil Law.

References

Banuri S (2005) The macroeconomic impact of natural disasters. Macroeconomic theory for applications, Department of Economics, Dr. Edward Day, Texas. http://utdallas.edu/~sxb030200/sb_cp3_nat_disasters_122005.pdf. Accessed 1 June 2011

Ehrmann H (2005) Risk management. Friedrich Kiehl, Ludwigshafen

Harmeling S (2010) Global climate risk index 2011. Who suffers most from extreme weather events? Weather-related loss events in 2009 and 1990 to 2009. Germanwatch, Bonn

Kemfert C (2005) Deutsches Institut für Wirtschaftsforschung (DIW) Berlin Wochenbericht Nr. 12–13/2005: pp. 209–215

Kirchner M (2002) Risikomanagement. Rainer Hampp, München

Krause L, Borens D (2009a) Strategisches Risikomanagement nach ISO 31000. ZRFC Part 1 04/09:180–186

Krause L, Borens D (2009b) Strategisches Risikomanagement nach ISO 31000. ZRFC Part 2 05/09:227–231

Mills E, Roth RJ Jr, Lecomte E (2006) Availability and affordability of insurance under climate change: a growing challenge for the US. J Insur Regul 25(2):109–149

Munich Re (2009) Klimafreundliche Projekte – Hoch im Kurs. Pressnote by October, 9th 2009, Munich

Nell M, Richter A (2004) Improving risk allocation through indexed cat bonds. The Geneva Papers on Risk and Insurance 29:183–201

Pollner JD (1999) Using capital markets to develop private catastrophe insurance. The World Bank Group, Note No. 197

Renn O, Schweizer P-J, Dreyer M, Klinke A (2007) Risk. How society deals with uncertainty. oekom, Munich

WEF (2007) Global risk report 2007. World Ecology Forum, Cologny, Geneva

WEF (2008) Global risk report 2008. World Ecology Forum, Cologny, Geneva

WEF (2009) Global risk report 2009. World Ecology Forum, Cologny, Geneva

WEF (2011) Global risk report 2011. World Ecology Forum, Cologny, Geneva

WWF Deutschland, SAM Group (2006) Carbonizing valuation – climate risks for the company's value. 11/2006

Chapter 14
The Cumulative Impacts of Climate Change on Subsistence Agriculture in the Sudano-Sahel Zone of Cameroon: Enhancing Adaptation Policies

Prosper Somah Techoro and Michael Schmidt

14.1 Introduction

With irrefutable evidences from around the globe, climate change impacts are being felt and are adding significantly to the development challenges in ensuring food security and alleviating poverty particularly to most Sub-Saharan African countries as a whole (Low 2005) and the Sudano-Sahel region of Cameroon in particular. This is strongly due to the fact that agriculture is the mainstay in the region where the majority of the population is very dependent on rain-fed agriculture, accounting for more than 90 % of staple food productivity (Calzadilla et al. 2008). Subsistence agriculture provides food, power, stability and resilience of rural livelihood. With the growing number of extreme weather events, the societal vulnerability to the risk associated with climate change and land degradation will exacerbate the present on-going socio-economic challenges in ensuring the food security and livelihood of ruralites.

The availability of food depends on agricultural production with food productivity inherently sensitive to climate. Seasonal and geographical crop yield availability depends on space, time and rainfall distribution patterns. Since climate is a primary determinant of agricultural production, any adverse changes in climate would likely have a devastating effect on this sector, thus threatening crop failures (Gregory et al. 2009). On the other hand, natural conditions have also made certain areas of Cameroon's Sudano-Sahel susceptible to soil degradation. Increased incidences of extreme events such as droughts and flash floods exacerbate the degradation processes with anthropogenic causes primarily related to high population pressure. Some of these human activities contributing to land degradation include the unsustainable agricultural land use, poor soils and water management practices, removal of natural vegetation, poor irrigation practices, inappropriate cultivation

P.S. Techoro (✉) • M. Schmidt
Brandenburg University of Technology (BTU) Cottbus-Senftenberg, P.O Box 101344, 03013 Cottbus, Germany
e-mail: ssomahs@googlemail.com

E. Albrecht et al. (eds.), *Implementing Adaptation Strategies by Legal, Economic and Planning Instruments on Climate Change*, Environmental Protection in the European Union 4, DOI 10.1007/978-3-540-77614-7_14, © Springer-Verlag Berlin Heidelberg 2014

techniques, deforestation and land conversion and land fragmentation (Ellis and Pontius 2006). Apart from impacting agricultural productivity of the rural population, the land use change and degradation are important sources of green house gases globally, responsible for about 20 % of emission (IPCC 2007). Land degradation leads to increased carbon emission both through loss of biomass when vegetation is destroyed and through increase soil erosion (Sivakumar 2007).

Climate change and land degradation could be major threats to agriculture when dealing with the survival and livelihood of the Sudano-Sahel population in Cameroon. Adaptation to both conditions is therefore very imperative and critical in dealing with some of the posed unavoidable impacts. Thus far, the enactment strategies to aid in the adaptation of subsistence farmers have been very slow. Adaptation is a process of deliberate change; integrative approach is imperative and has to be integrated, incorporated and mainstreamed into the policy apparatus of the government (Nelson et al. 2007). Therefore, identifying the precise drivers of these changes, whether environmentally, climatically, or socio-economically driven, is very important for policy implications.

This paper therefore is aimed at bringing into the limelight some of the current adaptation strategies used by the subsistence farmers in the Sudano-Sahelian region of Cameroon for policy enhancement. It primarily employs the Pressure-State-Response (PSR) indicator conceptual framework in incorporating the cause effect relationships (OECD 1994). This framework is based upon the premise that human activities exert pressure on the environment that results in changes in the state of the environment as a whole, as well as its individual parts. These changes often cause a societal response, which results in changing environmental policies or implementing management actions (OECD 1994). It therefore hypothesizes that cumulating the outcomes of the PSR indicator framework could play a crucial role in the development, implementation and enforcement of laws, regulations and policies for subsistence farmers' adaptation to climate change.

14.2 Materials and Methods

14.2.1 The Study Area Description

The Sudano-Sahel is a vast area just south of the Sahara. Cameroon's Sudano-Sahelian zone extends between Latitude 7° 30′ to 13° North and Longitude 9° to 15° East (Fig. 14.1) (Kenga et al. 2005). It has an estimate population of about 6.5 million inhabitants with 67 % constituting the rural population; that records the highest poverty level in the entire country (ECAM II 2001). From the political point of view, this zone constitutes three regions: the Far North, North and the Adamawa regions with their capitals being Maroua, Garoua and Ngaoundere respectively. The climate type is AW/BS according to Koeppen classification and is characterized by a pronounced dry season with the driest months having little or no precipitation

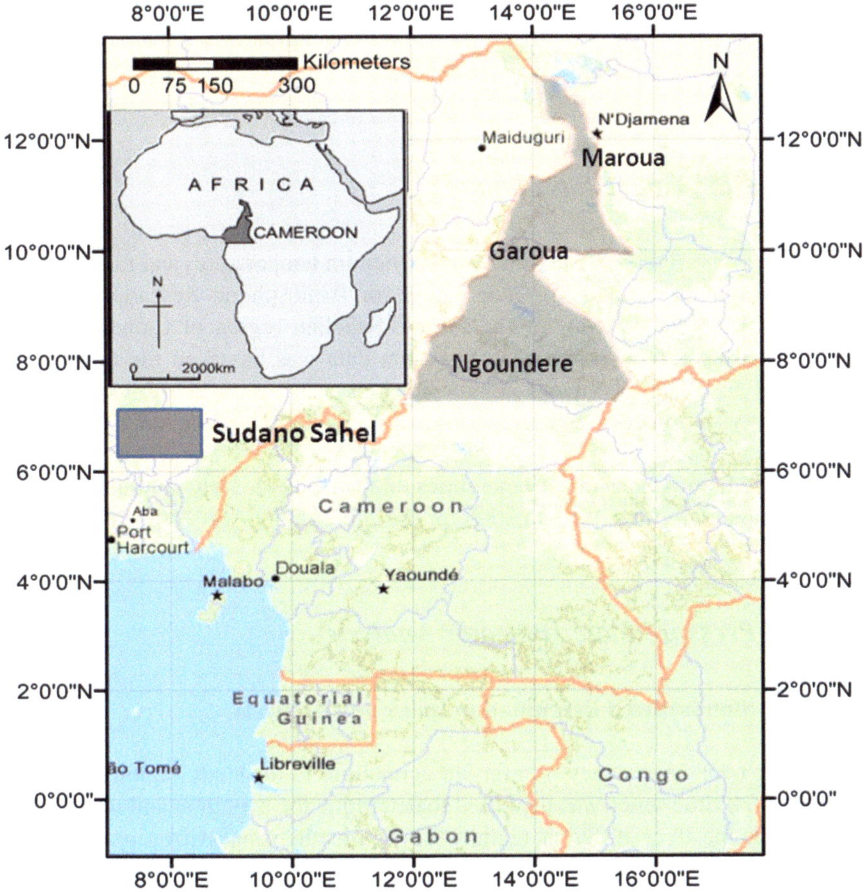

Fig. 14.1 Location map of Cameroon (shaded area shows Sudano-Sahelian study region)

(AW). The potential evapo-transpiration (PET) usually exceeds precipitation in its desert climate (BS) (Koeppen 1936). Rainfall distribution is monomodal or unimodol (Yengoha et al. 2011); and the region experiences a long dry season. Rainfall in this zone is highly variable with its onset very erratic and lasts for about five months (May–September) averaging about 300–800 mm of rain per year.

Over time, various soil formation processes have given rise to important pedological heterogeneity in the Sudano-Sahel. Based on the FAO soil classification, different major soil types have been distinguished. These include: the Vertisols, andosols, the fluvisols, the Planosols and the Lixisols (Kenga et al. 2005). The vegetation is semi-xerophytic and is transitional between the Sudan savanna woody grassland to the south and the open desert to the north. In the southern parts where the Sahel merges with the Sudan zone, savanna makes up the vegetation. Diversity is the norm in Sudano-Sahelian farming systems. Subsistence and commercial farming are practiced. Stephenne and Lambin (2001) put forward land-uses that generate basic resource of the population in the Sudano-Sahel. These include: food for subsistence, fuel wood in the natural vegetation areas, market needs in croplands, fallowing, and livestock in the pastoral land.

14.2.2 *Data Collection Method*

Daily climatic variables of maximum and minimum temperatures and rainfall were obtained from the University Cooperation for Atmospheric Research (UCAR) (http://dss.ucar.edu/datasets) for the Sudano-Sahelian region of Cameroon. The choice of using the observed weather station data was based on the findings of Tingem et al. (2008). To avoid inhomogeneities in the data, quality control and homogeneity test procedures were performed using ClimDex Version 1.3 software developed by Byron Gleason from NCDC/NOAA, USA. (http://cccma.seos.uvic.ca/ETCCDMI/software.shtml). Time series for the mean monthly rainfall and mean yearly temperature from 1961 to 2006 for the entire region was computed.

14.2.3 *Pressure/State Indicator Analysis*

14.2.3.1 Standardized Precipitation Index and Analysis

Due to the over-dependency on rain-fed agriculture in the study region, the computed rainfall time series was then used to determine the evolution and intensity of droughts in the three different regions. The 9-months standardized precipitation index (SPI) by McKee et al. (1993) was used in determining the agricultural drought analysis. The SPI is a common index used to detecting the change in droughts. It is an effective tool for regional climate monitoring and represents abnormal wetness and dryness over a given period of time. It offers a better understanding of the deficit in precipitation. Agricultural drought implies water shortages during the crops' growing period. It is calculated on the basis of selected periods and its equivalent to the Z-Score used in statistics and represents deviation in precipitation totals (McKee et al. 1993). Positive values in the SPI indicate greater than mean precipitation while negative values indicate less than mean precipitation.

14.2.3.2 Remote Sensing Method Analysis

Spatio-temporal land use and land cover dynamics were studied using remote sensing techniques for the study area. The image data used in the study were Landsat Thematic Mapper (TM) and Enhanced Thematic Mapper Plus (TM+). The TM data are one of the most frequently used data for environmental assessment and monitoring. The images acquired were between 1987 and 1990 from the United States Geological Survey (USGS 2012). The date of acquisition of the TM + images on the other hand was for the year 2000 and was lastly modified in 2009. The bands selected in generating false color composite images were 4, 3 and 2.

The data from Landsat imageries were then processed by ERDAS Imagine 9.1 software for spatio-temporal analysis of land use and land cover (ERDAS IMAGINE 2006). Supervised digital image technique was employed and complemented with the field surveys that provided on-the-ground information about the land use types and land cover dynamics as a result of human and natural influences (Eastman 2009). Calculation of the percentage area of the resulting land use and land cover types each of the study area and subsequent result comparison was done. These helped in the identification of change in land use types particularly based on the different classes. Given the homogeneity of subsistence farming in the Sudano-Sahel of Cameroon, the Maroua study area was selected as a hotspot area for the Sudano-Sahel. This was aided by purposive sampling techniques used in the administered questionnaire surveys.

Also considered was the national data on agricultural land availability and areas under cultivation for the Sudano-Sahel region. A comparative analysis was done to show the evolution of the population with agricultural farm sizes and land distribution over the periods of 1996 and 2010.

14.2.4 Response Indicator Analysis

14.2.4.1 Purposive Sampling/Key Informant Interview

The sensitivity and vulnerability of subsistence farmers were investigated by employing a participatory research approach in exploring the adaptation behaviors and patterns perceived by local communities in the face of variable climatic conditions via purposive sampling. Two key informant interviews were first conducted from each of the three different regions of the Sudano-Sahel zone of Cameroon. The key informants were mainly Agricultural Extension Service workers from IRAD (Institut de Recherche Agricole pour le Développement). The key informant interviews aided in the selection of the neighborhoods where subsistence farming was the order of the day.

Snowball sampling was the second process of purposive sampling used in this research and involved the identification of someone who met the criteria in the study. Snowballing sampling is especially useful when trying to reach population that is inaccessible or hard to find (Trochim 2006). This actually worked well with the household questionnaires whereby the subsistence farmers in the Sudano-Sahel villages were the target population groups, no more no less.

14.2.4.2 Household Questionnaires

The household questionnaires involved semi-structured open-ended, closed-ended questionnaires conducted across villages in the three regions of the Sudano-Sahel with Mayo Tsanaga (Far North region), Mayo-rey (North) and Gbaya and Mbum (Adamawa) being the most surveyed areas due to their subsistence farmers' population pool. These were the easily accessible villages chosen for the interviews. Questions covered a number of topics ranging from household information, agricultural dependency and crop types produced, the state of knowledge of subsistence farmers on climate change, currents adaptation practices, constrains and ways forward. A total of 300 questionnaires each were administered for each region. In some questions, respondents were given the options of multiple responses. The study was conducted at a face-to-face level and a respondent rate of about 81 % was obtained. Quantitative data were organized and facilitated for qualitative analysis. Analysis was based mainly on descriptive analysis and presentation. The statistical package (SPSS) 17.0 was used for analysis. Analysis of variance (ANOVA) was first used to check if there was significant difference between the mean responses to climate change perceptions and adaptation strategies towards curbing the impacts at a 95 % confidence interval ratio in the three different region of the Sudano-Sahel.

14.3 Results and Discussion

14.3.1 Drought Analysis

The drought vulnerability in the three study areas of the Sudano-Sahel was assessed by reconstructing historical occurrence of droughts at a 9-months time scale, employing the SPI approach, starting from March and ending in November, corresponding to the past 9 months of observed precipitation totals respectively. This time scale is a reflection of the agricultural period from planting to harvesting in the region of study including the likelihood where an iota of rainfall could be observed. The positive values in the SPI indicate greater than mean precipitation while negative values indicate less than mean precipitation.

The patterns of SPI values for Maroua, Garoua, and Ngaoundere stations are quite similar but the magnitude, intensity and occurrence for year of drought are different for the different stations and are also difficult to differentiate among the SPI time series. It is obvious that drought phenomena will create more vulnerable environment for the subsistence farming sector. While Ngaoundere experienced less extreme drought, Maroua and Garoua were more prone to random droughty conditions as several moderate and severe droughts were detected between the covered periods of 1961–2006. Based on the similar patterns shown in the curves, major events have been observed in the years of 1966, 1970, 1972–1973, 1987–1990, 1997, 2000, and 2006 (Fig. 14.2).

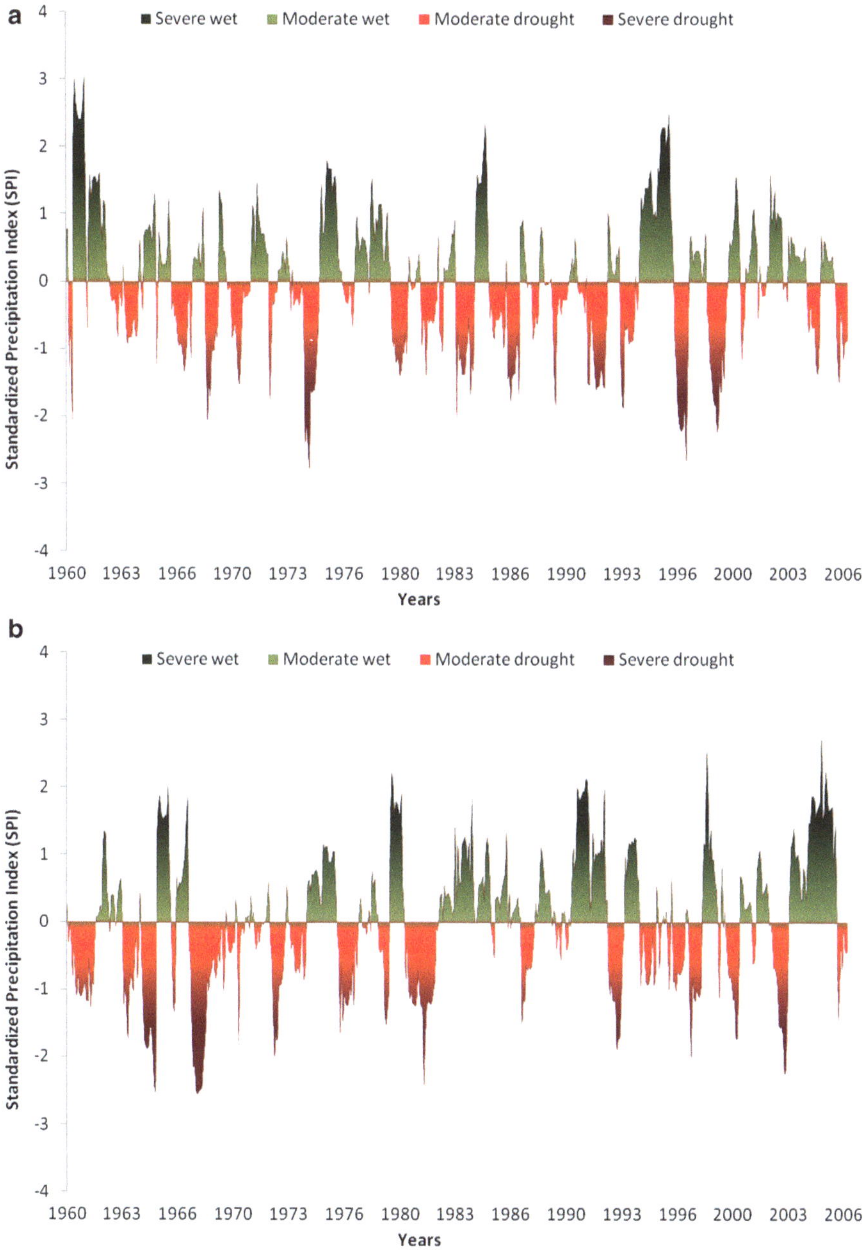

Fig. 14.2 (continued)

Droughts have a very pronounced impact because of the relative importance in the agricultural sector. Drought incidences have been prevalent in the Sudano Sahel based on the results of the Standardized Precipitation Index (SPI). It does not take

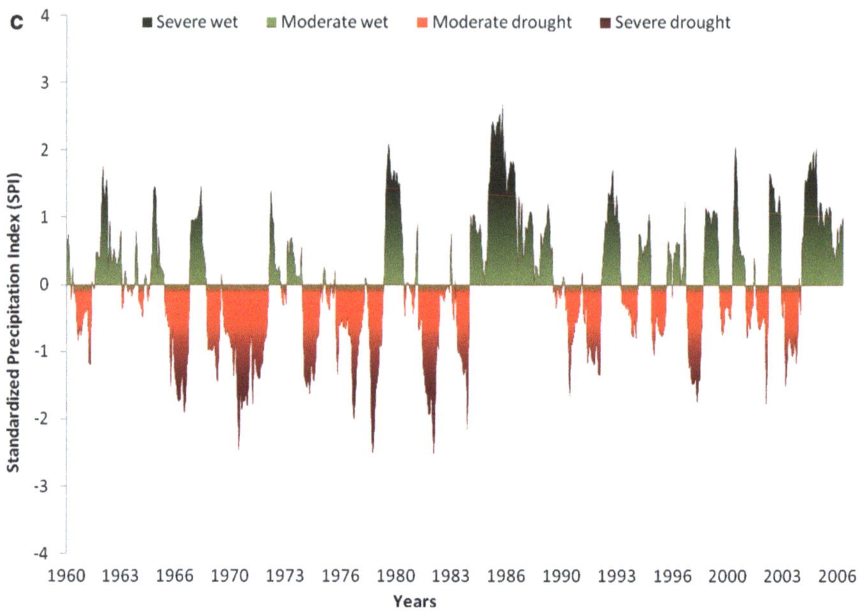

Fig. 14.2 (**a**) Nine-months SPI for Maroua from 1961 to 2006 (**b**) Nine-months SPI for Garoua from 1961 to 2006. (**c**) Nine-months SPI for Ngaoundere from 1961 to 2006

severe drought to affect crop yields. Even moderate lack of available water can drastically reduce crop yields. These types of droughty periods usually cause damage to rain-fed agriculture and the consequences are usually dramatic in that they render the agricultural soils very unstable, prone to crusting and soil hard-setting (Valentin 1995) thereby making crop production very difficult. The possibilities of desertification are also increased. The SPI droughts analyzed were in accordance with the response of the subsistence farmers claims that climate has been changing with many incidences of dry periods (Techoro and Schmidt). In trying to understand why people may declare one year as a drought, it is important to recognize their needs in terms of rainfall—their dependency on rain-fed agriculture. This falls within the premise that people's needs are used as a kind of benchmark when they compare individual years. Drought per se is a very diffuse concept and the threshold identified for defining it is somewhat arbitrary (Agnew and Chappell 1999). The mid-1960s, mid-1980s, early 1990s were droughty years characterized by severe famine in the Sudano-Sahelian. This coincides with the findings of (Molua 2008; Tingem et al. 2008) who both referred to these periods as years of the Sahelian drought.

Moreover, plethora of hypotheses have been put forward as to the causes of the Sahelian drought, with Charney et al. (1975) attributing it to the systematic and irreversible land degradation and desertification caused by the farmers and pastoralists of the region. Nicholson (2001) on the other hand offers one of the most convincing hypotheses today in arguing that during the second half of the twentieth

Table 14.1 Description of each land use and land cover type in the study area

Land use and land cover classes	Description of each class
Settlements	Areas allotted to settlements patterns.
Farmlands	Allotted areas of subsistence farming and other /rain-fed cultivation.
Light vegetation	Grasslands and savannah vegetation with bushes mixed with patch trees, usually used for grazing and browsing.
Bare soils	Are parts of the land surface which is mainly covered by bare soil and exposed rocks.
Water bodies	Allotted to wadis or streams or rivers.

century the warming of the South Atlantic and the Indian Ocean in contrast to the cooling of the North Atlantic reduced the land-ocean temperature differential. This in turn caused the monsoon to weaken. Rainfall in the Sahel is characterized by the great variations from year to year and from decade to decade, governed by the motions of the Inter-tropical Convergence Zone (ITCZ). The ITCZ and its associated deep convection migrated southwards thus depriving the Sahel of rainfall.

14.3.2 The Dynamics of Land Use and Land Cover Types in the Hotspot Study Areas of the Sudano-Sahel

By using the application of image classification methods, five major land use and land cover types were identified in the hotspot chosen in the Sudano-Sahel study area. Each of these classes has been described in Table 14.1 and includes: settlements, farmlands, light vegetation, bare soils and water bodies.

As illustrated in Figs. 14.3a and 14.4, in the Mayo Tsanaga hotspot zone in the Maroua study area, the greatest share of land use and land cover from all classes are farmlands, covering an estimate of 32 % of the total land area. Light vegetation and bare soils cover an aerial size representing 29 % and 27 % respectively of the total area in 1987. Settlements covered a rough estimate of 12 %.

On the other hand, the Enhanced thematic mapper (ETM+) portrayed that for the Mayo Tsanaga hotspot zone in the Maroua study area, light vegetation instead accounted for the greatest share of the aerial size of land use and land cover (51 %). The farmlands dropped to 22 %, with settlements increasing to 16 %, while bare soils also significantly decrease to just 10 % to the total surface area, and finally water bodies increasing by a percentage in Figs. 14.3b and 14.4.

These highlighted results were quite interesting as the 10 % drop in subsistence farmers' farmlands could be attributed to the rapid growth in the population of the Grand Nord (Stephenne and Lambin 2001). It therefore indicates that farmlands were no longer productive and there was the tendency for the rural population to leave these areas they cultivated and move to other fertile area (Fotsing 2009). Settlement increased by 4 %, an indication of the rapid population growth leading

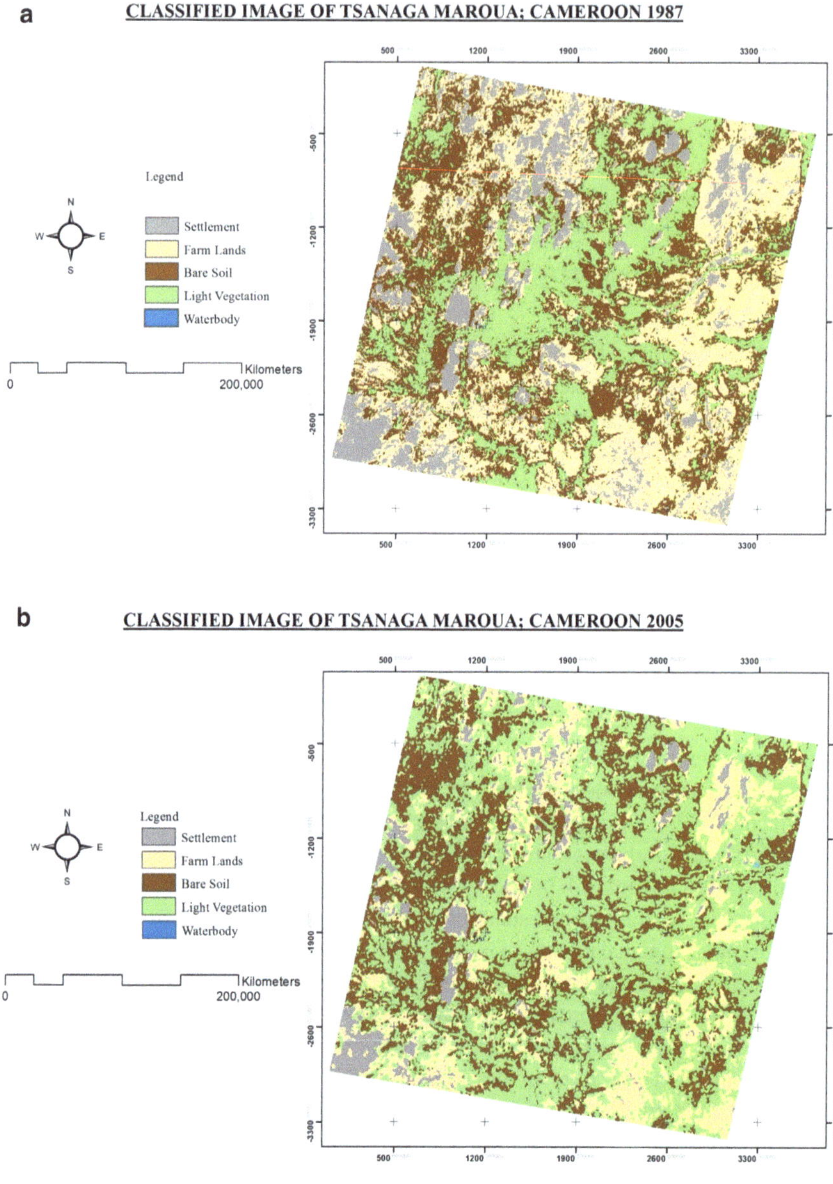

Fig. 14.3 (**a**) Land use and land cover of Mayo Tsanaga (Maroua) 1987. (**b**) Land use and land cover of Mayo Tsanaga (Maroua) 2005

to the development of settlement colonies in the study area. Rapid population growth tends to intensify the stresses that humans place on the ecosystem as the population dynamics in the Sudano-Sahel show a 5 % increase, implying the driving force in the observed land use and land cover dynamics. The drop in bare

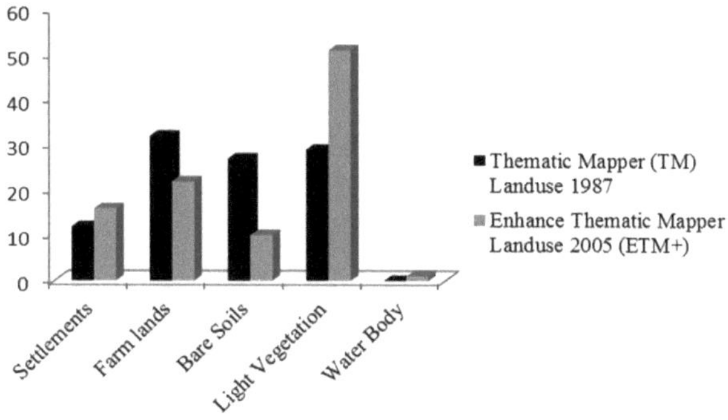

Fig. 14.4 Aerial coverage and percentage land use types of the Mayo Tsanaga hotspot zone

soil areas reduced by 7 % coupled with a very significant increase in light vegetation for up to 24 % was very conspicuous. These two changes could be accounted for by the fact that the images of 1987 were taken during the era of the Sudano-Sahelian droughts during the mid-1980s; drought affected the area causing severe hunger and famine. The findings corroborate with those of Tingem et al. (2008) and Molua (2008) who both described the Sahelian drought periods.

The improved vegetation cover on the other hand could be attributed to the Operation Green Sahel Initiative that was launched by the government in line with the United Nations Environment Program (UNEP) Billion Tree Campaign on the Sahara belt (Tchindjang et al. 2012). This afforestation campaign substantially contributed to reducing the farmlands and bare soils. Most recently, the organizational chart of the Ministry of the Environment and Planning (MINEP) created the Department of Restoration of Nature that so far has been working in habitat restoration measures and have galvanized the attainments of the motives of the "Operation Green Sahel". With the Intergovernmental Panel on Climate Change (IPCC) recognizing afforestation as one of the viable options of Carbon sequestration in terrestrial ecosystems (Gorte 2007), this Operation Green Sahel Initiative could make enormous contributions to soil carbon storage in the Sudano-Sahel drylands.

The slight increase in water bodies could be attributed to the presence of ephemeral streams or wadis. These are caused by the variability in the rainfall patterns, whereby intense wet spells have the likelihood of leaving behind patches of water (wadis), locally called Kouri in the local Hausa language.

14.3.3 Analysis of Land Distribution and Trends in Farm Sizes

Small farm sizes dominate the share of total household farms. As indicated in Table 14.2 from the 34,260 km^2 land area of the far north region (Maroua), 4,117 km^2 were cultivated in 1996. As the population increased by 5 %, the cultivated agricultural land area also increased to about 5,764 km^2. A similar

Table 14.2 Agricultural land area distribution and evolution of average farm sizes

Region/total area (km^2)	Year	Population	Poverty index of population (%)	Cultivated agricultural land (km^2)	Percentage change in agricultural cultivated area (km^2)	Average farm size (ha)
Far North (Maroua) 34,260	1996	2,467,000	49	4,117	4.8	<2
	2010	3,480,414		5,763.8		0.5
North (Garoua) 67,808	1996	1,000,000	44	1,500	1.8	1.5
	2010	1,800,000		2,700		0.5
Adamawa (Ngaoundere) 61,992	1996	500,000	37	850	1	<1
	2010	850,000		1,445		1

increase was also observed for the North and Adamawa regions as farm sizes were reduced for all three Sudano-Sahel regions.

Although agricultural cultivated land areas have increased, the average farm sizes tend to decrease considerably. This is a clear indication that with increase in population there is the tendency for bare soils to be converted into farming lands. Land holdings are becoming increasingly sub-divided, parceled and as a consequence fragmented. In the Maroua study region alone, the average farm size shrank from less than 2 ha in 1996 to 0.5 ha, in 2010. It is worth mentioning that land fragmentation is an obvious and common feature of agriculture in the Sudano-Sahel. In this situation a single farm has a number of parcels of land averaging less than 2–0.5 ha.

Fragmented lands are commonly seen where due to the increasing population and the expansion of settlements with families, lands are shared amongst the number of households. The parceling of these lands is especially seen in polygamous homes where land fragmentation practices are exacerbated. These findings corroborate with those of Jayne et al. (2012) who after reviewing nationally representative farm surveys in Sub Saharan Africa found a tendency of declining mean farm size over time within densely populated smallholder farming areas. At the household level, the inheritance from parents to sons, sons getting married and sharing the lands with wives, individuals equally sharing resources including land when divorcing, and land redistribution play significant roles such that small-sized farms are likely to be more fragmented.

Land fragmentation thus has a negative impact on the degradation of land and concomitantly crop productivity, thus increasing the pauperization of the population. The above findings concur with those of Nagayets (2005) who found a similar reduction in the average farm sizes in small holders in some selected developed countries with increasing population sizes. Therefore agricultural activities and livelihood options for the rural poor are affected not only by climatic conditions, markets, infrastructure, physical conditions, but also by human population impacts pertaining to the pressures on land.

14.3.4 Subsistence Farmers' Perceived Adaptation Responses

In instances where climate change is perceived, farmers perform a plethora of practices in order to cope with the changing environment. Quite a number of adaptation strategies have been adopted by subsistence farmers in the wake of the varying climate and are shown in Fig. 14.5, with about 21 % of the respondents saying they changed planting dates as adaptation strategies. Some 20 % of the respondents cultivated different crop varieties. Another adaptation strategy perceived by the farmers was the migration from the rural suburbs to urban areas in search of greener pastures. Some 17 % of the farmers have increased their cultivation areas of land while about 16 % use local indicators. Some 14 % have perceived switching from crops to livestock as an option, while on the other hand just 1 % switched the opposite way i.e. from livestock to crops. Some 6 % of respondents perceived the search for off-farm jobs as another adaptive means, 7 % used fertilizers with 8 % seeing soil conservation strategies as an adaptation measure. The use of irrigation was opined by 4 % of the respondents while 7 % of farmers have even hinged on daily prayers for better climatic conditions.

Perception is a necessary prerequisite for adaptation. Perceptions are considered to be important as farmers' perceptions are regarded as critical determinants of necessary preconditions for adaptation (Koch et al. 2006). From the findings, subsistence farmers' perception on climatic variability and change are based on assessments of mainly temperature and rainfall as they are experienced within the localities. Although they appear to be well aware of climate change, few seemed to actively take steps toward adjusting to the changing climate. Subsistence farmers in the Sudano-Sahel to an extent have been able to constantly cope with and adapt to the challenging environmental conditions (Techoro 2012). They have developed innovative responses to difficult or changing conditions. It must be reiterated that the strategies pointed out by the subsistence farmers are based on lessons learned from previous climatic stresses and have been handed down from generations to generations, with most of the strategies generally having some similarities for most *SSA* (Sub-Saharan Africa) countries (Reij and Waters-Bayer 2001).

As an ex ante adaptation strategy, subsistence farmers have developed the tendency of switching crops varieties as climate changes. They seek to grow crop varieties that have different sensitivities to climate. The reason for the positive farmers' response that crop switching is a viable option for adaptation could be associated to the conspicuous presence of the agronomic institute in the Grand North of Cameroon, although at times the supplies of these crops are limited. Shifts in planting dates are usually aimed at minimizing the effect of temperature-induced spikelet sterility that can be used to reduce yield instability by avoiding coincidence of the sensitive flowering stage with the hottest part of the growing season. Farmers' perception in the changing of planting dates as an adaptation option has also been reported in the finding by Bradshaw et al. (2004), Deressa et al. (2009), Tingem et al. (2008) and Molua (2008).

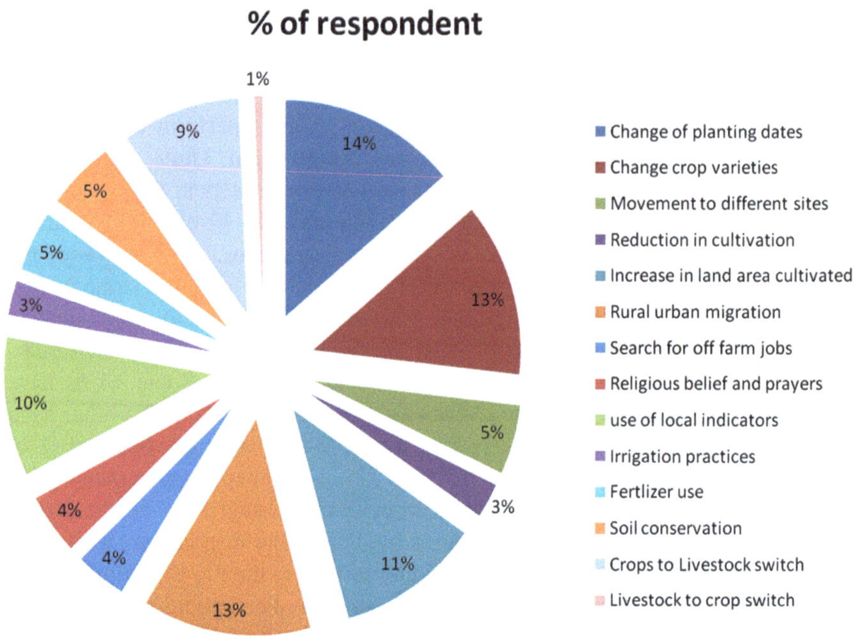

Fig. 14.5 Farmer's perceived adaptation practices to climatic variability and changed in the Sudano-Sahel of Cameroon

The use of local indicators is a traditional indigenous knowledge that enables farmers to live in a changing climate. Drought forecast indicators used have been the flower and fruit production of certain local trees—when less productive, it gave reasons for apprehension of poor climate. A good example of a drought forecast indicator plant were the shea tree Vitellaria Paradoxa Gaertn (Techoro and Schmidt 2012).

Since rain-fed crops are constrained by the harsh climate, further expansion of cultivated areas is seen at the expense of livestock, which were more or less sensitive to climate factors than crop production. Therefore, there is a shift from crop cultivation to livestock. To supplement the recent shifts from crop to livestock as a perceived adaptation strategy, the increase use of cotton cake as feed for cattle has increased. The presence of para-statal organization Société de développement du coton du Cameroun operating in the North and the Far North regions (SODECOTON) provides a constant supply of these feeds (Techoro and Schmidt 2012). As an addendum to the above-mentioned, since 1994, the exportation of cotton seed cakes to the Scandinavian countries stopped, implying that local cotton cake consumption produced by SODECOTON has been wholly local.

Soil conservation practices are perceived by farmers as another adaptive strategy. They use fallow periods and manure to maintain the integrity of the soil. Composting is also practiced, with cow dung after decomposition serving as source of nutrients for the household crops. Other practices include the use of farm yard manure, crop rotation, green manure, crop covers, mulching, agro-forestry residue retention, just to name a few. It is worth mentioning that these improved agricultural practices or better agronomic practices are inherent in ecological agriculture and can reduce the negative effects of droughts and degraded farmlands and concomitantly have enormous potential in climate change mitigation (Niggli et al. 2009).

Of the many factors that hinder subsistence farmers' adaption to climatic variability and change, most if not all could be coined under the banner of poverty. Adapting to climate change is a costly process (Mendelsohn 2006). Availability of credit facilities would therefore imply that farmers would be able to afford the necessary equipment, seed varieties and fertilizers and indulge themselves in other off-farming activities. Studies by various researchers (Adger et al. 2007; Brooks et al. 2004; Downing et al. 2005; Ziervogel et al. 2006) found out that wealthier households in subsistence farming communities are better able to act quickly to offset climate risk than poorer households. Loss of labor and man-power highlighted by subsistence farmers as an obstacle to adaptation could be attributed to the mass rural migration within the Sudano-Sahel of Cameroon. Young locals leave the villages and migrate into towns searching for greener pastures or better living conditions. The HI-Virus have also ravaged the working population and account for the loss of labor and man-power.

14.4 Conclusion

This study was based on bringing into the limelight the current strategies used by the subsistence farmers in Sudano-Sahel of Cameroon that could be enhanced in the present day climate change adaptation strategies. By employing the pressure-state-response approach, agricultural droughts and population dynamics were considered as the pressure indicators. They were analyzed via the standardized precipitation index (SPI) and remote sensing respectively. Comparative analyses of the current state of agricultural land distribution and trends of farm sizes were also done. Response indicators were the subsistence farmers themselves and were achieved based on structured questionnaires to deduce their perceived strategies and constraints encountered coping with the changing climate.

The frequency, duration and intensity of droughts have increased in the Sudano-Sahel during the past decades based on the SPI results. These are indications that subsistence agriculture to an extent has been impacted by droughty conditions and concomitantly the livelihoods of the ruralites. These droughty conditions, coupled with the ever-increasing population, have exacerbated the degradation of the land as portrayed by the remote sensing analyses. These have also impacted agricultural productivity. Notwithstanding, the improved vegetation cover shown in the remote sensing analyses are better indications that dryland agricultural management practices could lead to greater carbon sequestration. The enhanced storage of carbon in dryland could have direct environmental, economic and social benefits as it provides a win–win situation since improved agronomic productivity may also help in climate change mitigation.

Analyses of the response indicators show that subsistence farmers, no doubt, have been operating in marginal lands and have a repertoire of coping strategies to the extreme climate conditions that have prevailed over the past decades. Some of these indigenous adaptation strategies range from changing planting dates, and crop varieties, movements from rural to urban areas, the use of local indicators, switching from crops to livestock, increment in cultivated lands, and soil conservation practices.

Based on the foregone account, the adaptation by subsistence agriculture to climate change should go beyond the above mentioned practices; else climate change will continue to impact subsistence farmers in the Sudano-Sahel of Cameroon. With a combination of a multitude of factors, policy driven measures should be cumulative in the pretext of analyzing past, present and foreseeable future actions and also taking into cognizance the synergistic interactions of different effects. This approach could play a crucial role in the determination and promotion of adaptation policy options for subsistence agriculture in Cameroon's Sudano-Sahel.

Biography Prosper Somah Techoro is a Ph.D. holder at the Brandenburg University of Technology Cottbus-Senftenberg. His current interests are focused on the human dimensions of climate change, understanding the impacts of climate change and how they relate to the vulnerability of the subsistence farmers particularly in Sub-Saharan Africa. The title of his Ph.D. research was 'Climatic Change Impacts on Subsistence Agriculture in the Sudano-Sahel Zone of Cameroon—Constraints and Opportunities for Adaptation'

Michael Schmidt is the head of the Department of Environmental Planning at the Brandenburg University of Technology (BTU) Cottbus-Senftenberg, Germany. His scientific research and lecturing fields include environmental planning, environmental assessment, strategies for sustainable development, techniques for combating desertification as well as monitoring and evaluation. He promotes long-term cooperation in research and education with environmental experts in Syria, Yemen, Jordan and Vietnam.

References

Adger N, Agrawala S, Mirza MMQ, Conde C, O'Brien K, Pulhin J, Pulwarty R, Smit B, Takahashi T (2007) Assessment of adaptation practices, options, constraints and capacity. In: Parry ML, Canziani OF, Palutikof JP, van der Linden PJ, Hanson CE (eds) Climate change 2007: impacts adaptation and vulnerability. Contribution of working group ii to the fourth assessment report of the intergovernmental panel on climate change. Cambridge University Press, Cambridge, UK, pp 717–743

Agnew CT, Chappell A (1999) Drought in the Sahel. GeoJ 48:299–311

Bradshaw B, Dolan H, Smit B (2004) Farm-level adaptation to climatic variability and change: crop diversification in the Canadian prairies. Climatic Change 67:119–141

Brooks N, Adger WN, Kelly PM (2004) The determinants of vulnerability and adaptive capacity at the national level and the implications for adaptation. Glob Environ Change 15:151–163

Calzadilla A, Zhu T, Rehdanz K, Tol SJR, Ringler C (2008) Economic-wide impacts of climate change on agriculture in Sub-Saharan Africa. University of Hamburg Working Paper FNU-170, Hamburg, Germany

Charney J, Stone PH, Quirk WJ (1975) Drought in the Sahara: a biophysical feedback mechanism. Science 187:434–435

Deressa TT, Hassan RM, Ringler C, Alemu T, Yusuf M (2009) Determinants of farmers' choice of adaptation methods to climate change in the Nile Basin of Ethiopia. Global Environ Change 1:248–255

Downing TE, Patwardhan A, Klein RJT, Mukhala E, Stephan L, Winograd M, Ziervogel G (2005) Assessing vulnerability for climate adaptation. In: Lim B, Spanger-Siegfried E, Burton I, Malone E, Huq S (eds) Adaptation policy frameworks for climate change: developing strategies, policies and measures. Cambridge University Press, Cambridge

Eastman JR (2009) IDRISI taiga guide to GIS and image processing (manual version 16.02) [software]. Clark Labs: Clark University, Worcester, MA

Ecam II (2001) Deuxieme Enquête Camerounaise auprès des Ménages (ECAM II): pauvreté et genre au Cameroun. NIS, Yaounde, Cameroon

Ellis E, Pontius Jr RG (2006) Land-use and land-cover change—encyclopedia of earth. Available at http://www.eoearth.org/article/land-use_and_land-cover_change. Accessed 10 Feb 2013

ERDAS IMAGINE (2006) 9.1. On-line help system; leica geosystems geospatial imaging. LLC, Norcross, GA

Fotsing E (2009) SMALL Savannah: an information system for the integrated analysis of land use change in the far north of Cameroon. Ph.D. thesis. Department of Conservation Biology, Institute of Environmental Sciences (CML), Faculty of Science, Leiden University, 2009-12-08

Gorte RW (2007) Carbon sequestration in forests. CRS Report for Congress RL31432. Congressional Research Service

Gregory PJ, Johnson SN, Newton AC, Ingram JSI (2009) Integrating pests and pathogens into the climate change/food security debate. J Exp Bot 60:2827–2838

IPCC (2007) Climate Change 2007: impacts, adaptation and vulnerability. Contribution of working group II to the fourth assessment report of the intergovernmental panel on climate change. In: Parry ML, Canziani OF, Palutikof JP, Van der Linden PJ, Hanson CE (eds). Cambridge University Press, Cambridge

Jayne TS, Chamberlin J, Muyanga M (2012) Emerging land issues in African agriculture: implications for food security and poverty reduction strategies. Available at http://foodsecurity.stanford.edu/publications/emerging_land_issues_in_african_agriculture_implications_for_food_security_and_poverty_reduction_strategies/. Accessed 10 Feb 2013

Kenga R, M'biandoun M, Njoya A, Havard M, Vall E (2005) Analysis of Constraints to Agricultural Production in the Sudan-Sahelian Zone of Cameroon Using a Diagnostic Survey Actes du Colloque, 27–31 Mai 2002. Garoua, Cameroun

Koch IC, Vogel C, Patel Z (2006) Institutional dynamics and climate change adaptation in South Africa. Springer Science + Business Media B.V, Heidelberg, pp 1323–1339, 12

Koeppen W (1936) Das geographische system der climate, Berlin, Gebr. Borntrager. Part C of Joeppen-Geiger, Handbuch der Klimatologie, Vol 1, 44 pp

Low PS (2005) Climate change and Africa. Cambridge University Press, Cambridge, 412

McKee TB, Doesken NJ, Kliest J (1993) The relationship of drought frequency and duration to time scales. In: Proceedings of the 8th conference of applied climatology, 17–22 January, Anaheim, CA. American Meterological Society, Boston, MA 179–184

Mendelsohn R (2006) Climate change impacts on agriculture. In: Evenson R, Pingali P, Schultz P (eds), Handbook of agricultural economics: agricultural development, vol III, Chapter 19

Molua EL (2008) Turning up the heat on African agriculture: the impact of climate change on Cameroon's agriculture. Afr J Agricult Resour Econ 2(1):45–64

Nagayets O (2005) Small farms: current status and key trends. In: The future of small farms: proceedings of a research workshop, Wye, UK June 26–29 IFPRI Available online at http://www.ifpri.org/publication/future-small farms. Accessed June 2012

Nelson DR, Adger WN, Brown K (2007) Adaptation to environmental change: contributions of a resilience framework. Annu Rev Environ Resour 32:395–419

Nicholson SE (2001) Climatic and environmental change in Africa during the last two centuries. Climate Res 17:123–144

Niggli U, Fließbach A, Hepperly P, Scialabba N (2009) Low greenhouse gas agriculture: mitigation and adaptation potential of sustainable farming systems. April 2009, Rev. 2 - 2009. FAO, Rome

OECD Organisation for Economic Cooperation and Development (1994) Environmental indicators, core set. OECD, Paris

Reij C, Waters-Bayer A (2001) Farmer innovation in Africa: a source of inspiration for agricultural development. Earthscan, London

Sivakumar MVK (2007) Interactions between climate and desertification. Agric Forest Meteorol 2–4:143–55

Stephenne N, Lambin EF (2001) A dynamic simulation model of land-use changes in Sudano-sahelian countries of Africa (SALU). Agricult Ecosyst Environ 85:145–161

Tchindjang M, Amougou JA, Abossolo SA, Bell SB (2012) Challenges of climate change, landscape dynamics and environmental risks in Cameroon. In: Runge J (ed) Landscape evolution, neotectonics and quaternary environmental change in southern Cameroon palaeoecology of Africa, vol 31. CRC Press, Boca Raton, FL, pp 237–286

Techoro PS (2012) Sustainable Agriculture in the Light of Climate Change in Sub-Saharan Africa. In: Fongwa EA (ed) Sustainability Assessment: Practice, method and emerging socio-cultural issues for sustainable development. Südwestdeutscher Verlag für Hochschulschriften (SVH), Germany, pp 105–115

Techoro PS, Schmidt M (2012) The impacts of climatic variability and change on subsistent crops in the Sudano-Sahelian zone of Cameroon-options for adaptation. In: Behnassi M, Pollmann O, Kissinger G (eds) Sustainable food security in the era of local and global environmental change. Springer, Heidelberg

Tingem M, Rivington M, Bellocchi G, Colls JJ (2008) Crop yield model validation for Cameroon. Theor Appl Climatol doi: 10.1007/s00704-008-0030-8

Trochim W (2006) The research methods knowledge, 3rd edn. Atomic Dog Publishing, Mason, OH

USGS United States Geological Survey (2012) U.S. federal region III land cover set. U.S. geological survey, EROS data center, Souix Falls, SD. Available online at http://www.usgs.gov/climate_landuse. Accessed 12 Mar 2012

Valentin C (1995) Sealing, crusting and hardsetting soils in Sahelian agriculture. In: Sealing, crusting and hardsetting soils: productivity and conservation, So HB, International Symposium on Sealing, Crusting, and Hardsetting Soils (eds) Australian Society of Soil Science, Brisbane 53–76

Yengoha GT, Hicklerb T, Tchuintec T (2011) Agro-climatic resources and challenges to food production in Cameroon. Geocarto Int 26(4):2011. doi:10.1080.10106049.2011.556756

Ziervogel G, Bharwani S, Downing TE (2006) Adapting to climate variability: pumpkins, people and policy. Nat Resour Forum 30:294–305

Chapter 15
Climate Change in Cameroon and Its Impacts on Agriculture

Cornelius M. Lambi and Sunday S. Kometa

15.1 Introduction

Today, climate change has become a global problem. Because of the extreme climatic manifestations in many parts of the world, climate change has been drawn to the centre of most discussions. So perhaps the most spectacular and worrying issue which is very poorly understood is the aspect of climate change. As the short- and long-term implications of the phenomenon could have negative impacts on man and his environment, it has logically become an issue of global concern.

Since the dawn of the twenty-first century, dramatic climatic events with dire consequences on the environment, human safety, and economic livelihoods have been so preponderant and so highly intensified that global warming and climate change today have become hot currency among scholars around the world. And amongst all environmental problems that have plagued the planet since the 1980s, it has been estimated that more than 70 % of them are climatically related. In the wake of these phenomenal events like droughts, floods, tropical depressions as hurricanes, storms and heat waves which have had an overwhelming negative impact on humankind, the environment and economic livelihoods, the most recurrent questions are: What is wrong with our weather? Has our weather gone crazy?

These are all salient indicators that the present climatic manifestations have deviated significantly from the normal patterns or trends. Another popular recurrent arena of discussion is the problem of global warming which has been attributed to the effect of greenhouse gases like carbon dioxide, methane, chlorofluorocarbons (CFCs) and nitrous oxides which have occurred from intensified human activities

C.M. Lambi (✉)
University of Buea, Box 63, Buea, Cameroon
e-mail: lambimc@yahoo.com

S.S. Kometa
Geography Department, Higher Teachers Training College, University of Bamenda,
P.O. Box 39, Bambili, Cameroon
e-mail: shendek@hotmail.com

E. Albrecht et al. (eds.), *Implementing Adaptation Strategies by Legal, Economic and Planning Instruments on Climate Change*, Environmental Protection in the European Union 4, DOI 10.1007/978-3-540-77614-7_15,

since the dawn of the Industrial Revolution. Their cumulative influence has been climatic variability which also negatively impacts on the existing natural systems.

15.2 Conceptual Framework

While we largely subscribe to the concept of global warming today, one important question remains: Is our climate experiencing a warming or a cooling trend? In this respect, fierce debates based on different schools of thought have emerged.

Firstly, the Global Warming School holds that there is a general warming trend on the earth's surface due to a combination of natural and anthropogenic factors with the latter contributing more significantly to the warming process. Evidences in support of this school of thought are:

- Increased ice melting in Antarctica
- The occurrence of floods of unprecedented magnitude in much of South East Asia, the Eastern Coast of Ecuador and Peru
- Changing rainfall patterns which have been heralded by the increased frequency of hurricanes and tornadoes in some parts of South East Asia, Central and North America
- Greater frequencies of hurricanes and storms have been witnessed on the North Eastern Coast of the United States and have killed thousands of people
- Floods from tropical depressions (typhoons) have periodically swept along the coast of India, Bangladesh and other parts of South East Asia causing widespread death and destruction to economic livelihood
- The occurrence of droughts in Sub-Saharan Africa
- The occurrence of heat waves as well as bushfires in Australia and some parts of Southern Europe

Secondly, the Global Cooling School of thought claims that the recent winters have proven to be colder than is expected in some parts of Western Europe and Asia. As a pointer to the above, they mentioned the increased accumulation of ice in some parts of the Polar Regions.

Thirdly, the Climate Change Neutralists believe that the earth has been changing since the existence of man. They see climate change as a process of evolution that moves from warmer to colder periods and *vice versa*. Hence, they claim that the climate is changing. It always has been and always will be. So perhaps the climate change scenario that has taken the world by storm today is merely respecting its evolutionary phase of warm and cold periods as has been the case throughout the geological history of the earth.

15.3 The Study Area

In North Cameroon, Ndenecho and Lambi (2010) showed that the desertification process is provoked by anthropic activities but emphasized that the climatic factor remains the trigger mechanism. So while anthropic activities play a significant role in the aridification process, the climatic factor remains the essential motive force behind the mechanism. This paper examines the various elements of climate change in Cameroon and how climate change could affect the future agricultural systems and livelihoods in Cameroon. Furthermore, the paper looks at some adaptation strategies which could be put in place in order to minimize the adverse effects of climate change especially in the hot, dry and the low precipitation zone of North Cameroon (Fig. 15.1).

From 2003 to 2005, we (Lambi & Molua) were part of the World Bank Project on Climate Change and Agriculture in Africa. This was a Global Environmental Facility (GEF) under the Centre for Environmental Economics and Policy in Africa (CEEPA) that was involved in the impact assessment and adaptation strategies as an insurance against the negative impacts of climate change in Africa. The main goals of the eleven country project were:

(1) to assess quantitatively how climate affects current agricultural systems in Africa,
(2) to predict how the agricultural systems may be affected in the future by climate change under various global warming scenarios, and
(3) to suggest what role various adaptations could play in the wake of climate change.

So the analyses in this respect focused mainly on quantitative assessments of the economic impacts of climate change on agriculture and the farming communities in Africa, based on both the cross-sectional (Richardian) method and crop response simulation modelling (Dinar et al. 2008). The cross-sectional analysis made possible the assessment of the probable role of adaptation. This project used the river-basin hydrological modelling to generate climatic attributes for the impact assessment and climatic scenario analyses such as surface run-off and stream flow for Cameroon as well as for all districts of the selected countries.

15.4 Regional Settings

Cameroon runs from the southern coastal equatorial zone through the tropical savannah to the Sudano-Sahelian region in the north (Fig. 15.2). While the humid southern part is well-watered, the Sudano-Sahelian zones are dry. The later are water-deficient and are noted for droughts and the vagaries of a capricious climate and increasing aridification. Located between Latitude 10° and 11.30°N, the water

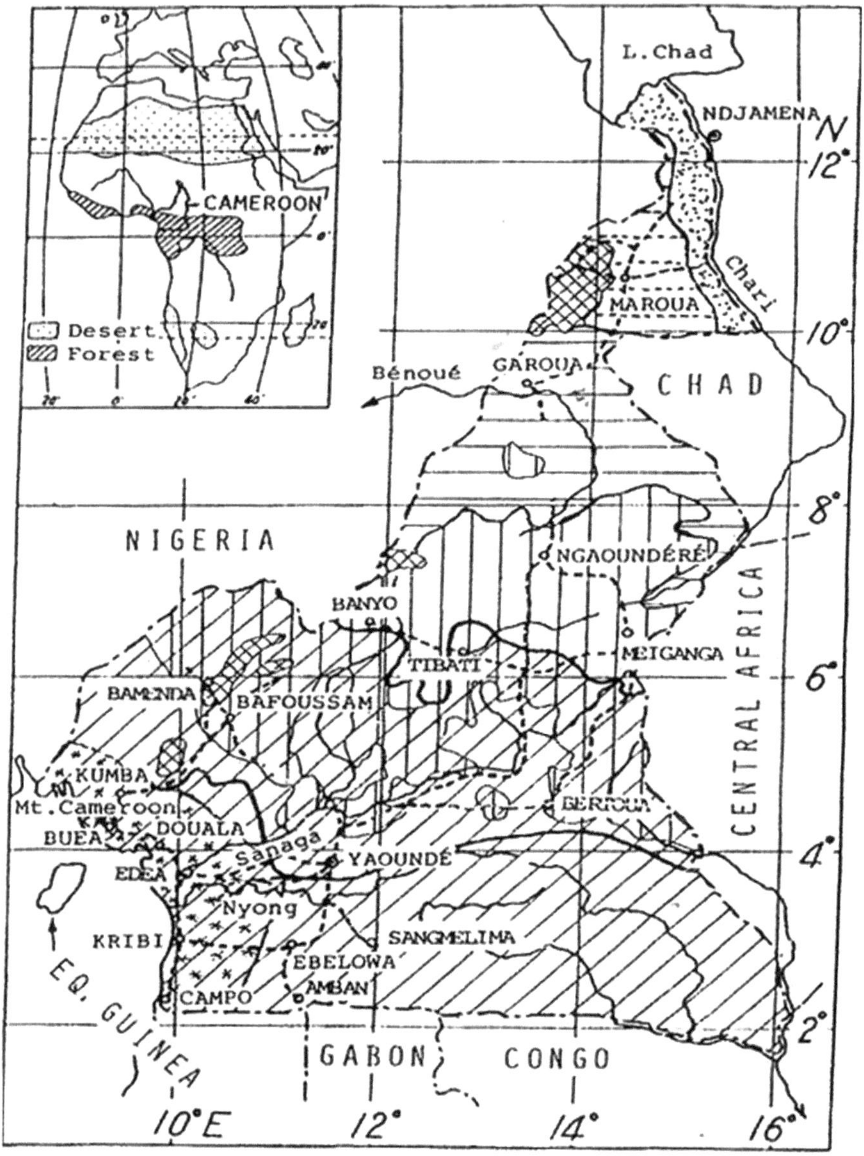

Fig. 15.1 Map of Africa and Cameroon showing the relief, the main drainage basins, the vegetation and rainfall distribution

stressed Sahelian zone passes for Cameroon's arid lands in transition (Fig. 15.3). So two major agro-climatological zones are evident in Cameroon. These are:

- the humid tropical well watered part, south of the Adamawa Highlands which depends on rain-fed agriculture, and

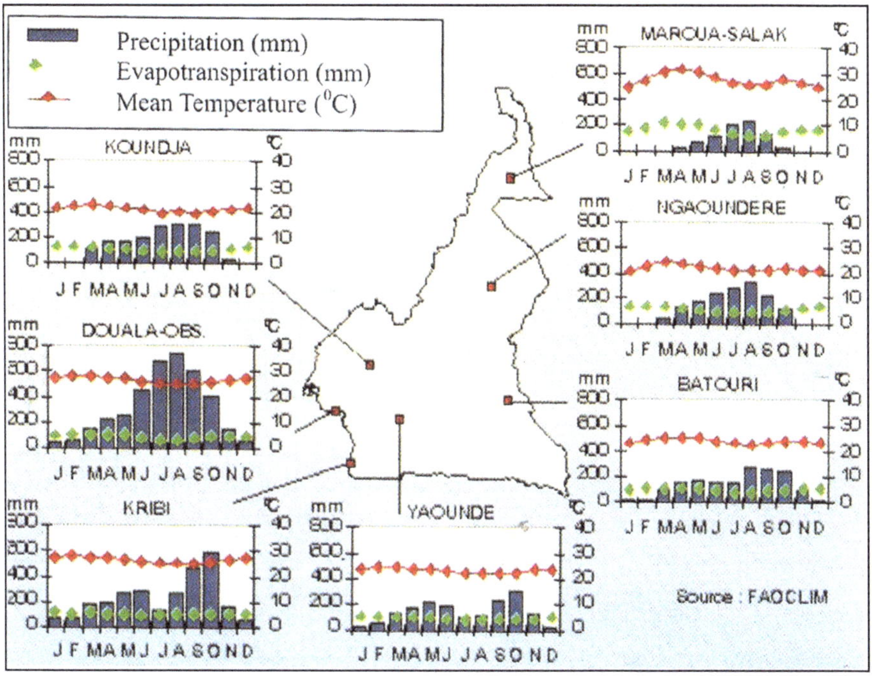

Fig. 15.2 Temperature and rainfall distribution across Cameroon. *Source*: CEEPA Policy Note no. 17 (2006)

- the northern dry Sudano-Sahelian region with high temperatures, low rainfall representing a water-stressed environment which calls for new technologies and other adaptations for successful outcomes in agricultural production.

The following climatic manifestations are some of the evidences which geographers and agro-climatologists for Cameroon have drawn upon in support of climate variability and climate change:

- In 1984, Cameroon's Sahelian region was hit by drought. Animals died, the northern grain crop failed and food aid was given to people of the Sahelian zone.
- According to Kadomura (1986), the sequence of superficial deposits in northern Cameroon shows it coincides with the arid-humid-arid climatic oscillations ranging from Late Pleistocene to Mid-Holocene.
- The Mid-Holocene aridification formed the Palaeo sand dunes in the Yagoua-Limani-Waga axis.
- Short-term and long-term climatic variations have had a great influence on the history of plant communities in the Sahelian region, a climatically sensitive part of Cameroon at the southern fringes of the Sahara Desert.

Here, rainfall variation from year to year and the occurrence of sporadic floods and droughts are significant. These inter-annual variations have been illustrated by

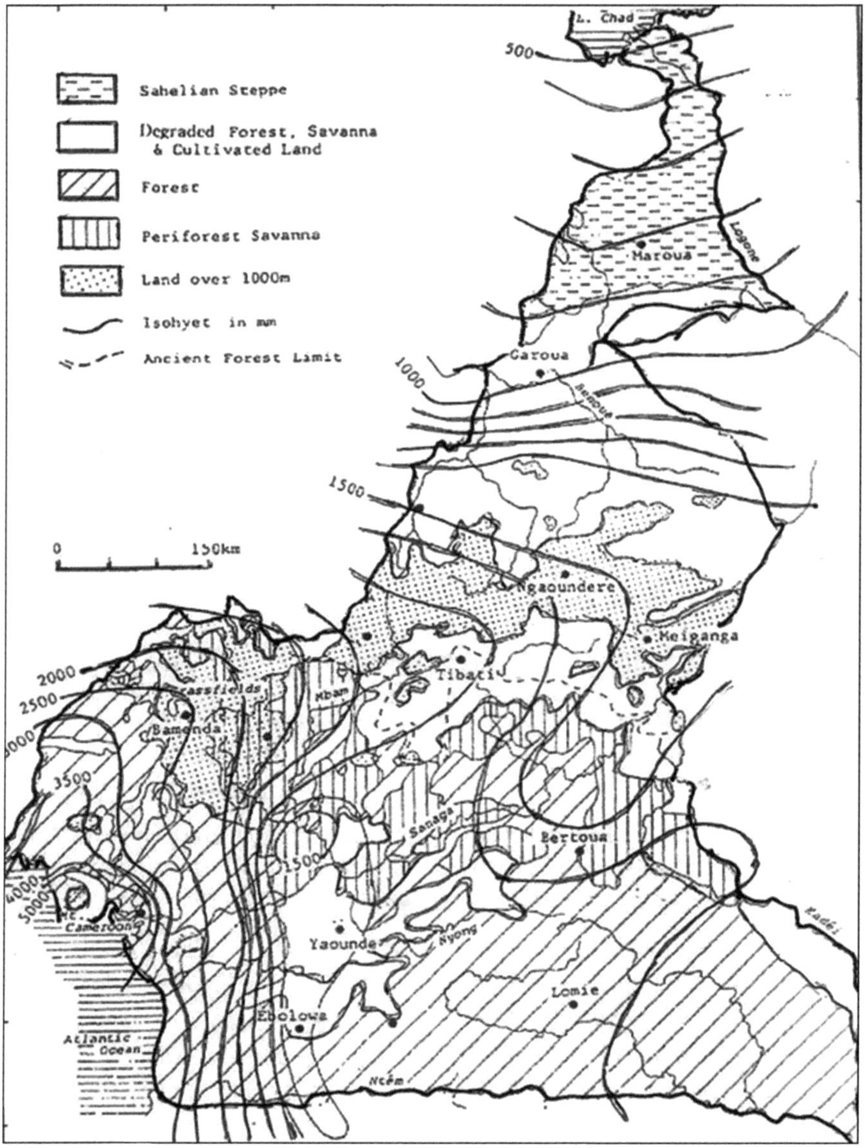

Fig. 15.3 Cameroon's arid lands in transition (N. Cameroon) showing the geomorphic units and isohyets (*Source*: Ndenecho and Lambi (2010)

Table 15.1, and Figs. 15.4 and 15.5. Variability decreases southwards, westwards and altitudinally (Table 15.1). This table shows that there were droughts in the Sahelian north between the 1960s and 1981–1982.

- Africa experienced great changes in climatic conditions during the Quaternary Period (Kadomura 1980; Hamilton 1982). Grove (1970) estimates the size of the

Table 15.1 Rainfall variability between 1934 and 1984

	Lowest rainfall		Highest rainfall		
Location	Year	Amount (mm)	Year	Amount (mm)	Variability (%)
Guider	1946	682.6	1982	1,181.0	73.0
Lam	1967	615.6	1978	1,280.3	108.0
Maroua	1944	548.1	1981	1,032.0	88.3
Mora	1941	390.0	1938	1,117.0	186.4
Waza	1969	381.9	1970	879.3	130.2
Kaele	1945	380.3	1955	1,162.9	195.1
Doukoula	1983	558.2	1953	1,261.7	126.0
Bago	1984	400.6	1975	933.6	133.0
Yagoua	1934	480.9	1953	1,213.5	152.3

Source: Beauvelin (1985)

Mega Chad to be about the size of the Caspian Sea some 5,000–10,000 years ago.

- Drier conditions prevailed and this is illustrated by the ancient sand dunes which blanket extensive areas far beyond the limits of the present day Sahara. Thus, the desert limits must have been fluctuating following the climatic fluctuations between arid and the pluvial phases (Fig. 15.6).
- Lake Chad has shrunk to nearly one tenth of its original size today between 1963 and 2006, as shown by Fig. 15.7.
- Lake Fianga completely dried up in December 1984.
- Extensive and intensive drying up of vegetation and the lack of natural pastures in December 1984.
- Total crop failure necessitating food aid distribution in 1984.
- Installation of a thermal station due to a short fall in electricity supply from the Lagdo Hydro-Electricity Power Plant.
- The 1968–1972 droughts in the Cameroonian Sahel were regarded as indications of long-term trends towards greater aridity in the arid and semi-arid zones of West Africa and similar zones of Sub-Saharan Africa.
- Agro-climatologists in Cameroon's arid lands agree that the total amount of rainfall has dropped significantly over the past 20 years (Fig. 15.4).
- While some streams and springs dry up, the volumes of some rivers have significantly decreased over time.
- The volume of the Bamendjin Dam which was supposed to augment and sustain electricity supply from the Songlulu and Edea HEP Stations failed since 1998, thereby necessitating an erratic distribution of electricity in urban areas.
- Another climatic unwelcome visitor is the occurrence of heat-waves which affect mankind in the major cities during the dry season.
- Short-term climatic variations are superimposed on long-term Palaeo-climatic variations. Dresh (1973) noted that the northern Sudano-Sahelian Zone was subjected to wet and dry years.

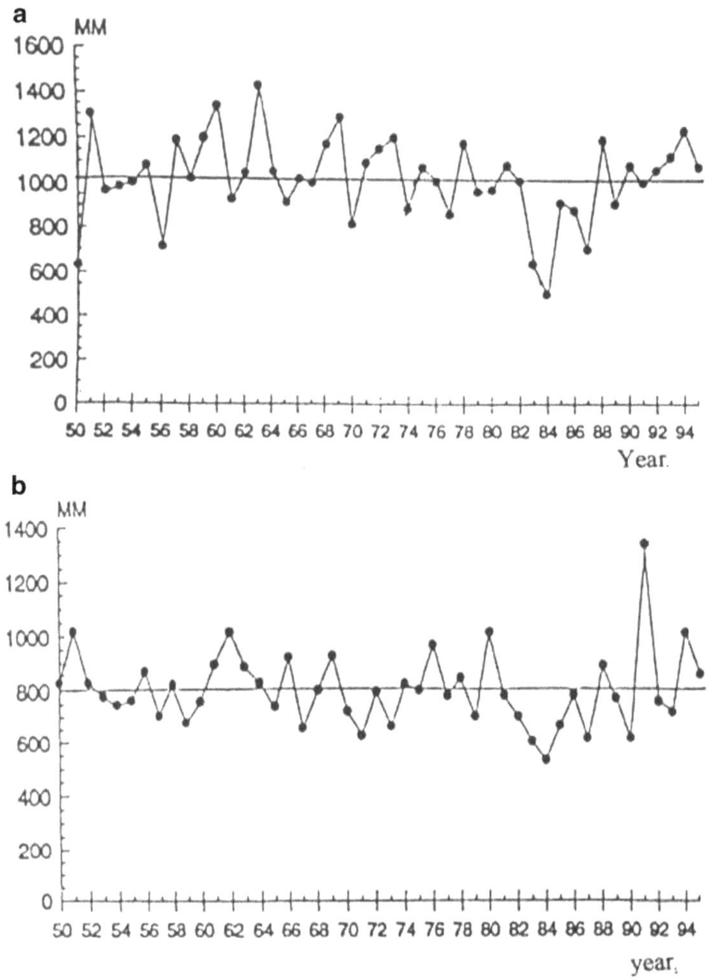

Fig. 15.4 Evolution of annual rainfall in some stations. (**a**) Garoua from 1950 to 1995 and (**b**) Maroua from 1950 to 1995 (after Donfack et al. 1996)

There were thus marked dry periods between 1910 and 1914 and from 1941 to 1942. From rainfall data, Ayonghe (2001) established a high rainfall from 1930 to 1950, 1968 to 1976 and 1981 to 1988. Rainfall data for the Chad Basin of Cameroon and Chad Basin in Nigeria were reported to be decreasing. The overall change in the total amount of rainfall in Cameroon was a drop of 283 mm between 1960 and 1990. On the other hand, the net decrease in the number of rainy days over the same period was minus 7 days. When the data is projected to 2060, there is a net rainfall decrease of 559 mm and a net decrease of 16 rainy days. Carter and Alkali (1996) reported a similar rainfall decrease in the Chad Basin in Nigeria.

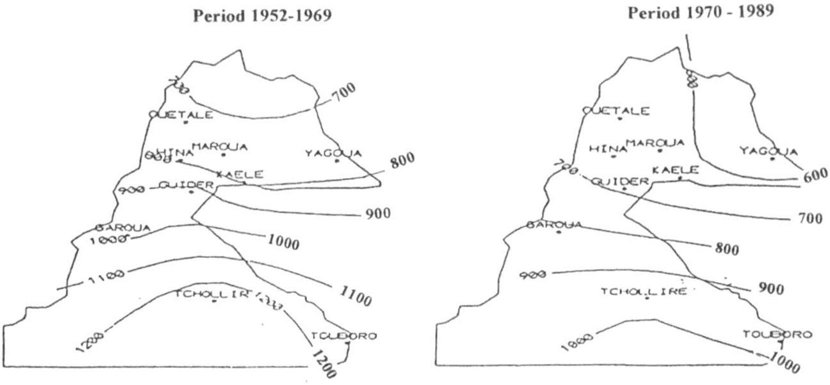

Fig. 15.5 Frequency of annual rainfall (mm) 8/10 years (after Donfack et al. 1996)

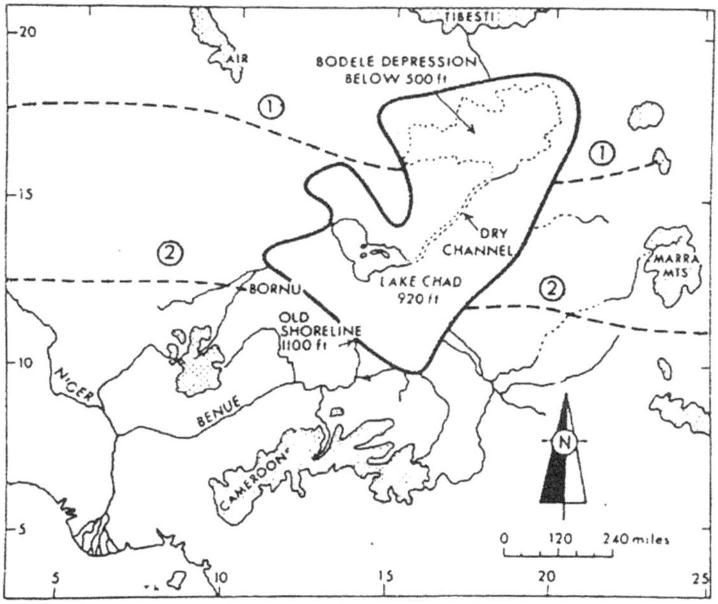

Fig. 15.6 Shifting lake shores and desert limits in the Chad Basin (*Source*: Ndenecho and Fonteh 2012, pp 187)

Nwagbara et al. (2009) from a study on decadal variations in stressed and non-stressed vegetation in neighbouring Northern Nigeria, showed that there is a steady increase in temperature over the decades from the 1970s to 2000 while rainfall varied over the decades in response to the droughts of the 1970s and 1980s. Tume (2009) showed a rainfall variability of approximately 18 % for the Bui Plateau of the Bamenda Highlands of Cameroon. While Amawa (2009) in analyzing stream flow discharge for the Bamenda Highlands and the Adamawa

Fig. 15.7 The decreasing trends in the volume of Lake Chad between 1963 and 2006 (*Source*: FAO 2006 in Lambi 2010)

Plateau showed a high flow discharge for the Bamenda Highlands and a lower flow discharge for the latter, indicating decreasing amounts of rainfall as one moves northwards from the south. Ayonghe (2001) analyzed the temperatures for Cameroon and reports that the net trend of temperature variation per decade from 1930 to 2000 is an increase of 0.14 °C (0.94 °F). Maroua in the Chad Basin showed a decadal increase of 0.28 °C. When projected to 2060, there is a net increase in temperature of 1.8 °C for Cameroon and 1.4 °C for Cameroon's Chad Basin.

15.5 Implications for Agriculture

More than 70 % of the Cameroonian population depends on agriculture and about 35 % of Cameroon's GDP comes from agriculture and related activities. Thus as the Cameroonian economy is predominantly agrarian and agriculture remains the driving force of the country's economic development, the role of climate cannot be overemphasized. The agrarian system is highly dependent on climate as temperatures and rainfalls (water) are the main drivers of crop production.

With the increasing evidence of global warming today, there is a growing concern that this phenomenon could leave a significantly negative impact on crop cultivation. The low rainfall of 1997 in Northern Cameroon caused low yields and cattle deaths necessitating food aids. In 2005, Cameroon was again hit by food shortages. The decreased rainfall in northern Cameroon was thought to contribute to increasing desertification leading to shifts in the ecological zones and the increasing exploitation of marginal lands. Global climate change therefore remains a major challenge to Cameroon's agricultural policy makers. The northern region of Cameroon is an example, *par excellence* of a harsh, hostile and fragile climatic region for crop production. Nearly 80–90 % of all the rainfall comes within 3–4 months (June to September). At the same time, evapo-transpiration is very high. Maroua, for example, has a rainfall of 804 mm whereas the rate of evapo-transpiration is 3.576 mm/year.

Within this climatic context, the possibilities of reconstituting ground water reserves and the availability of water to plants are very limited. Thus, there are large deficits between evapo-transpiration and rainfall. And all of this rainfall comes in 3–4 months (Fig. 15.8) of the year. These remain strong and limiting factors for agricultural activities. This scenario gives rise to a system of agricultural uncertainties as capricious climatic variabilities in the past have made nonsense of crop production and ushered in spells of crop failure followed by food aid for the victims of the Sudano-Sahelian region. Agricultural activities in this region undertaken to ensure human livelihoods are bewildering as there is a delicate balance between nature and the human race in this fragile climatic environment. The small variety of local grain crops does not serve as an effective insurance against the vagaries of the sporadic climatic uncertainties within the water deficient northern zone of Cameroon.

Any climate change scenario in Sub-Saharan Africa would, in general, affect the vigour and rhythm of crop production. In this respect, increased rainfall variability may decrease the number of plants that would withstand the changing rainfall amounts.

On the other hand, increasing temperatures with varying amounts of water could perhaps account for the production of greater biomass. However, as water becomes a limiting factor, biomass production would be significantly reduced. Tchatat (2009) attempted an economic analysis of the consequences of climate change for rice cultivation around the Yaguoa Sub-basin of North Cameroon and indicated that there would be vulnerability in crop production with weather changes. This drop in

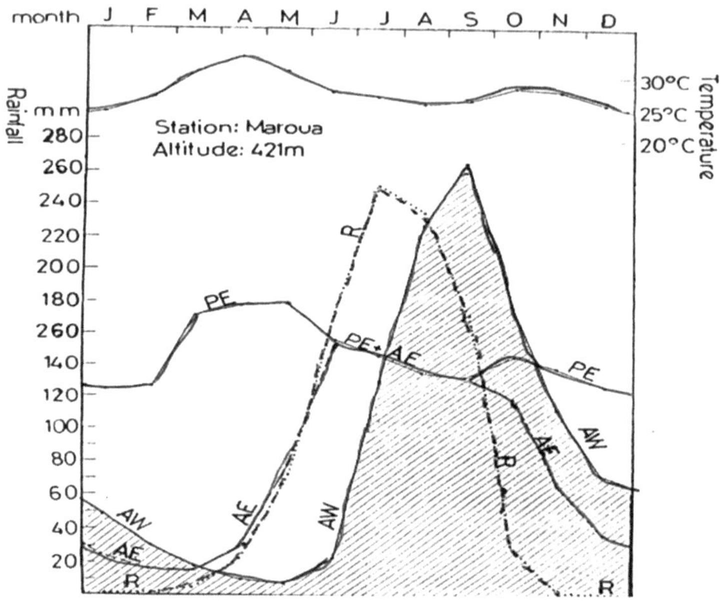

Fig. 15.8 Climatic and soil water balance for Maroua

agricultural productivity indicates a drop in the GDP and in the welfare of the agrarian northern population which depends on crop cultivation as a major source of rural livelihoods (Fig. 15.9).

In the face of this climatic uncertainty, there is a need for greater crop biodiversity as high biodiversity gives room for local adaptations to the changing climatic process. Moreover, the indigenous people or local farmers should be supported in maintaining their local varieties as an insurance against future global warming and the occurrence of new pests and diseases. In this drier northern milieu, livelihood sustenance would be largely based on the help of elaborate irrigation systems (Fig. 15.10a, b).

In earlier analyses of climatic data on climate change and agriculture in Cameroon, the results show that as precipitation decreases or temperatures increase, the net revenue from crop production falls. The net revenue that would be obtained from crop production was regressed on climate, water flow, soil and other economic variables. The overall results show that agricultural production in Cameroon is strongly dependent on the seasonality and the amount of available soil moisture. The following scenarios were arrived at from an empirical analysis (CEEPA 2006).

1. An increase in temperature of 2.5 °C will cause a net fall in crop revenue of 0.5 billion US dollars,
2. while a 5 °C increase would cause net revenue to fall by 1.7 billion US dollars,
3. a 7 % decrease in precipitation would cause net revenues from crops to fall by 1.96 billion US dollars, and

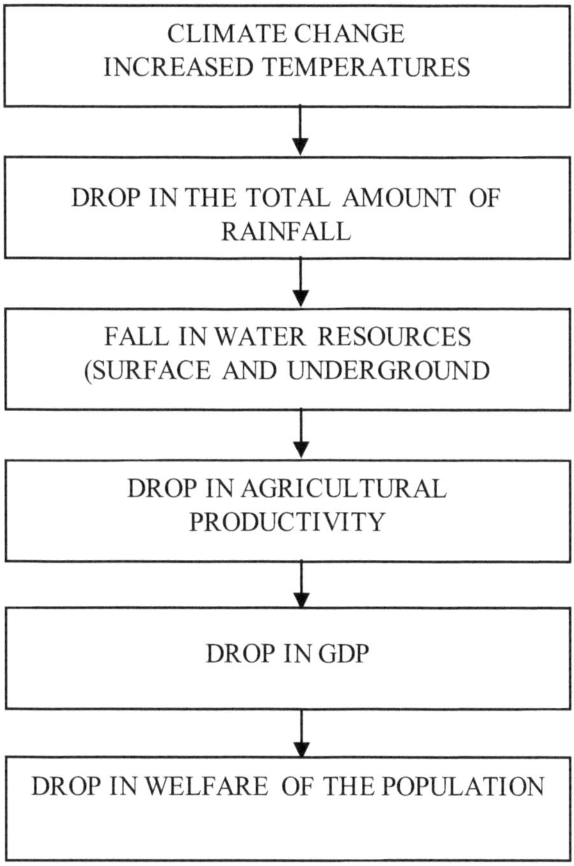

Fig. 15.9 Climate change, water resources and population welfare in Yagoua

4. a 14 % decrease in precipitation would cause a net fall in crop revenue by 3.8 billion US dollars.
5. The study showed that an increase in precipitation would have positive effects on net agricultural revenues.

In concrete terms, the crop maps of Cameroon (Fig. 15.11a, b) show tree and root crops in the humid south and grain crops in the dry and semi arid north. With increasing temperatures, it means that there could be a northward shift of ecological belts and consequently, such tree crops as cocoa, rubber, oil palms, coffee, tea, and bananas. The limiting factor, however, will be the amount of rainfall available. The north is generally synonymous with water-stress or water deficiency which will accommodate such short-cycle grain crops such as sorghum, millet, maize, groundnuts, rice and cotton. Molua and Lambi (2009) remarked that in the drier Sudano-Sahelian zone, moisture stress is particularly important at the onset of the farming season. With the higher temperatures, the heat stress compounds the water stress. Consequently, the water stress problems have important implications for

Fig. 15.10 (**a**) Indigenous polderisation and irrigation of cultivated fields in the Lake Chad Basin (*Source*: FAO 2006). (**b**) Extensive irrigation for crop cultivation during the dry season in the Lake Chad Basin (*Source*: FAO 2006 in Lambi 2011)

Cameroon's agrarian policy. With a view to protecting the agricultural sector by mitigating future climatic changes associated with global warming and hydrological droughts, they recommended the following: In order to ease the problems of water constraints so as to enhance agricultural productivity in the Sudano-Sahelian region of Cameroon, there is, first and foremost, a need to consider improving crop patterns and a need to cultivate crops with lower water requirements. Secondly, there is an urgent need to improve the irrigation efficiency by altering the widely-practised traditional irrigation systems to the modern drip irrigation and pipe irrigation. In such water-stressed environments, we cannot afford to waste water when it is the essential ingredient for plant growth.

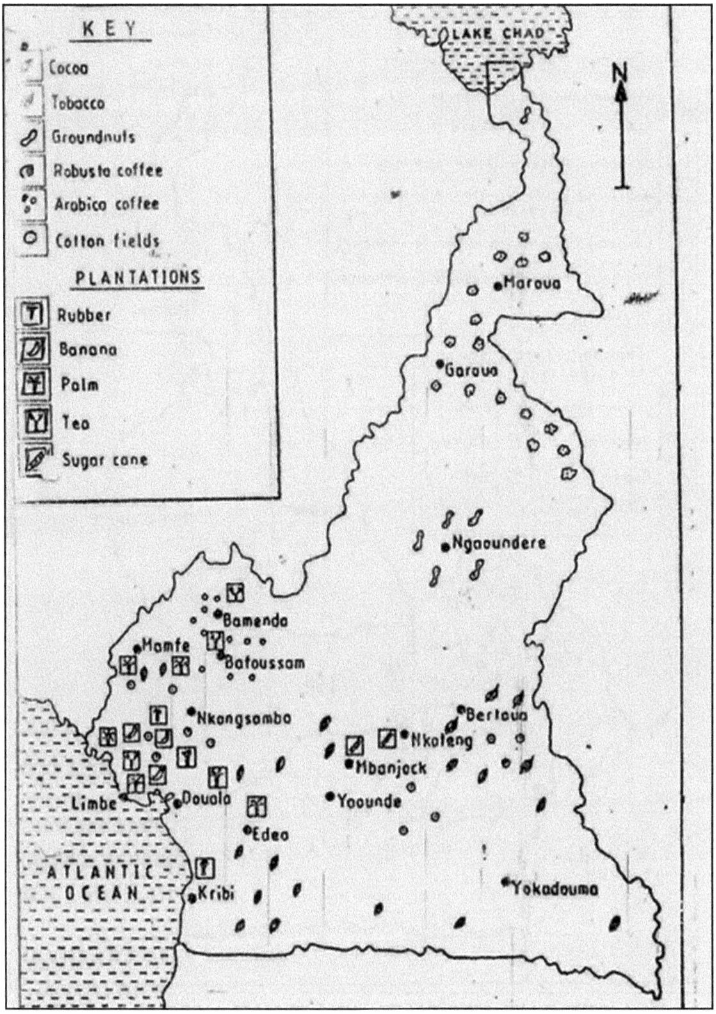

Fig. 15.11 (continued)

Fig. 15.11 Spatial patterns of crop production in Cameroon

15.6 Conclusion

Since the agricultural sector in Cameroon employs more than 70 % of the population, long-term climatic changes which could bring unpleasant consequences should be forestalled. As the farmers are continuously adapting to climatic variations, government should aid the adaptation process by providing the necessary resources and irrigation infrastructure especially in the drier north to cushion the impact of low soil moisture arising from high temperatures and high evapotranspiration. The future negative effects of climate change could thus be effectively neutralized if we think and plan ahead before the arrival of the unpleasant climatic scenarios in our Cameroonian Sudano-Sahelian Region.

As the mainstay of rural livelihoods, agricultural production must be carefully planned and managed by taking into cognizance the climate-related problems that confront agriculture in Cameroon. With a view to maximizing the returns from this primary economic sector, the future climatic scenarios must remain the focal point of all agricultural planning for better outcomes. In this respect, perhaps the application of water technologies should be called to play as the increasing aridification holds bleak prospects for mankind in this harsh or inhospitable Sudano-Sahelian zone of Cameroon.

Biography Cornelius M. Lambi is Professor of Geography and Environmental Science at the University of Buea. He is the former Vice-Chancellor of the University of Buea and a visiting Professor in a number of Universities in Africa, Europe and the United States of America. He is a Fellow of the Royal Geographical Society of London, Fellow of the Geological Society of France, and a Fellow of the Cameroon Academy of Science.

Dr. Sunday Shende Kometa is Head of Department at the University of Bamenda. He gives lectures in Physical Geography, Hydrology, Climatology and Environmental Issues both at the Department of Geography Universities of Bamenda and Buea. He obtained a PhD in Geography from the University of Buea, Cameroon. In 2010, Dr. Kometa was a visiting Research Fellow at McGill University particularly at the Brace Centre for Integrated Water Resource Management. His research interests are in the areas of Applied Geomorphology, Environmental Impact Assessment, Climate Change and Agriculture, Climate Change and Health and Integrated Water Resource Management. He has published a number of articles in both local and international journals and he is an Associate Editor in some local journals.

References

Amawa SG (2009) Stream discharge, seasonality, and variability along the Bamenda and Adamawa Highlands (Cameroon). J Appl Soc Sci 8(1 & 2):89–120

Ayonghe SN (2001) A quantitative evaluation of global warming in Cameroon from 1930–1995 and projections to 2060: effects on the environment and water resources. In: Lambi CM, Eze EB (eds) Readings in geography. Unique Printers, Bamenda, Cameroon, pp 142–156

Beauvelin A (1985) Remarques sur la situation pluviometrique actuelle au Nord Cameroun. Cameroon Geogr Rev 5(1):47–61

Carter R, Alkali A (1996) Shallow groundwater in the northeastern arid zone of Nigeria. Quart J Eng Geol 29:341–355

CEEPA (2006) Climate change and African agriculture. Policy note no. 17

Dinar A, Hassan R, Mendelsohn R, Benhin J (2008) Climate change and agriculture in Africa: impact assessment and adaptation strategies. Earthscan, London. ISBN 13: 978-1-84407-547-8

Donfack P, Boukar S, M'Biandoun M (1996) Les grandes caractéristiques du milieu Physique. In: Boukar S, Donfack P, M'Biandoun M (eds) Agricultures des savannes du Nord Cameroun. Actes de l"atelier d" échange. CIRAD/CA, pp 29–40

Dresh J (1973) Drought over Africa. The United Nations Economic, Social and Cultural Courier, New York, NY

Grove AT (1970) Africa south of the Sahara. Oxford University Press, Oxford, 280 pp

Hamilton AC (1982) Environmental history of East Africa: a case study of the quaternary. Academic, London

Kadomura H (1980) Reconstruction of palaeoenvironments in tropical Africa during the Wurm glacial maximum: a review. Environ Sci Hokkaido 3(2):147–154

Kadomura H (1986) Geomorphology and environmental changes in Africa. Case studies of Cameroon and Kenya. Laboratory of Fundamental Research, University of Aokkaido, Sapporo, Japan

Lambi CM (2010) The environment and development frontier in Sub-Saharan Africa: some global lessons. Agwecam Printers, Bamenda, Cameroon

Lambi CM (2011) The environment-development frontier in Sub-Saharan Africa: some global lessons. Nab Ventures, Bamenda, Cameroon

Molua EL, Lambi CM (2009) A cropwat experiment on maize water requirements in diverse ecological zones in Cameroon: some implications for climate change. J Appl Soc Sci 8(1 & 2): 8–32

Ndenecho EN, Fonteh ML (2012) Fresh water and coastal resource management in Cameroon: building resistance and resilience to climate change. Agwecams Printers, Bamenda, Cameroon

Ndenecho EN, Lambi CM (2010) Cameroon arid lands in transition: a case study of a fragile environment at the ante-room of desertification. J Appl Soc Sci 9(1):6–22

Nwagbara MO, Ijioma MA, Chima GN, Idoka MA (2009) Decadal variations in stressed and non-stressed vegetation in Northern Nigeria. J Appl Soc Sci 8(1 & 2):141–154

Tchatat G (2009) Climate change impact on rice production in the Yagoua Sub-Basin of Lake Chad: Cameroon. In: Lambi CM (ed) Cameroon: a country at crises crossroads: an anthology in the social sciences. NAB Ventures, Bamenda, Cameroon, pp 273–290

Tume SJP (2009) Incompatible land uses: a source of conflict at the watershed scale of the Bui Plateau, North West Region of Cameroon. In: Conflict prevention, management and resolution, II Postgraduate Seminar, Faculty of Social and Management Sciences, University of Buea, pp 38–56

Chapter 16
Cameroon's Sustainable Forest Management Initiatives with Potentials for Climate Change Mitigation and Adaptation

Terence Onang Egute and Eike Albrecht

16.1 Introduction

Climate change, caused primarily by increasing concentration of greenhouse gases (GHGs) in the atmosphere which stem from both natural and anthropogenic activities, is one of the most profound contemporary global environmental problems (IPCC 2007). It is an overriding environmental crisis with economic, social and ecological dimensions. The impacts of climate change in developing countries, especially in Sub-Saharan African countries, will be devastating (Nkem et al. 2007) and a country like Cameroon is highly vulnerable partly because of its high dependence on natural resources and limited capacity to adapt to the changing climate (CEEPA 2006; Molua and Lambi 2007; Fuo and Semie 2011). The impacts of climate change are expected to deepen poverty, food insecurity, risk of infrastructure destruction, degradation of environmental resources and unsustainable development (Nkem et al. 2007). Given that climate change is a severe environmental challenge facing humanity, countries are designing and implementing strategies for mitigation and adaptation as required by the United Nations Framework Convention on Climate Change (UNFCCC) (ILM 31 (1992), 849) adopted in New York on 9 May 1992 and the Kyoto Protocol to the United Nations Framework Convention on Climate Change adopted in Kyoto, Japan on 11 December 1997 (ILM 37 (1998), 22). Under these international environmental agreements, the international community is committed to combating climate change. One of the sectors that could contribute greatly in combating climate change is the forest sector given that forests are a gigantic carbon sink and as such play a vital role in regulating the planet's GHGs.

Forests more than ever before are seen as important in the struggle to alleviate poverty, mitigate and adapt to climate change (Agrawal et al. 2008). The forests,

T.O. Egute (✉) • E. Albrecht
Department of Civil Law and Public Law with References to the Law of Europe and the Environment, BTU Cottbus-Senftenberg, Cottbus, Germany
e-mail: eguteter@tu-cottbus.de; albrecht@tu-cottbus.de

E. Albrecht et al. (eds.), *Implementing Adaptation Strategies by Legal, Economic and Planning Instruments on Climate Change*, Environmental Protection in the European Union 4, DOI 10.1007/978-3-540-77614-7_16, © Springer-Verlag Berlin Heidelberg 2014

especially tropical rainforests such as those of the Congo Basin, could serve as a sink for absorbing CO_2 emissions, store carbon for a long period and are an essential component of climate change adaptation strategies. The Congo Basin is the world's largest tropical forest after the Amazon and its forests are estimated to contain between 25 and 30 billion tons of carbon and therefore could play a significant role in the global efforts to combat climate change (Yadlapalli 2010). The Intergovernmental Panel on Climate Change (IPCC) estimates that about 65 % of the total climate change mitigation potential in the forest sector is located in the tropics and about 50 % of this total could be achieved by reducing deforestation which account for about 20 % of global CO_2 emissions (IPCC 2007; Yadlapalli 2010). Cameroon is one of the six countries in the Congo Basin area and has a significant portion of the Congo Basin rainforest which has a great potential for carbon uptake and storage. The country's forests cover a surface area of about 21.2 million ha i.e. 45 % of the national territory and constitute 11 % of the Congo Basin's rainforests (Essama-Nssah and Gockowski 2000; MINFOF and FAO 2005). Because the forests suffer from deforestation and degradation which are the primary drivers of carbon emissions, conservation and Sustainable Forest Management (SFM) are essential for securing the forests' economic, social and ecological services such as climate regulatory potential. In addition, conservation and SFM practices may contribute to reducing the vulnerability of the social system and equally maintain the mitigation capacity of the forests (Locatelli et al. 2008). Cameroon has taken some conservation and SFM initiatives that could potentially contribute to climate change mitigation and adaptation. Based primarily on thorough review and analysis of policy and law documents as well as relevant literature, this paper presents Cameroon's forest sector conservation and management initiatives, especially efforts to practice SFM, and the potentials of these initiatives to contribute to climate change mitigation and adaptation. It identifies the problems encountered in practicing SFM beneficial for climate change mitigation and adaptation, and proffers recommendations for improvement.

16.2 Overview of Cameroon and Climate Change Observed

Cameroon is located in the central African region. The country lies between latitude 2° and 13° North and longitude 8° and 16° East and shares a border with Nigeria to the West, Chad and the Central African Republic to the East, the People's Republic of Congo, Gabon and Equatorial Guinea to the South. It covers a surface area of 475,440 km^2 and has a coastline of 402 km. It is administratively divided into ten regions and five of these regions, i.e. South, East, Centre, Littoral and Southwest, are located in the moist forest zone. The Far North and North regions are located in the Sahelian zone while the West, Northwest and Adamawa regions are located in the moist Savannah zone (Foahom 2001). The country contains almost all the fascinating aspects of Africa which range from culture, landscape, vegetation and climate. That is why it is often referred to as "Africa in Miniature", meaning all of what Africa has can be found in Cameroon (Lambi 2001). Cameroon's climate varies with terrain and ranges from equatorial in the South, to tropical above 8° North latitude and to Sahelian in the North. The climate is hot and humid in the forested South and Western regions. Temperatures in the Southern region of the country depend largely on altitude and range between 20–25 °C and vary little with seasons. It is cooler in the highland Grassfields region of the West and Northwest regions and hotter and drier in the savannah and Sahel of the North. The semi-arid Northern region i.e. North of 6°N is the hottest and driest part of the country and experiences average temperatures between 25–27 °C during cooler seasons and 27–30 °C during warmer seasons (McSweeney et al. 2008). The country has two main seasons i.e. a wet and a dry season. The wet season runs in the South from March to October and the dry season from November to February. The Northern part of the country has a different weather pattern. The wet season runs from May to October while the dry season runs from November to April. In the humid Southern region of Cameroon, annual rainfall often averages 1,500 mm, while in the North it averages 500 mm (Molua and Lambi 2007).

Cameroon is experiencing diverse climate change impacts which vary between different regions of the country. Since 1930, average temperatures in the country have been increasing with a net increase of 0.95 °C between 1930 and 1995. It is predicted that Cameroon's temperature rise will reach as high as 2 °C by 2060 (CEEPA 2006). Average rainfall has decreased by about 2 % per decade since 1960 and it is predicted that the average rainfall will continue to decrease (CEEPA 2006; Molua and Lambi 2007). These changes have been attributed to extreme weather conditions across the country. The Northern Sudano-Sahelian region is increasingly experiencing incidences of drought and desert advancement that have scorched large expanses of land (Brown et al. 2010), while the Southwest coastal and rainforest regions have experienced increased periods of prolonged rainfall. The persistent rainfall has led to disastrous floods that have claimed lives and resulted in widespread loss of property (Norrington-Davies 2011). Furthermore, Lake Chad which is the major source of water in the Northern region of Cameroon and the surrounding countries is increasingly shrinking as a result of changes in global climate and it is predicted that the lake may completely dry up by 2060 (CEEPA 2006; Ehode 2011). Farmers, fishermen and herders near the lake have been relocating to other areas as a result of continuous shrinking of the lake where they make their livelihood. In addition, continuous rise in temperatures and desertification in the region have forced some people to migrate to the South in search of water and arable land (Norrington-Davies 2011). Agricultural production which contributed 20 % to Cameroon's GDP in 2008, is expected to decrease due to increased desertification in the North and higher incidence of flooding in the South (Norrington-Davies 2011). Agriculture and livestock production will be most affected by changes in temperature with agricultural production projected to decrease by between 10 and 25 % depending on the warming situation (MINEF 2005).

Slow agricultural productivity, sea level rise and health risk are the key vulnerability problems identified by Cameroon in its First National Communication submitted to the UNFCCC in 2005 in accordance with Art. 4 and 12 of the convention (MINEF 2005). The climate change impacts will likely affect all of Cameroon's forest landscapes (Brown et al. 2010). As Sonwa et al. (2012) opine, the rise in atmospheric concentration of CO_2, increase in temperature, precipitation changes, floods and prolonged drought will severely affect tree growth. Pressure on forest resources will intensify with increasing population, declining productivity of agricultural land caused by climate change and dependence of the rural population on Non-Timber Forest Products (NTFPs) to meet their nutritional, medicinal and cultural needs. The consequences for forests which serve as carbon sequester and sink, will be severe as rising energy prices and continuous electricity shortages in the country, further concentrate dependence on fuelwood and charcoal as a source of energy (Sonwa et al. 2012). The 2007 Fourth Assessment Report of the IPCC reveals that a 2 °C rise in global temperature will result in a sea-level rise of between 69 cm and 1 m across the world (IPCC 2007). Given its location along the coast, Cameroon is expected to experience the impacts of sea level rise. As Molua (2010) asserts, the 5.8 million Cameroonians living in coastal regions will be at risk of the anticipated sea level rise. A rise in the sea level will adversely affect mangrove forests and its biodiversity, cause flooding, coastal erosion, sedimentation and increased salinity. Considering the risk on coastal infrastructure, Cameroon estimated in its 2005 National Communication to the UNFCCC that just its coastal industrial properties were at risk of experiencing about 2.74 billion CFA Franc (€3,111,000) in damages as a result of climate change by 2100 (MINEF 2005). An increase in temperature is expected to lead to the proliferation of pests, diseases, crop stress, livestock strain and increased outbreak of diseases such as malaria, dysentery and cholera (MINEF 2005; Bele et al. 2010; Norrington-Davies 2011). With these adverse impacts, the government of Cameroon has taken some SFM initiatives which could potentially contribute to climate change mitigation and adaptation. Before examining the initiatives, it is imperative to provide an overview of Cameroon's forest.

16.3 Overview of Cameroon's Forest

Cameroon's forest is one of the most important natural resources in the country and plays a major role in sustaining life. The forests which cover a surface area of about 21.2 million ha are a reservoir of biodiversity and have exceptionally rich and diverse flora and fauna (Burnham 2000). It is estimated that the forests constitute between 8,000 and 12,000 species (Liviu 2005). There are about 9,000 species of plants in which at least 156 are endemic. The country contains 409 species of mammals and 925 species of birds. In Africa, it is ranked fifth in biodiversity richness with a high degree of endemism after South Africa, Democratic Republic of Congo, Madagascar and Tanzania. Cameroon's forests just like other tropical forests have multiple functions and are of great significance for the wellbeing of humanity. In addition to inhabiting most of Cameroon's biodiversity, the forests serve as agent of erosion control, provide the country with wood and non-wood forest products, serve as watershed and above all regulate the climate (Essama-Nssah and Gockowski 2000; MINFOF and FAO 2005; Liviu 2005; Topa et al. 2009). The forests which are one of the important sources for the development of the national economy account for 6 % of Cameroon's GDP (Bele et al. 2010). They suffer from deforestation and there is increasing degradation and loss of biodiversity. The factors which lead to deforestation and loss of biodiversity include population growth and undesirable agricultural practices such as slash and burn, agro-industry, fuel wood collection, bushfire and logging by timber companies (Essama-Nssah and Gockowski 2000; Neba and Lotsmart 2003; MINFOF and FAO 2005; MINEP 2008; Egute and Albrecht 2011).

The economic crisis that gripped Sub-Saharan African countries in the mid-1980s also severely affected Cameroon. In order to ensure economic development, the government embarked on incessant exploitation of its forest resources with little or no regard for sustainable management (Fombad 1997; Njoh 2007). The forest area that has been logged or allocated as logging concession is about 76 %. Due to a poor management system and intensification of forest use, about 2 million ha of forest was lost between 1980 and 1995 i.e. about a tenth of the forest cover of 1980. During that period, while permanent cultivable land increased by 25 %, forest cover decreased by 4.8 % (Neba and Lotsmart 2003). The present annual average deforestation rate is about 1 % (Alemagi 2011). Continuous deforestation and forest degradation leads to increasing emission of CO_2 into the atmosphere, thereby contributing to global warming and climate change. Climate change will not only be a risk to the forest ecosystem composition, health and vitality; it will threaten forest continuous provision of the wide range of ecosystem services that are essential for the well-being of humans (Locatelli et al. 2008). This necessitates the practice of SFM which is worth elucidating before proceeding with the government's forest management measures.

16.4 Concept of Sustainable Forest Management

One of the appropriate approaches for mitigating and adapting to climate change is SFM. The concept of SFM can be traced to "the Non-legally Binding Authoritative Statement of Principles for a Global Consensus on the Management, Conservation, and Sustainable Development of all Types of Forests" (ILM 31 (1992), 881) commonly known as the Forest Principles and Chapter 11 "Combating Deforestation" of Agenda 21 (U.N. Doc. A/CONF. 151/26 (Vol. I–III) [1992]) adopted at the United Nations Conference on Environment and Development that was held in Rio de Janeiro in 1992. Principle 2(b) of the Forest Principles specifically states that "Forest resources and forest lands should be sustainably managed to meet the social, economic, ecological, cultural and spiritual needs of present and future generations". The UN Resolution 62/98, establishing the Non-legally Binding Instrument on All Types of Forests describes SFM as a dynamic and evolving concept that aims at maintaining and enhancing the economic, social and ecological values of all types of forests for the benefit of the present and future generations.

The concept gained prominence from 1992 and a number of international and regional initiatives have attempted to provide an explicit definition of it. The Ministerial Conference on the Protection of Forests in Europe (MCPFE) defines SFM as "the stewardship and use of forests and forest lands in a way, and at a rate, that maintains their biodiversity, productivity, regeneration capacity, vitality and their potential to fulfill, now and in the future, relevant ecological, economic and social functions, at local, national, and global levels, and that does not cause damage to other ecosystems" (MCPFE 2011). From the above definition, it can be deduced that SFM aims at managing forests in accordance with the principle of sustainable development as it encompasses and seeks to achieve a balance between all the three pillars of sustainability which are environmental, economic and social values (Eddington and Howel 2006). Criteria and indicators which provide a framework for assessing, monitoring and reporting on progress towards the implementation of SFM have been developed for tropical and other forests types. The appraisal of SFM is based on thematic areas defined by the United Nations Forum on Forest and these include the extent of forest resources; biological diversity; forest health and vitality; productive, protective and socio-economic functions; and the legal, policy and institutional framework. As Locatelli et al. (2010) assert, these thematic areas are similar to the key elements that should be considered in climate change adaptation plans. As climate change is expected to affect forests, all the SFM elements are relevant for climate change mitigation and adaptation. SFM, which helps in the maintenance and resilience of forest ecosystem functions, has a great potential to serve as a tool in mitigating and adapting to climate change. It is one of the most important tools to achieve reduced CO_2 emissions and includes measures for the effective conservation and management of forest resources in order to meet the present and future needs of especially the local population. SFM framework guides the implementation of forest-based adaption which encompasses adaptation for forests and forest-dependent people (Locatelli et al. 2010).The United Nations has been working to promote the widespread use of SFM and Cameroon has made an effort to implement this forest management tool.

16.5 Cameroon's Sustainable Forest Management Initiatives

The SFM initiatives that could potentially contribute to climate change mitigation and adaptation that the government of Cameroon has taken constitute the following sections.

16.5.1 Forest Policy Initiative

The government of Cameroon has taken some policy measures to promote and implement SFM in order to secure the environmental services of the forest including climate change regulation. The government recognised the importance to develop a national program for SFM especially after Cameroon's participation in the United Nations Conference on Environment and Development held at Rio de Janeiro in 1992. In 1993, a forest policy document i.e. the National Forestry Action Program was adopted. It sets the objectives of the forestry and wildlife sector which include: the protection of Cameroon's forest heritage and participation in safeguarding the global environment, preservation of biodiversity in a sustainable manner, improvement of participation of the local populations in forest conservation and management in order to enable forestry to contribute to raising their living standard, development of forest resources with a view to improving the contribution of forestry production in the GDP while preserving the production potential, ensure the regeneration of forest resources by plantations in order to perpetuate their potentials and revitalise the forest sector by setting up an efficient institutional framework (Government of Cameroon 1993). The Forest-Environment Sector Program (FESP) was also put in place to address biodiversity conservation, SFM and land degradation. Even though this policy document does not contain the phrase climate change, it indirectly deals with climate change (Bele et al. 2010). In 1994, the government put in place policy tools for promoting and implementing SFM i.e. Law No 94/01 of 20 January 1994 to lay down Forestry, Wildlife and Fishery Regulations (hereafter 1994 law) and its decrees of implementation, Decree No. 95-531-PM of 23 August 1995 to determine the conditions of implementation of forestry regulations and Decree No. 95-466-PM of 20 July 1995 to lay down the conditions for the implementation of wildlife regulations (MINEF 1999). As stipulated in section 1, the 1994 law was enacted with a view to attain the general objectives of the forestry, wildlife and fisheries policy within the framework of an integrated management ensuring sustainable conservation and use of the said resources and of the various ecosystems. The 1994 law is the first law in Cameroon which provides for an integrated approach to natural resource use and management as it decentralises forest resource governance by recognising the local communities' use and management of forest resources. The law is widely cited as the first attempt to decentralise forest resource governance and address issues of forest sustainability and inequity in the Congo Basin (Alemagi and Kozak 2010).

One of the most important aspects of the 1994 law is the classification of the national forest estate into two main categories based on the 1993 zoning plan drafted for Southern Cameroon. Section 20 of the law classifies the national forest estate into Permanent Forest Estate (PFE) and Non-permanent Forest Estate (NPFE). While the PFE and NPFE are under some form of state control, they differ with respect to how the state can allocate for use. The PFE category is made up of forests belonging to the State and to rural councils, and comprise of forestlands used solely for forestry or as wildlife habitat. The PFE is designated to permanently remain forested and protected as well as sustainably managed. The NPFE comprise of forestlands used for other purposes other than forestry and are made up of community forests which are forests whose management is delegated to communities by the responsible forestry officials, given that well-managed forests offer a range of livelihood benefits that could assist them in adapting to climate change (Seppälä et al. 2009). For instance, the forests which serve as safety nets can help communities to cope with climate shocks. Many products from the forests are more resilient to climate variability as well as extreme situations than crops and therefore, these forests products are essential for the local livelihoods. When drought reduces the productivity of agricultural land or assets are lost as a result of floods, communities can sell timber and NTFPs for income. The communities can also consume NTFPs such as eru (Gnetum africanum), cashew nut (Tetracarpidium concophorum), bush mango (Irvingia gabonensis), mushrooms and bush meat as food. The forests also protect soil from erosion, agricultural lands and community's watershed (Locatelli et al. 2008; CIFOR 2012). By ensuring good management of forests, the adaptive capacity of forests dependent local communities will be improved. The NPFE also comprise of private forests which are forests planted by natural persons or corporate bodies, orchards, agricultural plantations, fallow land, wooded land, pastoral land and agroforestry land (Sec. 35, 37, 39 of 1994 law). The PFE accounts for the majority of Cameroon's forests, covering 80 % of the total forest area, with the NPFE accounting for the remaining 20 % (De Wasseige et al. 2009). The enactment of the 1994 law was a laudable measure taken by the government to ensure the protection and sustainable management of Cameroon's forests. Deforestation and forest degradation are a threat to climate stability and the putting in place of such policy instrument to halt forest disappearance has become a priority and even though not explicitly stated in policy documents, this is one of the components of government's strategy to mitigate and adapt to climate change. One of the 1993 forest policy objectives is the protection of Cameroon's forest heritage and participation in safeguarding the global environment which of course includes climate stability. SFM is a mechanism used by the state to fight against deforestation and forest degradation given that failure to manage forests sustainably reduces their capacity to provide ecosystem services especially to forest dependent communities in the long term. That is why Cameroon identified SFM in its First National Communication as a management and policy measure for adapting forests to climate change (MINEF 2005). SFM has the potential for carbon sequestration and safeguard of the climate even though forest management in Cameroon does not explicitly include carbon sequestration as a purpose.

One of the specific objectives of the 1994 law is to guarantee sustainable management of forest resources through the preparation and implementation of forest management plans for each Forest Management Unit (FMU) granted to logging companies in order to ensure ecological, economic and social sustainability in forest management (Sec. 23, 29, 31, 32 and 37(2) of 1994 law). After a FMU is allocated to the highest bidding logging company, it must within 3 years prepare a management plan and once approved, exploitation of the FMU must be in accordance with the approved management plan in order to ensure the sustainability of forest resources (Sec. 23 of 1994 law). In 2001, Decree No. 0222/A/MINEF of 25 May 2001 regulating the preparation, adoption and implementation of forest management plans was passed. The decree clarifies the procedure for the selection of the set of commercial tree species on which management decisions must be taken. The number of selected species for sustainable management must be at least 20 from a list of 60 species that are the most harvested in the country. To ensure regeneration, the selected species must represent at least 75 % of the trees inventoried in the FMU as laid down in Order No. 0222/A/MINEF of 25 May 2001 establishing procedures for the preparation, approval, monitoring and control of the implementation of management plans of permanent forest production. These measures contribute to ensuring reduced-impact logging which is an important SFM activity relevant for climate change adaptation. As of mid-2009, 67 FMUs covering an area of about 4.5 million ha had approved forest management plans and that represented about 65 % of a total of 101 FMUs in Cameroon (Cerutti et al. 2011). For community forests, the use of a simple management plan to promote replanting of harvested tree species is recommended in the manual of the procedures for the attribution of community forests in Cameroon. Section 7.9 of the manual stipulates that while implementing a management plan, the community must undertake operations to maintain the ecological potential of their forest and such actions i.e. conservation and regeneration activities must be in line with the management plan (MINFOF 2009). Approved forest management plans and their effective enforcement could contribute to the sustainability of Cameroon's forest heritage and this is beneficial for the country's forest-dependent people and the local as well as global climate.

Despite government's efforts to sustainably manage the forest, illegal practices which include illegal exploitation of NFTPs, approval of forest management plans that do not comply with legal prescriptions, issuance of fake concession permits, timber exploitation beyond concession boundaries, illegal transportation and export processes that do not comply with the law are still common in Cameroon (Fombad 1997; Essama-Nssah and Gockowski 2000; MINFOF and FAO 2005; Njoh 2007; Cerutti et al. 2008; Sonwa et al. 2012). The government has instituted enforcement measures like verification checkpoints, regular auditing and monitoring of FMUs and sawmills to combat illegal logging operations (Alemagi 2011). In addition, an international non-governmental organization (NGO) i.e. Resource Extraction Monitoring has been established to serve as an independent observer to monitor forest exploitation activities. There is presently a quarterly publication of illegal logging cases and fines even though only a small proportion of the fines are actually paid due to negotiations with culprits that are not in compliance with the law. This negatively affects the work of the monitoring organization (Dkamela 2011).

A partnership of the Ministry of Forestry and Wildlife (MINFOF) and World Resource Institute (WRI), which is a non-profit organization that focuses on the intersection of the environment and socio-economic development, has also contributed to the development of information to monitor compliance with logging standards and the state of forest resources. It will be important to integrate climate change adaptation into monitoring initiatives because this will be important for early detection of changes in forest status and health; maintaining forest health, vitality and diversity: removing invasive species; and addressing pests as well as disease threats. The government put in place a Special Fund for Forestry Development (FSDF) for funding activities in the forestry sector but there is no adequate disbursement of funds to replenish it. For community forests, the local communities and MINFOF cooperate in fighting against illegal activities. Article 31(3) of Decree No. 95-531-PM of 23 August 1995 to determine the conditions of implementation of forestry regulations provides that in case of infringement of community forests rights, it should be up to the ministry in charge of forests to bring legal actions against the culprits when notified by the official in charge of the community forest.

All these measures have been put in place to ensure the continuous availability of the forests with its social, economic and ecological benefits. In addition, if climate change mitigation and adaptation needs are adequately integrated in the measures, this will contribute to reducing the threats to be posed to forests in the face of climate change by invasive alien species, fire, insects and diseases.

Concerning benefit sharing which serves as an incentive for the local people to participate in forest conservation and management, there exists an Annual Forestry Fee which allows the local communities to benefit from taxes on forest exploitation. The taxes from forestry activities are shared as follows: 50 % goes to the national treasury, 20 % to the rural council where the concession is located, 20 % to the Special Council Support Fund for Mutual Assistance known in its French acronym as FEICOM and 10 % to the neighbouring village communities where the logging concession is situated for their own development projects (Freudenthal et al. 2011). This benefit-sharing mechanism ought to serve as an incentive for forest resource protection in the sense that it can encourage the local communities to develop a sense of custodianship of the forest and cooperation with the forestry authority in ensuring that logging companies are practicing sustainable forest exploitation.

But the 10 % of Annual Forestry Fee meant for the development of the local communities is often embezzled by the local elites (Alemagi 2011). In order to further promote the socioeconomic sustainability of this model, the provision of social amenities like schools, hospitals, good roads and electricity to the communities in which logging concessions are located is part of a contract that is supposed to be negotiated by concessionaires and communities after a logging contract has been awarded to a concessionaire by the state. The concessionaires are expected to meet their corporate social responsibility. Concessionaires are also obliged to set up timber processing units in their area of operation to partly ensure the creation of jobs in the forestry sector (Alemagi 2011). Under the 1994 law, with the exception of protected areas, the local population has the right to harvest forest products for their household use but all commercial use is subject to licensing. Cameroon has also established a community forestry model that transfers forest management rights from the government to the village communities even though the forest area is limited to 5,000 ha. These measures aim at enabling the local communities to use forest resources in a way that does not jeopardize the forests' resilience to future climate impacts. As the forests serve as a source of livelihood for especially forest communities, the forests remain an indispensable asset for supporting poverty reduction strategies and climate change adaptation.

16.5.2 Protected Areas Initiative

Forest conservation is an important strategy for mitigating and adapting to climate change. Forest conservation in the form of protected areas was identified in Cameroon's First National Communication as a management and policy measure for adapting forests to climate change (MINEF 2005). Due to the importance of protected areas in situ conservation of biodiversity and securing environmental services such as climate change mitigation, the 1994 law sets aside a network of protected areas. As of 2008 about 3,482,741 ha of forest was protected under the category of forest reserves, wildlife sanctuaries, botanical gardens, zoological gardens and national parks. The total area under protection excluding 18 % safari hunting zones is 11 % of the national territory (Republic of Cameroon 2009).

Protected areas are classified into 15 categories and divided into forest protected areas and wildlife protected areas. Some of the protected areas correspond to the IUCN categories II, IV and VI. This implies that their legal status especially pertaining to conservation and management is clearly stated.

These protected areas include: national parks (II), wildlife and related reserves (IV), production forests (VI) (Foahom 2001). Wildlife protected areas are mainly created for the conservation of fauna. National parks which are under wildlife protected areas are created primarily for the conservation of biodiversity. Presently, there are 18 national parks in Cameroon. The law prohibits habitat destruction and exploitation of especially protected areas. Lighting of fire that could cause damage to the vegetation of the national forest estate is forbidden and prior study of environmental hazard is required for any project that can lead to the destruction of forest and aquatic environment. It is forbidden to dump toxic product or waste that could destroy or modify plant and animal life in the national forest (Sec. 14, 17 and 18 of 1994 law). Even though these measures contribute to enhancing forest and biodiversity conservation, increase resilience and reduce climate change vulnerability risk, the protected areas still face the problems of encroachment and poaching (Neba and Lotsmart 2003; MINFOF and FAO 2005; MINEP 2008; Egute and Albrecht 2011). However, all the forest protected areas have high carbon uptake potential and contribute to mitigating climate change.

16.5.3 REDD Program

The Bali Action Plan adopted by the 13th Conference of the Parties (COP) to the UNFCCC in 2007 proposes that forests in developing countries should be considered an important tool for climate change mitigation. The 15th Conference of

Parties (COP) of the UNFCCC in 2009 in Copenhagen, Denmark then decided upon the implementation of a mechanism for Reducing Emissions from Deforestation and Forest Degradation (REDD) in developing countries given that deforestation of the tropical forest account for 20 % of CO_2 in the atmosphere which is more than the emissions contributions of the whole transport sector and seconded only by the energy sector (UN-REDD Program 2009).

The REDD program represents adaptation in the sense that it is designed to create a financial value for the carbon stored in forests and as a result provides incentives for developing countries to reduce emissions from forested land by way of investing in a low-carbon path to sustainable development. Under the compensated reduction mechanism of the REDD program, developing countries that are willing to reduce their emissions from deforestation using their 1980–1990 annual average rate of deforestation as baseline would be permitted to sell carbon certificates to governments and the private sector. The protection of the forests will contribute to alleviating poverty through revenue that will be generated from the eventual sale of carbon stocks (Bellassen and Gitz 2008). REDD has the potential to contribute to climate change adaptation in the sense that, through it, local livelihoods can be improved, local institutions strengthened and forest ecosystem services conserved (Locatelli et al. 2010).

The REDD program has been upgraded to REDD+ which comprises not only deforestation and forest degradation goals but also the goals of conservation, SFM and enhancement of the forest carbon stock (UN-REDD Program 2009). The main goal of any carbon mitigation project is to achieve permanent emission reduction as sequestered carbon is beneficial for climate change mitigation if and only if it remains sequestered and any anthropogenic or natural activity causing the release of stored carbon do not undermine previously stored carbon (Streck and Scholz 2006). REDD+ can best succeed by encouraging SFM. Forests that are sustainably managed can better adapt to climate change. SFM can reverse forest loss as well as degradation and enhance resilience of forest to climate change (Seppälä et al. 2009).

Since the establishment of the REDD program, some developing countries have taken some initiatives to carry out pilot projects which aim at testing the methodologies required for the successful implementation of the program. In this light, Cameroon is a host of REDD pilot projects such as REDD PP initiated in 2007 by an international German based remote sensing company (GAF AG) and phased out in 2012 (Justice and Bih 2009). The REDD PP project was designed to develop tools to account for national deforestation and forest degradation emissions, identify opportunities for national incentive schemes and strengthen dialogue between stakeholders, and facilitate the regional and international exchange of learning experiences.

Cameroon presently hosts more than eight sub-national REDD pilot projects (Freudenthal et al. 2011). Since 2005, the country has been engaged in international negotiations on REDD and it is an active member of the Central African Forest Commission (COMIFAC) as well as the Working Group on Climate Change. In this context, Cameroon contributed to preparing and communicating five Congo Basin country submissions to the UNFCCC that helped to highlight the role of forest degradation and sustainable forest management (SFM) in the REDD discussions. In its proposal, the country endorses a two-phased approach, with funds and markets, a historical reference emission level with development adjustment factors and national and sub-national implementation (UN-REDD Program 2009). Several analyses on forest cover change have been carried out in Cameroon, in different periods and regions. For instance, in 2005, as part of its First National Communication to the UNFCCC, the total GHG emissions of Cameroon was estimated based on the 1994 national statistics and the guidelines of IPCC. The country's Readiness Plan Idea Note (R-PIN) document outlines the limitations of the estimates. This includes a lack of reliable data, national competence, reliable cartographic information as well as GIS data, and the lack of reliable national statistics that is reliable (MINEP 2008; UN-REDD Program 2009). Decree No. 2009/410 of 10 December 2009 established the National Observatory on Climate Change which is in charge of providing official data including forest emissions data to enhance knowledge and understanding around climate change at the national level.

16.5.4 Forest Sector International and Regional Initiatives

The government's forest sector response to climate change is evident at both international and regional levels. At the international and regional levels, Cameroon is active in a number of processes. One of these processes is the COMIFAC established in 2005 as part of a commitment to SFM in the Congo Basin. Through COMIFAC the country has played a significant role in contributing to the development of the REDD+ option in particular with respect to avoiding deforestation and forest degradation and also the enhancement of carbon stocks in protected areas (De Wasseige et al. 2009). Under the COMIFAC, a trilateral agreement was established among Cameroon, Gabon and Congo to protect 14.6 million ha of forest in the region. Cameroon hosted the 1999 Central African Heads of State Summit on SFM and put in place an Emergency Action Plan as an implementation mechanism of that summit with provisions for the sustainable management of the forest resources as well as recognition of the role of forest products in local livelihoods and national development (Ndibi and Kay 1997 in Sonwa et al. 2012).

The country also ratified the UNFCCC in 1994 and the Kyoto Protocol in 2008. Forest-related mitigation measures in developing countries including Cameroon are confined under the Kyoto Protocol's Clean Development Mechanism (CDM) to afforestation and reforestation.

Even though there is no afforestation and reforestation CDM project in Cameroon, the country in 2008 renewed a 1960 tree planting project i.e. Operation Green Sahel in the Northern region to mitigate global warming and combat desert encroachment (Tchindjang et al. 2012). The trees planting initiative is important for climate change adaptation. For instance, with livestock production as one of the important economic activities in Northern Cameroon, the trees planted will help in protecting livestock from climate variability and fodder from the trees will help in sustaining livestock for months during periods of drought. With the semi-arid Northern region i.e. North of 6°N experiencing average temperature between 25–27 °C during cooler seasons and 27–30 °C during warmer seasons (McSweeney et al. 2008), the trees planted will play an important role in adaptation to climate variability and change by reducing temperatures when there is heat waves (CIFOR 2012).

This project which contributes to carbon sequestration is in compliance with Cameroon's commitments under the climate change agreements, the Convention on Biological Diversity (ILM 31 (1992), 818) which it ratified on October 19th 1994 and the United Nations Convention to Combat Desertification in Countries Experiencing Serious Drought and/or Desertification, particularly in Africa (Paris) (ILM 33 (1994), 33480) which it ratified on May 29th 1997.

Cameroon is one of six countries in the Congo Basin Region participating in the project "Climate Change and Forests in the Congo Basin: Synergies between Adaptation and Mitigation" (COBAM). COBAM is designed to ensure that policy makers, practitioners and local communities have the information, analysis and tools they need in order to implement policies and projects for adaptation to climate change and reduction of carbon emissions in the forests of the Congo Basin. This involves equitable impacts and co-benefits which include poverty reduction, enhancement of ecosystem services and the protection of local livelihoods and rights. The project is implemented by the Centre for International Forestry Research (CIFOR) under the African Development Bank grant to the Economic Community of Central African States in view of financing the Congo Basin Ecosystems Conservation Support Program (CIFOR 2012). Cameroon is also involved in a global World Wide Fund for Nature project examining its approach to building resilience to climate change in tropical mangroves and associated coral reefs. As Cameroon identified sea level rise as one of the key vulnerability problems within its First National Communication submitted to the UNFCCC, such projects are very important for adaptation because mangroves help to reduce the risk from disasters relating to sea level rise such as coastal flooding and erosion (CIFOR 2012). As a member of the COMIFAC, the country will also benefit from COMIFAC's current project on climate change scenarios for the Congo Basin which is expected to enable decision makers to adapt and prepare their natural resource management strategies to meet the regional challenges of climate change.

16.5.5 Voluntary Forest Certification

Forestry laws are not however the only instruments through which SFM can be implemented and forest management improved for climate change mitigation and adaptation. Forest certification, which is a market-based tool for third party auditing of sustainability in forest management, is important for achieving SFM. Third-party forest certification standards such as the Forest Stewardship Council (FSC) international standard is largely considered as one of the best international voluntary standards for SFM (Auld et al. 2008). The FSC certification is intended as a tool to enhance good forest management practices given that it assures consumers of forest products that the products they purchase are produced sustainably, equitably and with appropriate management (CIEFE/ICEFS and IMAFLORA 2006). There is proliferation of such voluntary sustainability standards schemes in Cameroon. Some logging companies operating in a socially and environmentally responsible manner have already voluntarily sought and obtained the FSC certification. As of 2009, 13 FMUs were certified in accordance with FSC requirements and cover about 870,000 ha (Cerutti et al. 2011). The FSC standard FSC-STD-20-001, section 19.1 requires a certification body to issue a certificate only if there is compliance with the FSC principles and criteria. Forest certification enables logging companies or forest exploiters to comply with forestry laws that were otherwise largely unenforced in order to be certified.

National forest sectorial codes of practice based on the rules of FSC certification such as the FSC Standard for Community Forests and SLIMFs in Cameroon have been developed and need to be promoted to ensure SFM. FSC requirements for certified forests such as the identification and conservation of High Conservation Value Forests and habitats for threatened and endangered species contribute to biodiversity conservation. The economic, social and environmental benefits of certified forests are glaring and need not be overemphasized. Compliance with the FSC certification requirements in Cameroon could contribute to SFM and biodiversity conservation, which is what the country's pieces of forestry legislation were designed to achieve. SFM can best be promoted in Cameroon through legal frameworks and market-oriented instruments such as the FSC standard. For instance, the adoption of forest management plans require logging companies to reduce their annual allowable cut by 11 % on average while certified FMUs undergo an average reduction of about 18 % (Cerutti et al. 2011). This is a clear positive impact of certification allowing a better recovery of valuable species for the next rotation and reducing the damages to the residual stand by lowering the harvest intensity. SFM as such retains forest cover over time and thus sustains or increases carbon stock. As Cerutti et al. (2011) assert, certification has enabled the companies to improve upon their management plans, something they would not have done if they continue operating only within the management system as provided by the national forestry legislation. The government could significantly enhance its forest sustainability goals by supporting companies that have adopted or are interested in adopting forest certification management systems.

16.6 Institutional Framework Initiatives

The institutional framework of Cameroon plays a crucial role in the way SFM management is implemented. The institutions in charge of SFM in Cameroon constitute the following sections.

16.6.1 Ministry of Forestry and Wildlife

In order to implement its forest conservation and management policies, the Ministry of Forestry and Wildlife (MINFOF) was created in 2004 by Decree 2004/320 of 8 December 2004. As stipulated in the decree, it is responsible for the development, implementation and reviewing of the Cameroon's forest heritage policy. In addition, the ministry is responsible for forest conservation and management, development and monitoring of the implementation of regeneration programs, afforestation, reforestation and inventory. MINFOF has taken action to advance ecological sustainability in Cameroon's forest sector. In this light, it has established an Urgent Action Program and the National Brigade for the Control of Forests and Wildlife which is in charge of monitoring and regulating illegal logging in forest concessions (Tieguhong and Betti 2008). The ministry ensures compliance with the 1994 law by the forest stakeholders. For instance, it recently suspended the licenses of 27 companies that had failed to comply with the law (Ngalame 2012). The MINFOF also collaborates with the international development agencies to combat illegal logging in Cameroon. An example is the cooperation with the German Agency for Technical Cooperation (GTZ), now German Agency for International Cooperation (GIZ), to ensure the application of satellite imageries and Geographic Information System (GIS) to identify illegal operations within forest concessions in the Southeast Region of Cameroon (Alemagi 2011).

16.6.2 Ministry of Environment, Nature Protection and Sustainable Development

The Ministry of Environment, Nature Protection and Sustainable Development (MINEPDED) is in charge of the implementation and coordination of national, regional and international environmental policies. It also plays a role in conserving and managing Cameroon's forest heritage. Its main functions include development, coordination and monitoring of the national environmental policy. MINEPDED is in charge of negotiating international environmental agreements and their implementation. Activities carried out by MINEPDED include implementation and coordination of regional and international environmental policy, ensuring the sustainable use of the country's natural resources while liaising with organizations involved in the exploitation of natural resources, and increases public awareness of environmental issues.

The MINEPDED and MINFOF collaborate in coordinating REDD activities nationally. In 2009, the government passed Decision No. 09/MINEP of 15 January 2009 creating a Steering Committee for the REDD Cameroon pilot project and the Prime Ministerial Decree of 13 June 2012 spell out the functions of the Steering committee. The committee serves as REDD focal point in Cameroon. It is in charge of matters concerning reducing emissions resulting from deforestation and environmental degradation, and technical as well as strategic orientations of the REDD+ Pilot Project in Cameroon. It is also responsible for formulating policies and strategies to meet the REDD+ process in Cameroon and put in place criteria for the selection of projects which will lead to the achievement of the REDD+ objectives. The steering committee is placed under the tutelage of the MINEPDED which chaired it and is assisted in this task by MINFOF.

The Climate Change Unit, Ecological Monitoring Unit and a designated national authority for the Clean Development Mechanism (CDM) are in different departments in MINEPDED. These UNFCCC focal points contribute to facilitating government direct engagement in climate change which focuses on mitigation through the REDD Readiness Preparation Proposal (RPP) process and also preparation of the country's first National Adaptation Plan of Action (NAPA) which will increase understanding of the need for adaptation. Cameroon already submitted its Readiness Plan Idea Note (R-PIN) to the Forest Carbon Partnership Facility (FCPF) through MINEPDED. Some fundamental weaknesses experienced include many non-functioning REDD coordinating committees established within MINEPDED, logistical problems in the ministry, overlap in functions and collaboration challenges between MINEPDED and MINFOF (Brown et al. 2010; Dkamela 2011).

16.7 Recommendations

Cameroon's SFM initiatives with potentials for climate change mitigation and adaptation and also major challenges of SFM have been discussed. This section of the paper proffers recommendations for policy makers to consider in their effort to foster SFM that contributes to climate change mitigation and adaptation.

16.7.1 Review Forest Policy and Ensure Effective Law Enforcement

The government should review the country's forest policy and programs to ensure that they address current and predicted climate change impacts. Forest conservation and sustainable management including REDD should be an integral part of the new policy. As the Non-legally Binding Instrument on All Types of Forests highlights, governance is a crucial component of successful conservation and SFM as is law enforcement. Though regulation can be a highly effective means to combat deforestation and promote the use of SFM, lack of capacity in monitoring and lack of resources for law enforcement are inhibiting factors.

Adequate monitoring is highly recommended if SFM is to be achieved in Cameroon. Indeed, in the absence of sufficient monitoring, enforcement of laws geared at promoting SFM is difficult. The government should ensure effective monitoring and enforcement of the law. It should adequately protect protected areas and be more proactive in ensuring that the social and ecological obligations of the logging companies are fulfilled. It should not relend its effort in suspending or revoking forestry licenses granted to logging companies that violate forestry laws. This will serve as a deterrent to defaulters and contribute to forest sustainability.

Within the country's forest policy, broad afforestation and reforestation programs should be incorporated in order to increase the peoples' resilience in the face of changing climate. In line with the country's forest policy, the government should support international environmental agreements that target deforestation. The government should improve upon the country's notion of community forest to ensure that the local communities benefit from it. It should also improve on the management of protected areas and enforce the practice of SFM to maintain and restore healthy forest ecosystems.

Most importantly, as adaptation remains a viable option to cope with climate change impacts, the government should finalize sooner than later the drafting of the National Adaptation Plan on Climate Change launched on the 15th of June 2012 to ensure that the country has an instrument which defines priority adaptation activities with practical modalities for their execution. Above all, it should mainstream forest-climate change mitigation and adaptation into the National Adaptation Plan on Climate Change, new and existing policies, laws and national development plans.

16.7.2 Government Should Be in Charge of Preparing Management Plans

The 1994 law requires the preparation of forest management plans for each forest management unit auctioned. As Cerutti et al. (2008) point out, the intention was for the government to be developing the management plans, but due to insufficient financial and human resources it delegated the task to logging companies which consequently focus on economic than ecological and social considerations. But given that logging companies unilaterally prepare management plans because of the government's inability to prepare them, the management plans approved by the ministry lay emphasis on economic than ecological and social aspects. This implies that the significant number of approved management plans does not mean that the forests are sustainably managed given that logging companies often harvest the most traded species as if management plans do not exist. The government represented by MINFOF should assume the responsibility of preparing forest management plans to ensure that they are in line with the legal prescription laid down in Decree No 0222/A/MINEF of 25 May 2001 regulating the preparation, adoption and implementation of forest management plans. In addition, it should make sure that logging companies comply with their management plans to ensure sustainability in forest management.

16.7.3 Encourage Voluntary Forest Certification

As discussed above, legal instruments are not the only tools to implement SFM and achieve improved forest management. Market-oriented instruments such as voluntary forest certification are some of the tools for achieving SFM. The certification standards often required logging companies to apply stricter silvicultural parameters than the requirements of approved management plans to ensure sustainability in forest management. National codes of practices based on the rules of certification should be developed and promoted by the government. It should also provide direct and indirect incentives to logging companies voluntarily applying forest certification standards and effectively practicing SFM to enable them to remain competitive.

16.7.4 Build Technical and Institutional Capacity

The conflicting responsibilities in MINFOF and MINEPDED should be streamlined in order to eradicate the overlaps. Adequate financial resources should be put in place and the institutional capacity to monitor emissions from the forest sector should be strengthened. The government should build the technical and human capacity of institutions such as National Observatory on Climate Change and Steering Committee for REDD to make them effective. Existing non-functioning climate change or REDD coordinating committees established within MINEPDED should be dissolved.

16.7.5 Eradicate Corrupt Practices in the Forestry Sector

Corruption is one of the main obstacles to effective enforcement of the 1994 law. In addition to local elites that manage the local communities' share of the Annual Forest Fee as earlier mentioned; it is practiced by personnel in charge of law implementation and enforcement. According to an eminent Cameroonian lawyer i.e. Nico Halle in Ndifiembeu (2006, p. 10), "No matter how good the laws are, if those implementing them are not honest, fair, equitable and patriotic, such laws will represent nothing". The government should fight against corruption within the law implementation and enforcement institutions, and the civil society organizations should be involved. As watchdogs, the non-governmental organizations can play a meaningful role in fighting against corruption and ensuring enforcement as well as compliance with environmental laws to ensure forest sustainability essential for climate change mitigation and adaptation.

16.8 Conclusion

In conclusion, the legal and institutional frameworks governing the forest sector in Cameroon have contributed significantly in the sustainability of forest management in the country and this plays a pivotal role in climate change mitigation and adaptation. By conserving and sustainably managing Cameroon's forests, there is protection and maintenance of the carbon in them thereby reducing carbon emissions from the forests. By tackling the causes of deforestation and forest degradation, there is reduction in the rate and amount of loss of forest cover and this also protects the ecosystem services including climate regulation provided by Cameroon's forests. Strengthening forest management can contribute to enabling Cameroon to effectively practice SFM which is an effective instrument for forest-based climate change mitigation and adaptation. Improved forest management practices lead to the survival of forest ecosystems and enhances their environmental, socio-cultural and economic functions. In addition, improved forest management practices secure forests' contribution to climate change mitigation and enhance the adaptive capacity of both forests and forest-dependent people. SFM in Cameroon should thus be a critical component of any policy and action program that seeks to address the growing concern about the impact of climate change.

References

Agrawal A, Chhatre A, Hardin R (2008) Changing governance of the world's forest. Science 320:1460–1462

Alemagi D (2011) Sustainable development in Cameroon's forestry sector: progress, challenges, and strategies for improvement. Afr J Environ Sci Technol 5(2):65–72

Alemagi D, Kozak RA (2010) Illegal logging in Cameroon: causes and the path forward. Forest Pol Econ 12(8):554–561

Auld G, Gulbrandsen LH, McDermott CL (2008) Certification schemes and the impacts on forests and forestry. Annu Rev Environ Resour 33:187–211

Bele MY, Somorin O, Sonwa DJ, Nkem JD, Locatelli B (2010) Forests and climate change adaptation policies in Cameroon. Mitig Adapt Strat Glob Chang 16(3):369–385

Bellassen V, Gitz V (2008) Reducing emissions from deforestation and degradation in Cameroon: assessing costs and benefits. Ecol Econ 68(1):336–344

Brown HCP, Nkem JN, Sonwa DJ, Bele MY (2010) Institutional adaptive capacity and climate change response in the Congo Basin forests of Cameroon. Mitig Adapt Strat Glob Chang 15:263–282

Burnham P (2000) Whose forest? Whose myth? Conceptualizations of community forests in Cameroon. In: Abramson A, Theodossopoulis D (eds) Land, law and environment: mythical land, legal boundaries. Pluto, London, pp 31–58

CEEPA (2006) The economic impact of climate change on agriculture in Cameroon. Policy Note prepared by the Centre for Environment Economics and Policies Africa

Cerutti PO, Nasi R, Tacconi L (2008) Sustainable forests management in Cameroon needs more than just forests management plans. Ecol Soc 13(2):1–13

Cerutti PO, Tacconi L, Nasi R, Lescuyer G (2011) Legal vs. certified timber: preliminary impacts of forest certification in Cameroon. Forest Pol Econ 13(3):184–190

CIEFE/ICEFS, IMAFLORA (2006) FSC forest certification: simple guide for forest stakeholders. Icefs/Ciefe-Imafl ora. http://www.congobasinprogram.com/site/getfile.php?i. Accessed 25 Jan 2013

CIFOR (2012) Forests and climate change adaptation: what policymakers should know. http://www.cifor.org/fileadmin/factsheet/RIO+20_Factsheet%20-%20Adaptation.pdf. Accessed 18 Jan 2013

De Wasseige C, Devers D, De Marcken P, Eba'a Atyi R, Nasi R, Mayaux P (eds) (2009) Les forêts du Bassin du Congo. Etat des forêts 2008. European Union Publication Office, Brussels

Dkamela GP (2011) The context of REDD+ in Cameroon: drivers, agents and institutions. Center for International Forestry Research

Eddington MM, Howel CI (2006) Criteria and indicators of sustainable forest management: using global thinking to act locally. Veg Future 06: the conference in the field. http://ga.yourasp.com.au/vegfutures/pages/images/Colloquium%20B4_Howell.pdf. Accessed 25 Oct 2012

Egute TO, Albrecht E (2011) Environmental law and local traditional practices as potential tools in combating forest biodiversity loss in Cameroon. In: Forbid GT, Albrecht E, Egute TO (eds) Risk assessment and management for environmental protection in Sub-Saharan Africa. Lexxion Verlagsgesellschaft mbH, Berlin, pp 113–123

Ehode G (2011) Effects of climate change on water resources in Sub-Saharan Africa: case study of the Lake Chad Basin. In: Albrecht E, Egute TO, Forbid GT (eds) Risk assessment and management for environmental protection in Sub-Saharan Africa. Lexxion Verlagsgesellschaft mbH, Berlin, pp 113–123

Essama-Nssah B, Gockowski J (2000) Cameroon: forest sector development in a difficult political economy-evaluation country case study series. The World Bank, Washington, DC

Foahom B (2001) Biodiversity planning support program: integrating biodiversity into the forestry sector. Cameroon case study. Paper prepared for an international workshop on integrating biodiversity into the forestry planning program held in CIFOR Headquarters, Bogor, Indonesia on 13–16 August 2001

Fombad CM (1997) The effectiveness of environmental protection measures in Cameroon's 1994 law laying down forestry, wildlife and fisheries regulations. J Env Law 9(1):43–58

Freudenthal E, Nnah S, Kenrick J (2011) REDD and rights in Cameroon. A review of the treatment of indigenous peoples and local communities in policies and projects. Forest Peoples Program

Fuo ON, Semie SM (2011) Cameroon's environmental framework law and the balancing of interests in socio-economic development. In: Faure M, Du Plessis W (eds) The balancing of interests in environmental law in Africa. Pretoria University Law Press (PULP), Pretoria, South Africa, pp 75–94

Government of Cameroon (1993) La politique forestière du Cameroun. Ministère de L'Environnement et des Forêts Yaoundé, Cameroun

IPCC (2007) Climate change 2007: mitigation. In: Metz B, Davidson OR, Bosch PR, Dave R, Meyer LA (eds) Contribution of working group III to the fourth assessment report of the intergovernmental panel on climate change. Cambridge University Press, New York

Justice NS, Bih ET (2009) Case study Cameroon. In: Legal frameworks for REDD, design and implementation at the national level. IUCN, Gland, Switzerland

Lambi MC (2001) Environmental issues: problems and prospects. Unique Printers, Bamenda

Liviu A (2005) Legal compliance in the forestry sector: case study Cameroon. http://www.fao.org/forestry/195800b266e69b6880da012e1a1875f2319a15.pd. Accessed 19 Oct 2012

Locatelli B, Kanninen M, Brockhaus M, Colfer CJP, Murdiyarso D, Santoso H (2008) Facing an uncertain future: how forests and people can adapt to climate change. Forest Perspectives No. 5, Centre for International Forestry Research, Bogor, Indonesia

Locatelli B, Thompson I, Brockhaus M, Buck A (2010) Forests and adaptation to climate change: challenges and opportunities. Int For Rev 12(5):436

MCPFE (2011) Sustainable forest management. http://www.foresteurope.org/eng/What__for/Sustainable_Forest_Management/. Accessed 26 June 2012

McSweeney C, New M, Lizcano G (2008) UNDP climate change country profiles: Cameroon. United Nations Development Program and University of Oxford, Oxford

MINEF (1999) A compendium of official instruments on forest and wildlife management in Cameroon. National Printing Press, Yaounde, Cameroon

MINEF (2005) Cameroon. Communication nationale initiale du Cameroun sur les changements climatiques. Ministère de l'Environnement et des Forêts, Yaoundé. http://unfccc.int/resource/docs/natc/cmrnc1f.pdf. Accessed 25 Jan 2013

MINEP (2008) The forest Carbon Partnership Facility (FCPF) Readiness Plan Idea Note (R-PIN). http://www.forestcarbonpartnership.org/fcp/sites/forestcarbonpartnership.org/files/Documents/PDF/Cameroon_R-PIN_07-31-08.pdf. Accessed 5 Nov 2012

MINFOF and FAO (2005) Evaluation des ressources forestières nationales du Cameroun. Yaoundé, Cameroun

MINFOF (2009) Manual of the procedures for the attribution and norms for the management of community forests in Cameroon (revised version). Ministry of Forest and Fauna, Yaounde, Cameroon

Molua EL (2010) Climate and location vulnerability in southwestern Cameroon: assessing the options and cost of protection to property in coastal areas. CEEPA Discussion Paper No 46, Centre for Environmental Economics and Policy in Africa, University of Pretoria, Pretoria

Molua EL, Lambi CM (2007) The economic impact of climate change on agriculture in Cameroon. Policy Research Working Paper 4364. The World Bank, Development Research Group, Sustainable Rural and Urban Development

Ndifiembeu BN (2006) A handbook on the criminal procedure code of the Republic of Cameroon. Yaounde, Cameroon

Neba EN, Lotsmart NF (2003) Actors, options and the challenges of forest management in Anglophone Cameroon. GeoJ 57:95–111

Ngalame EN (2012) Cameroon: government cracks down on illegal logging. http://allafrica.com/stories/201203050086.html. Accessed 20 Aug 2012

Njoh AJ (2007) Politico-economic determinants of forestry policy in Cameroon. GeoJ 70:109–120

Nkem J, Santoso H, Murdiyarso D, Brockhaus M, Kanninen M (2007) Using tropical forest ecosystem goods and services for planning climate change adaptation with implications for food security and poverty reduction. J Semi-Arid Trop Agric Res 4(1):1–23

Norrington-Davies G (2011) Climate change financing and aid effectiveness: Cameroon case study. http://www.oecd.org/dataoecd/23/11/48458409.pdf. Accessed 9 Sept 2012

UN-REDD Program (2009) About REDD+. The United Nations Collaborative Program on reducing emissions from deforestation and forest degradation in developing countries. http://www.un-redd.org/About REDD / tabid/ 582/ Default.aspx. Accessed 20 Sept 2012

Republic of Cameroon (2009) Cameroon's fourth national report to the convention on biological diversity. Yaounde, Cameroon

Seppälä R, Buck A, Katila P (eds) (2009) Adaptation of forests and people to climate change. A global assessment report. IUFRO World Series vol 22. IUFRO, Vienna, pp 1–224

Sonwa DJ, Somorin OA, Jum C, Bele MY, Nkem JN (2012) Vulnerability, Forest-related Sectors and Climate Change Adaptation: The Case of Cameroon. Forest Pol Econ 23:1–9

Streck C, Scholz MS (2006) The role of forests in global climate change: whence we come and where we go. Int Aff 82(5):861–879

Tchindjang M, Amougou JA, Asossolo SA, Bell SB (2012) Challenges of climate change, landscape dynamics and environmental risks in Cameroon. In: Runge J (ed) Landscape evolution, neotectonics and quaternary environmental change in southern Cameroon: an international yearbook of landscape evolution and palaeoenvironments. CRC Press Taylor & Francis Group, Boca Raton, pp 237–287

Tieguhong JC, Betti JL (2008) Forest and protected area management in Cameroon. ITTO Tropical Forest Update 18(1):6–9

Topa G, Karsenty A, Megevand C, Debroux L (2009) The rainforests of Cameroon: experience and evidence from a decade of reform. World Bank, Washington, DC
Yadlapalli L (2010) EU-Cameroon trade agreement ensures timber legality from one of Africa's biggest exporters of tropical timber. http://www.efi.int/portal/news___events/news?bid=510. Accessed 20 Aug 2012

Chapter 17
The Renewable Energy Sources Act (EEG) as German Way of a Future-Oriented Energy Policy Change

André Zschiegner and Emmanuel Wanki

17.1 Introduction

In our modern, technology-based society the constant—actually unlimited—availability of raw materials of energy at low economical costs is a central functional condition. Both the German economy and private households depend heavily on the sufficient supply of cheap energy. The continuous and very one-sided focus on conventional forms of energy is increasingly proving to be a problem because the unpredictability of long-term ecological consequences of the use of fossil energy carriers on the global climate on one hand and the uncontrollability of the risks of atomic energy generation on the other set tasks on our society that can hardly be solved. Furthermore, the limited availability of fossil energy carriers coupled with increasing global energy needs is proving to be a problem. The constantly increasing energy needs of established industrialized nations have, in recent years, met the rapidly increasing demand for energy of transition countries such as China, India or Brazil. In view of the increasing shortage of resources, this has lead to horrific price development in world markets. On a long-term basis, the goal of a conscious energy policy—national as well as international—can therefore only be the development and establishment of alternative energy generation concepts based on regenerative energy carriers. However, the implementation of this objective in practice proves to be very problematic.

A. Zschiegner
Department of Civil and Public Law with References to the Law of Europe and the Environment, Brandenburg University of Technology Cottbus-Senftenberg, Erich-Weinert-Str. 1, LG 10/524, 03046 Cottbus, Germany
e-mail: zschiegn@tu-cottbus.de

E. Wanki (✉)
Department of Civil and Public Law with References to the Law of Europe and the Environment, Brandenburg University of Technology Cottbus-Senftenberg, Erich-Weinert-Str. 1, LG 10/522, 03046 Cottbus, Germany
e-mail: ateghemm@tu-cottbus.de

E. Albrecht et al. (eds.), *Implementing Adaptation Strategies by Legal, Economic and Planning Instruments on Climate Change*, Environmental Protection in the European Union 4, DOI 10.1007/978-3-540-77614-7_17, © Springer-Verlag Berlin Heidelberg 2014

As a result, alternative technologies for the generation of electricity from renewable energy sources—with high divergences within the individual forms—are still at the beginning of their technological development (Maslaton and Zschiegner 2009). The efficiency of these installations can still not compete with those of conventional generation forms and use only a fraction of their actual existing potential. An energy policy which sets as objective the promotion of renewable energies cannot limit itself merely to already established technologies in the market but must at the same time provide incentives to consistently promote technological development.

In addition, the competitive situation between conventional energy generation forms and renewable energies is strongly distorted because price-determination for the remuneration of electricity from conventional energy generation forms is almost exclusively pure electricity generation cost, while so far potential external costs are usually not taken into consideration. The subsequent price development mechanism for electricity from fossil energy carriers does not take into consideration the economic and ecologic consequences of the threatening global warming. The direct costs and the costs of necessary measures to address global warming are not components of the electricity price in the market. They meet the consumer nevertheless in form of taxes, which are also used as a measure against global warming. The same applies to electricity from nuclear power plants. The costs of risk management and the disposal of nuclear waste are not directly incorporated into the electricity pricing policy, but will ultimately be co-financed through tax revenues. As a result, there exists a de facto artificial splitting of the electricity price into the pure production cost on one hand and the external costs on the other. Only the production cost is taken into account in the competitive price of fossil generated electricity.

With renewable energies,[1] however, there is no such artificial splitting of the price. The electricity price is rather the sum total of the costs of generation and necessary subsequent measures. This is due to the fact that the use of renewable energies is already designed in order to cause as few environmental effects as possible in the course of their generation. In this respect, possible negative environmental effects that could offset price intensity have already been eliminated. Furthermore, the permit requirements for renewable electricity generation projects provide that permission is granted only if the operator—through public easement or a similar protection law—takes over a decommissioning obligation.[2] The financial feasibility of an unproblematic removal of the installation at the end of its operation period for

[1] Renewable energies according to the statutory definition of Art. 3 No. 3 EEG are hydropower, including wave, tidal, concentration gradient and flow energy, wind energy, solar energy, geothermal and biomass energy including biogas, landfill gas and sewage gas as well as the biodegradable fraction of waste from households and industry.

[2] Regulated for open outdoor areas specifically, Art. 35 para 5, sent. 2 and 3, of the Building Code (BauGB). Large plants can regularly be built only in open outdoor areas. This requirement concerns the absolute majority of projects for alternative energy production. In the case of construction in the development plan area, corresponding obligations are regularly determined which should be considered with regard to the legal norms for the outdoor areas.

large renewable energy projects is a precondition for granting a permit. Therefore the economic operation of the installation must immediately consider these costs from the onset; this obviously influences the final price of this form of electricity. Unlike the conventional forms of energy generation, payment for electricity generated from renewable energy carriers is not limited only to the electricity generation costs, but also extensively covers the potential decommissioning cost from the onset.

This, however, leads to the fact that the actual ecological advantage of the use of renewable energies to generate environmentally friendly electricity is not reflected on the economic level, since the substantial external costs of electricity generation from conventional energy sources—which is actually their competitive disadvantage—do not directly appear as part of the electricity price (Maslaton and Zschiegner 2009), which gives them a significant competitive advantage.

It follows from the above-mentioned aspects that the generation of renewable energies—depending again on the specific generation type[3]—in Germany is not yet competitive compared to conventional forms of energy generation. However, in order to accelerate the development of ecologically necessary renewable energy technologies, a legislative regulatory intervention in the electricity generation market is inevitable. It can therefore be stated that a legislation which is aimed at increasing the use of renewable energy generation forms cannot be effective without market-regulating measures.

17.2 The Renewable Energy Sources Act (EEG): Purpose

The legislative attempt to establish a long-term use of renewable energies in the energy generation concept in Germany is found primarily in the Renewable Energy Sources Act (EEG). This has to create the necessary economic framework conditions in order to enable the success of renewable energies in the market even in their early development phase, so as to achieve the key climate policy goals with regard to sustainable energy supply (Salje 2008). The EEG is thus a concretisation of the country's constitutional Environmental Protection objectives according to Art. 20a of the Basic Law (Grundgesetz—GG) as well as a transposition of the international environmental legal framework, as derived from the UN Framework Convention on Climate Change (UNFCCC) (ILM 31 (1992), 849) and the Kyoto Protocol (ILM 37 (1998), 22) into German national law (Federal Ministry for the Environment, Nature Conservation and Nuclear Safety 2009).

The EEG came into force on 1 April 2000 (Salje 2008) and has to date been amended twice—on 1 August 2004 (FLG[4] I p. 1918) and on 1 January 2009 (FLG I p. 2074). The current version is based on the second comprehensive revision and

[3] The leader so far should be the use of wind energy, in particular the latest generation installations, which can exhibit significantly improved efficiencies.

[4] Federal Law Gazette (= Bundesgesetzblatt – BGBl).

entered into force on 1 January 2009. It has till date gone through five additional changes; the current applicable version of the EEG came into force on 1 January 2012.

The regulatory purpose of the Act is—according to the Article 1 para 1 EEG—'to facilitate a sustainable development of energy supply, particularly for the sake of protecting our climate and the environment, to reduce the costs of energy supply to the national economy, also by incorporating external long-term effects, to conserve fossil fuels and to promote the further development of technologies for the generation of electricity from renewable energy sources'.

17.2.1 Sustainability of Energy Supply

A central objective of the EEG is to ensure a sustainable energy supply. This is based on the legislator's knowledge that the current focus of primary energy supply on fossil and nuclear energy carriers in Germany does not meet the criteria of sustainability (Salje 2008). Neither in terms of actual resource consumption nor the emission of greenhouse gases is the prevailing energy supply in Germany in compliance with the principles of sustainability, whose distinctive criteria lies in extensive resource conservation with the long-term goal of preserving the natural resource base for future generations (ibid.). The sustainability principle is no specific concept of the EEG but is also found in international frameworks such as the Kyoto Protocol as well as at the national level where it is a key aspect of the constitutionally laid down goal of environmental protection.

17.2.2 Environmental Protection and Nature Conservation

Directly related to the goal of sustainable energy supply, aspects of environment protection and nature conservation are additional legislative purposes of the EEG (Federal Ministry for the Environment, Nature Conservation and Nuclear Safety 2009). According to the legislator, the renunciation of fossil and nuclear energy carriers on the basis of sustainability has reduced extensively the environmental impacts of conventional energy generation besides conserving resources. In addition to the primary objective of changing the energy concept, the EEG has as a direct secondary objective aspects of environmental and nature protection (Salje 2008). This concerns particularly the issue of implementing the most environmentally friendly energy change possible. Here, the EEG sets specific requirements to ensure the possible development of environmentally friendly facilities for the use of renewable energy source. For example, Art. 23 para 4 EEG makes the remuneration of electricity from hydroelectric power plants dependent on the achievement of a targeted minimum environmental status of the water according to the requirements of the Water Management Act.[5] Similarly for wind mills, the requirements for the

[5] Water Management Act, 31.07.2009, FLG I. p. 2585.

extension of higher remuneration rates for repowered wind mills under certain conditions has extensively been put in place. This may result in a possible long-term use of existing wind parks in order to minimize the need for entirely new areas for wind power utilization (Art. 30 EEG). The purpose of the environmentally motivated land planning also serves as a provision of Art. 32 para 3 No. 2, 3 and 4 EEG for electricity from solar energy.

17.2.3 Reduction of Economic Costs

The EEG also aims at reducing the economic costs of energy supply. This is particularly reflected in the fact that energy generation from fossil and nuclear energy carriers is associated with enormous external costs (Federal Ministry for the Environment, Nature Conservation and Nuclear Safety 2009) which are incurred in addition to the actual electricity generation costs. This is a burden on the whole economy. The energy change brought about by the increasing shift from fossil and nuclear energy carriers has to reduce the external costs resulting from their use in the future. In addition, the establishment of new generation technologies has to lead to an increased electricity flow in the market and an effective pricing mechanism because the price of electricity is determined significantly by the most unviable power plants, which are still connected to the grid system, based on the 'merit-order-effect'. If these unviable installations are shut down as a result of grid overload or excess supply of electricity, this will certainly have a long-term impact on the development of electricity prices (Salje 2008). Finally, one must also consider that renewable energies are an economic sector with enormous growth potential. In this respect, they are already and will continue to assert its lasting impact on the labour market (Maslaton and Zschiegner 2009). Through early technology development in this industrial sector, Germany has earned a leading global market position which has led to large international demand for technologies developed in Germany for the generation of electricity from renewable energy sources.

17.2.4 Advancement of Technologies for Generating Electricity from Renewable Energies

The aspect of technological advancement is not only a consequence of the economic cost-reduction objective, but also a legal purpose of the EEG. This ultimately enables the long-term competitiveness of electricity generation from alternative energy sources (Federal Ministry for the Environment, Nature Conservation and Nuclear Safety 2009). Here, the legislator does not limit its method to calculate remuneration rates solely in a way that only a technologically and effectively conceived installation can be economically operated. Rather, the EEG system guarantees in different ways the highest possible degree of efficiency of the installations (Salje 2008). For example, the digressive structuring of remuneration rates depending on the year the installation starts operating makes the continuous development of technologies for the efficient operation of the installation inevitable. In addition, certain bonus payments are granted when particularly innovative or future-oriented technologies are deployed.[6] In this context, the EEG therefore has a strong controlling character on technological development.

This direct control character is shown in practice repeatedly and where relatively short interval adjustments in the remuneration rates or bonus payments[7] for individual technologies are made in the EEG, as was recently done for the remuneration of electricity from photovoltaic generation by the first law amending the EEG,[8] which standardized a special reduction in remuneration of 13 % from 1 July 2010 and a further 3 % from 1 October 2010. This certainly controversial additional lowering of EEG compensation beyond the legal scale also represents a reaction of the legislator to the rapid economic and technological development of electro-energetic use of solar energy in Germany, which was already well above the underlying expectations initially stipulated as compensation rates in the EEG from 2009.

17.2.5 Avoidance of Conflicts over Fossil Energy Resources

Beyond the above environmental and economic policy objectives, and given the increasing scarcity of resources, the EEG has to help avoid international conflicts over fossil fuels. The rapid development of transition countries is accompanied by a sharp increase in energy needs, which cannot be sufficiently compensated by the use of new, energy-efficient technologies. A consequence of this development is therefore an ever-increasing global demand for energy, combined with a drastic shortage of resources. In this situation, an increasing shift to alternative energy sources is not only able to avoid negative economic effects but is also designed to prevent potential international conflicts over the remaining fossil fuels, as it reduces

[6] So for electricity from biomass, the technology bonus according to Art. 27 para 4 No 1 EEG (each analogue granted for the energetic use of landfill gas in accordance with Art. 24 para 3 EEG, for sewage in accordance with Art. 25 para 3 EEG and for mine gas in accordance with Art. 26 para 3 EEG), the NawaRo-Bonus acc. to Art. 27 para 4 No 2 EEG, and combined heat and power-bonus acc. to Art. 27 para 4 No 3 EEG. For electricity generated from geothermal heat use-bonus acc. to EEG Art. 28 section 2 and the bonus for the use petro thermal techniques ('hot dry rock method') acc. to Art. 28 para 3 EEG. For wind energy, increase of the compensation for Repowering acc. to Art. 30 EEG. For photovoltaic electricity, the increased compensation for building systems acc. Art. 33 EEG.

[7] Thus, the transitional arrangements for palm oil and cooking oil electricity plants in the third Act to amend the energy conservation law of 28 March 2009, FLG. I, p. 643 and the Act to avoid short-term market bottlenecks in liquid biomass of 31 July 2010, FLG I, p. 1061.

[8] First amendment to the Renewable Energy Law of 11 August 2010, FLG I, p. 1170.

import dependence and increases the statistical range of internal reserves (Salje 2008). In addition to these direct consequences, the EEG also shows indirect conflict prevention potential, in the fact that, through its mitigation of climate change, scarcity of water resources can be reduced (ibid.).

17.3 The Renewable Energy Sources Act: Regulatory Content

To achieve the aforementioned objectives, the EEG regulates the German energy market in many ways. Contrary to popular misconception, however, this is not done through the granting of subsidies. The claim that renewable energy sources are being subsidized by the EEG stands opposed, not only that the regulatory content of the EEG has no specific hallmark of subsidization. Furthermore, the EEG lacks the sovereign dominated regulation of the state-citizen-relationship, which by definition is a precondition for subsidization. The EEG actually governs only the relations between the grid operator and the feeder, each as legal entities of private law.

With respect to grid operators, it standardizes essentially two significant limitations of the civil law applicable to private autonomy[9] of each individual entity. On one hand, existing contractual obligation between the network operator and the feeder under certain circumstances is justified by virtue of actual conditions—regardless of the existence of a contract (see Art. 4 para 1 EEG). The obligation of the grid operator to connect installations generating electricity from renewable energy sources as well as receiving electricity generated by these installations and feeding it into the grid system results from this contractual obligation. Therefore, the grid operator should not make the connection of installations generating electricity from renewable energy sources dependent on prior conclusion of a 'connection contract'.[10] Secondly, the EEG contains significant limitations on the content of the legal relationship between the feeder and the grid operators, called feed-in agreements. As long as the determination of the minimum compensation to be paid for electricity fed into the grid by the grid operator is based on the type and size of the installation, the grid operator is to pay this minimum compensation, regardless of whether the parties base the feed-in solely on the existing contractual obligation or have specifically signed a feed-in agreement. Since the parties in principle have the freedom—overlapping the contractual obligation under the EEG—to regulate their contracts in their legal relationships,[11] however, this should not be at the detriment of the feeder with respect to the content according to the legal requirements of the EEG, Art. 4 para 2 EEG. In this respect, the EEG is a positive obligatory law that

[9] On the notion of private autonomy Ellenberger (2014), overview before Art. 104 marginal no. 1 ff.

[10] A demand which unfortunately is often to be seen against the legal regulations in practice.

[11] And for practical purposes, it is advisable to regularly base the feed-in on a comprehensive feed-in agreement.

predetermines each feed-in contract, therefore limiting the private autonomy at the expense of the grid operators.

Regulatory intervention by the EEG is thus merely in the civil legal relationship between the grid operator and the feeder. The legal minimum feed-in compensations are not government subsidies. Besides, the regulation is to be distinguished between the provision of grid access by the standardisation of a legal connection (1.), purchase obligation (2.) and the guarantee of competitiveness by the fixing of specific compensation conditions—in particular the minimum compensation (3.).

17.3.1 Connection Obligation, Art. 5 EEG

The central obligation of grid operators—regulated in Art. 5 EEG—is to enable access to the public grid to installations generating electricity from renewable energy sources.

17.3.1.1 Principle

According to Art. 5 para 1 EEG, with respect to installations that generate electricity from renewable energy sources,[12] the 'grid system operators shall immediately and as a priority connect installations generating electricity from renewable energy sources to that point in their grid system which is suitable in terms of the voltage and which is at the shortest linear distance from the location of the installation if no other grid system has a technically and economically more favourable grid connection point'. The existence of the actual requirements of Art. 5 para 1 EEG justify the contractual obligation between the potential feeder and the grid operators to meet the connection obligation. This is expressly made clear in Art. 4 para 1 EEG. The EEG guarantees via the standardization of a legal connection and acceptance obligation a high level access of renewable energies to the general electricity market.

Promptness of the Connection

The connection of the installation has to be done immediately, without culpable delay. This means that the grid operator should basically connect the installation—provided that it is technically feasible—as soon as possible to its grid system. If he culpably violates this obligation, this can result in damage claims from the contractual obligation according to Art. 280 et seq. of the German Civil Code (Federal Ministry for the Environment, Nature Conservation and Nuclear Safety 2009). For a

[12] A statutory definition of renewable energies is found in Art. 3 No. 3 EEG.

substitutable claim of damage, the resulting income deficit of the declining minimum remuneration over the life span of the installation will be considered, provided the delay in connecting the installation leads to the lapse of a purposeful relevant date (usually the 31st of December of a year). In addition, the assertion of further damages caused by default seems possible under certain circumstances.

Priority of Renewable Energies

Installations generating energy from renewable energy sources have priority connection to the public grid system, see Art. 5 para 1 sent. 1 EEG. The priority refers to the relationship to installations generating energy from fossil and atomic energy carriers. Renewable energies basically have priority over these energy carriers. Practically, this means that the grid operator should not reject a connection desire with reference to the grid load, unless the load is caused solely by the supply of electricity from renewable energy sources (Federal Ministry for the Environment, Nature Conservation and Nuclear Safety 2009).[13] In practice, it is frequently the case that there is a grid overload when there are still installations generating energy from conventional energy sources in the grid system. Therefore, the grid operator is obliged to remove these old installations from the grid as required to enable the connection of installations generating energy from renewable energy sources.

As a result, there is not just a simple connection obligation on the grid operator but also a corresponding pressure to keep the grid free of conventional energy sources to the extent necessary for the connection of renewable energy sources. This is ultimately an expression of the legislative purpose of the EEG to support a sustainable energy transition.

Feed-in Obligation Even Under Necessity of Grid Expansion

The connection obligation does exist even if such becomes technically possible only when the grid is expanded or improved. This is presented in Art. 5 para 4 EEG explicitly clear and complements in general the existing obligation of the grid operator according to Art. 9 EEG, to expand their grids in the manner required to feed-in electricity from renewable energy sources. This also expresses the main purpose of the EEG to achieve lasting, future-oriented transformation of the German energy market towards renewable energies. Because this makes it necessary to adapt the grid system to the consequences of increased decentralize feed-in of renewable energy generation.

[13] Loading the network just with renewable energies is still hardly relevant practice. Only in the areas of large wind farms—especially in Northern Germany—is this constellation possible.

17.3.1.2 Public Supply Network

The connection obligation refers to the public supply network, which is legally defined and conceptualized in Art. 3 No. 7 EEG as 'all the interconnected technical facilities used for the purchase, transmission and distribution of electricity for general supply'. Crucial for an assignment to the public grid system is that the individual lines are laid down according to their dimensions so that they can serve more than just the supply of a single end user (potentially).[14]

The public supply network also include so-called branch lines, that is, lines that (initially) only serve the connection of a particular subscriber in the supply network, provided the connection of additional customers without the extension cable is not possible and excluded from the outset.[15] Practically relevant in this constellation is the connection of construction areas by individual branch lines in which initially only one end-user actually existed. In this case, the branch line is made despite the fact that they initially served as a connecting cable to a single individual who is a part of the public grid system as defined by Art. 3 No. 7 EEG.

17.3.1.3 Connection Point

In a logical continuation of the connection obligation, the EEG also provides binding statements on the relevant connection point of generation facilities to the grid system, which raises the question of 'how' the connection is in spatial terms.

The Shortest Distance Between Installation and Grid as a Principle

For the fulfilment of the connection obligation, the EEG standardises as suitable connection point the location on the grid 'which is suitable in terms of the voltage and which is at the shortest linear distance from the location of the installation' Art. 5 para 1 sent. 1 EEG. The determination of the grid connection point according to the conception of the EEG is fundamentally based only on actual circumstances. The suitability of the voltage level of the grid connection is a necessary prerequisite, while the reference to the shortest linear distance expresses primarily economic aspects of cost minimisation. From this economic consideration follows then, that in some cases a connection point which is not closest according to Art. 5 para 1 sent. 1 EEG may be a possibility, if it transpires that in the general consideration, the connection costs of the installation are lower than in the case

[14] Court of Appeal Munich, Decision of 23.09.2003, Az.: 15 U 1772/03; Court of Appeal Stuttgart, Decision of 16.06.2003, Az.: 2 U 43/03; Court of Appeal Nuremberg, Decision of 30.04.2002, Az.: 3 U 4066/01; Federal Civil Court, Decision of 10.11.2004, Az.: VIII ZR 391/03; Also see Explanation to the EEG 2009, Part B, German Parliamentary Bulletin 16/8148, p. 40.

[15] Federal Civil Court, Decision of 10.11.2007, Az.: VIII 391/03.

of a connection based on the shortest distance. Besides, this economic approach does not depend on which cost share variations the installation operators and the grid operators have—depending on whether there are grid connection costs (Art. 13 EEG) or grid expansion costs (Art. 14 EEG). The total expenses to be expected are exclusively decisive.[16]

Special Provision for Small Installations

A practically important exception to the principle of Art. 5 para 1 sent. 1 EEG for the determination of the connection point is contained in Art. 5, para 1, sent. 2 EEG. For small installations with a total maximum installed capacity of 30 kW, which are located on a property with an existing electricity connection, it is assumed that this is the most economically favored connection point. This should allow significant minimization of additional connection costs by the small installation (in practice mostly photovoltaic systems) which otherwise would hinder the project, thereby making its construction and operation profitable. Additionally, a purpose of the EEG is to establish a lasting energy concept while at the same time to significantly reduce economic costs and promote development.

Special Rights for Feeder and Grid Operator

Furthermore, in Art. 5, para 2 and 3 EEG, the installation operator (para 2) as well as the grid operator (para 3) have been granted the possibility, based on Art. 5 para 1 sent. 1 EEG, to choose a different connection point. This supports the interests of stakeholders in the deviation from the principal criteria of economic efficiency to choose in certain conditions a favorable connection point in accordance with Art. 5 para 1 EEG. If the grid operators makes use of this law according to Art. 5 para 3 EEG, he has to carry the resulting extra costs in accordance with Art. 14 para 2 EEG.

17.3.1.4 The Connection Obligation Supports the Feeder's Information Claims

The construction and operation of installations for the generation of electricity from renewable energy sources are regularly linked with massive investments from the operator. In view of this investment volume, the processes of project planning and realization need thorough calculation. With regards to connecting the installation to the grid system, two constellations are considered: On one hand, the installation operator must know which costs will incur for the realization. In addition, an

[16] Federal Civil Court, Decision of 18.07.2007, Az.: VIII 288/05.

accurate determination of the possible connection period is required. Lastly, since the minimum rate of remuneration is subject to an annual digression according to the EEG, a late connection of the installation to the grid can lead to significant economic losses for the entire operation period of the installation.

In order to grant the operator maximum planning security, the law in Art. 5 para 6 EEG gives him the right to information concerning the precise connection period (No. 1), the actual determination of connection point (No. 2) and the associated costs of connection to the grid (No. 3). Moreover, as soon as the feeder informs the grid operator of his desire to connect to the grid system, the grid operator has the obligation to provide the feeder with information on the precise processing period of the connection (Art. 6, para 5, sent. 1 EEG). The feeder has to be informed about the processing steps the examination of documents would follow (Art. 6 par 5 No. 1 EEG) and the information as well as data that is needed for the examination (Art. 6 para 5 No. 2 EEG).

17.3.1.5 Establishment of the Connection

Concerning the concrete realisation of the connection, the EEG contains unique regulations concerning on one hand the question of who establishes the connection and on the other hand the technical issues of implementation.

Personnel

According to Art. 7 para 2 EEG, the installation operator has the right to decide if he wants the grid operator or a qualified third party to connect the installation to the grid. This regulation should serve to avoid unnecessary costs and at the same time to prevent cost disputes.[17] It also takes into account the fact that the installation operator will frequently pay for the cost of connecting the installation to the grid. Unless the installation operator commissions a third party to connect the installation, the grid operator is obliged to help in the realization of the connection to the required extent (Federal Ministry for the Environment, Nature Conservation and Nuclear Safety 2009). He has to provide the necessary access data and access rights needed for the connection.

Technical Realization

With regard to the technical realization of the grid connection, Art. 7 para 2 EEG states clearly that the necessary technical know-how must meet the requirements of

[17] So already for Art.13 para 1 sent. 4 former version, Salje (2008), Art.13 marginal number 39; and Explanation to the EEG 2009, Part B, German Parliamentary Bulletin 16/8148, p. 43.

the grid operator and those of Art. 49 of the Energy Industry Act, which refers to the generally approved rules of the technology. The need to guarantee these requirements is valid for the entire period of operation of the installation. Thus, if the installation no longer meets the preconditions in Art. 7 para 2 EEG after its connection to the grid, the grid operator has the right to disconnect the installation from the grid, provided it does not concern a short-term disturbance (ibid.).

Beyond this natural principle, the EEG standardizes specific requirements for certain installation categories and types. It is fixed in Art. 6 Nr. 1 EEG that installations with a feed-in capacity of more than 100 kW must be fitted with a technical or operational device for remote-control reduction of feed-in load when the grid is overloaded and for the reading of the first feed-in which the grid operator may access. The reason for this technical specification is to facilitate the practical realization of the so-called feed-in management of imminent grid overload, which is a departure from the formerly applicable priority principle (Salje 2008) that is now implemented on the basis of a so-called sensitivity matrix. Unlike the EEG of 2004, according to Art. 4 para 3 sent. 1 EEG, the establishment of such a technical equipment is meant only for installations whose connection was at first thought to optimally supply the grid solely with renewable energies (ibid.). Presently, all operators of facilities that generate electricity from renewable energies above the threshold limit of 100 kW have the responsibility to install corresponding remote-controlled throttling equipment.

17.3.1.6 Requirements of the Grid System

The increasing establishment of renewable energies in the German Federal Energy Concept also leads to a drastic change of the requirements of the grid system. If the previous energy concept was characterised in particular by an extensive centralisation of the generation locations and their proximity to potential consumers, the shift to renewable energy sources has led more and more to a decentralisation of the generation locations. Unlike energy generation from fossil and atomic energy sources, the utilisation of renewable energy sources strongly depends on the location of the generation facility.[18] As a result, a capacitive expansion of the grid over the existing main axes is also essential to effectively integrate renewable energies into the grid permanently.

[18] This applies almost exclusively to renewable energy installations. Only in the case of biomass electricity conversion is such a direct dependence on location not given. Nevertheless, for logistic reasons, the relative nearness to the production location of the biomass will also make sense in this respect. Besides, the avoidance of unnecessary distances to the supply of the installation can also be an essential criterion for the granting of the installation permit according to the Federal Imission Control Act (BImSchG).

Expansion Obligation

As required for the connection of installation generating electricity from renewable energy sources, Art. 9 para 1 EEG sees a fundamental obligation of the grid operator to expand its grid. Grid expansion describes every measure which leads to an increase in the feed-in and transmission capacity of the grid and remains in the grid internal area. A possible under-sizing of the grid according to Art. 5 para 4 EEG—a logical continuation of the legal principle of obligatory grid expansion by the grid operator—cannot be held as claim for inability to grant access to the grid, as far as the expansion obligation according to Art. 9 para 1 EEG exists. The EEG via the grid expansion obligation places the grid operator in the duty to provide the necessary technical conditions for implementation of the energy shift in Germany by adapting their grid systems. The responsibility to expand and develop the grid according to Art. 9 para 1 sent. 2 EEG, lies explicitly on the grid operators.

Limits of Unacceptability

The basic expansion obligation, however, is not without exception. Rather, it finds according to Art. 9 para 3 EEG a limit to expansion in the criteria of economic unacceptability for the grid operator in particular cases. The EEG thus assumes the developed dispensation[19] of the prevailing legal situation (Federal Ministry for the Environment, Nature Conservation and Nuclear Safety 2009). Therefore, to assess the commercial unacceptability of a macroeconomic approach, the necessary costs for grid expansion are to be compared with the costs for connecting the installation to another connection point. Only when this comparison proves that the costs of grid expansion exceed that of an alternative connection of the installation to the grid, the grid operator can avoid his obligation to expand the grid with reference to their economic unacceptability. Admittedly, this will seldom be the case in practice, because in the calculation of the cost of linking the installation to another connection point other external costs also come into play. For example longer transmission distances are associated with higher feed-in losses (Maslaton and Zschiegner 2009). As a result, the fundamental responsibility of the grid operator to expand the grid will regularly remain.

Liability for Damages

The expansion obligation of the grid operator is accompanied by a standardised compensation claim of the feeder in Art. 10 para 1 EEG for culpable violation of its obligations according to Art. 9 para 1 EEG. Therefore, the installation operator can

[19] Federal Civil Court Decision of 10.11.2004, Az.: VIII ZR 391/03.

regularly demand damage compensation which originates from the fact that the installation cannot operate to its full capacity due to insufficient grid capacity.

Besides, the installation operator according to Art. 10 para 2 EEG can demand information from the grid operator to ascertain whether and to what extent the grid operator has fulfilled its commitment to expand the grid. This right to information should ultimately ensure that the installation operator can obtain factual information necessary for the filing of compensation claims (Federal Ministry for the Environment, Nature Conservation and Nuclear Safety 2009).

17.3.1.7 Bearing the Cost

The connection of installations for the use of renewable energies to the grid system may be associated with considerable costs in individual cases. In practice, it is therefore necessary to regularly clarify the question of who bears these costs and to which extent. Besides, it is the responsibility of the legislator to properly mediate between the opposing interests. On one hand, the grid operator will regularly try to shift the cost obligation onto the installation operator, though the grid operator equally profits from the connection of the installation. On the other hand, a one-sided strain of the feeder should not lead to unprofitability of the installation. Additionally, the necessary expansion of the grid for the connection of the installation equally benefits the grid operator indirectly. The EEG tries to resolve the rooted conflict of interest by differentiating the cost burden between the grid connection and the grid expansion.

Cost Bearing Obligation by Distinguishing Between Grid Connection and Grid Expansion

The installation operator basically bears the grid connection costs, Art. 13 sent. 1 EEG. Here, connection costs entail all the costs which directly arise from the connection of the installation to the grid.

In contrast, the grid operator pursuant to Art. 9 para 1 EEG bears the costs for the necessary expansion of the grid, Art. 14 EEG. Grid expansion is understood as: all the measures taken in the optimization, reinforcement and expansion of the grid. This cost-bearing obligation of the grid operator proves to be appropriate with regard to the above interests, both in factual and in legal-political terms. The expansion of the grid system, regularly and indirectly benefits the grid operator economically, in the form of a better usability of the grid system. There could be an excessive economic demand on the feeder if he must also bear the costs of grid expansion, which would jeopardize the establishment of the installation. This would stand in the way of the basic purpose of the EEG to speed up energy change in Germany. Finally, the grid operator—in contrast to the installation operator—regularly has the opportunity to split the grid expansion costs over the grid charges.

All this makes the legislative decision to place the cost-bearing obligation for grid expansion on the grid operator to be politically appropriate.

Boundary Problems

Despite the conceptually very clear distinction between connection and expansion measures, their practical distinction in terms of the distribution of cost burden is often difficult. The connection point of the installation to the grid according to Art. 5 para 1 EEG refers to the point which is the de facto dividing line between the connecting installation regularly owned by the installation operator and the grid of the grid operator. In addition to Art. 9 para 1 EEG, the provision of Art. 9 para 2 EEG on grid expansion claim is also applicable. Then, the expansion required costs of the grid operator should include all technical equipments necessary for the operation of the grid as well as those facilities owned by the grid operator or passing into the ownership of the grid system operator. This is also true in the latter case, as far as the connecting installations are affected, see Art. 9 para 2 EEG.

If the installations are technically affected by the concerned structural measures which are necessary for the reliable functioning of the grid, these would also affect measures adopted for grid expansion due to the fact that these installations are owned by the grid operator or passing into his ownership. It is normally irrelevant whether already existing installations will require expansion measures or if entirely new installations or extensions are constructed, provided the extension will be part of the grid system. In addition to the branch lines mentioned in the context of the definition of the public supply network, so-called parallel lines pose in this respect practical classification problems. These are lines that are more cost-effective alternatives to the extension of an existing (new) parallel installed cable. This will be used often by the grid operator to deny affiliation to the general supply network and, in so doing, trying to justify the cost obligation of the feeder. According to Art. 9 para 1 EEG, however, the grid operator should regularly construct the parallel line only as a cheaper alternative in order to fulfil its expansion obligation. The parallel cable is thus a necessary part of grid expansion, that is, an integral part of the grid system. The cost-bearing obligation with regard to a parallel line is also carried by the grid operator according to Art. 14 EEG.

17.3.2 Purchase Obligation

The grid operator has the legal obligation to connect installations generating electricity from renewable energy sources to the grid and the obligation to purchase electricity generated from renewable energy sources according to Art. 8 para 1 EEG.

17.3.2.1 Principle

According to Art. 8 para 1 EEG, the grid operator is obliged to purchase, transmit and distribute the entire quantity of electricity from renewable energy sources as a priority. The priority principle described within the framework of the connection obligation is also represented here. The grid operator therefore cannot refuse to purchase renewable electricity based on grid overload as long as feeders of non-renewable energy carriers are still connected to the grid (Federal Ministry for the Environment, Nature Conservation and Nuclear Safety 2009). The existing purchase obligation of the grid operator is ruled out only when the grid capacity is loaded completely with electricity produced from renewable energy and/or from heat-power-co-generation. In these cases, the EEG regulation of the so-called feed-in management takes effect, allowing the (temporary) throttling or complete disconnection of installations under specific conditions to ensure grid security. In addition it is in the fundamental obligation of the grid operator to expand his grid immediately for the connection and complete purchase of electricity.

However, the principle of priority purchase of electricity from renewable energy sources is not safe with regard to weakening the grid security requirements. The priority principle should apply generally only if the separation of conventional energy supply installations does not compromise grid security.[20] The grid operator can thus rule out its purchase obligation as far as he is able to justify this with system-related requirements of his grid. Of course the verification of the actual existence of conditions of grid overload for installation operator is often not possible, which at least justifies the danger that the grid operator could also use this exception of the principle of priority to enforce its economic interest of a maximum possible feed-in of electricity from its conventional power plants and disconnect installations generating electricity from renewable energy sources earlier from the grid.

17.3.2.2 Area Grid Supply

The electricity generated by the installation operator must be purchased according to Art. 8 para 2 EEG, even if this is not directly fed into the public supply network, but first fed into the non-public area grid and conveyed from this by means of commercially accountable transmission to the public grid. Here it is sufficient to supply just the electricity which was previously fed from renewable energy carriers from the area grid into the supply network without considering if the passed on electricity actually comes from renewable energy sources. The background of this regulation is to prevent the creation of an economically absurd direct connection

[20] The Explanation to the EEG 2009, Part B, German Parliamentary Bulletin 16/8148, p. 43: As long as the priority given to renewable energy 'does not affect the security of the transmission lines.

from the installation to the public supply system, when an indirect supply is possible (Maslaton and Zschiegner 2009). The actual transmission of electricity through the area network is therefore not required. It suffices to provide the accounting technique for the detection of the electricity potentially transmitted, taking into account the expected transmission losses (Federal Ministry for the Environment, Nature Conservation and Nuclear Safety 2009).

17.3.2.3 Different Contractual Agreement

As for the connection obligation, the standardized principle in Art. 4 para 1 EEG also applies to the purchase obligation of electricity that grid operators should not make their purchase obligation relay on the signing of a (feed-in) contract. The purchase obligation is therefore considered a direct legal obligation simply because of the existence of an actual feed-in situation due to the EEG. However, the EEG opens in Art. 8 para 3 the possibility for the parties to derogate from the legally standardized purchase priority for electricity from renewable energies on individual contract[21] basis. Reason for this legal possibility of deviation, which is to be understood as an offer to the parties as evidenced by the legal reasoning (Federal Ministry for the Environment, Nature Conservation and Nuclear Safety 2009), is to achieve a better integration of installations into the public grid through the agreement of individual purchase obligation regulations. The aim of such an agreement may in particular be to ensure a constant feed-in—by fixing specific periods in which little or no feed-in takes place. The resulting economic benefit for the grid operator—for example through the possible waiver of the energy balance provision—can then be allocated proportionally among the feeders in addition to the feed-in tariff. As a result, both sides can benefit from the self-regulation of the purchase obligation. The feeders would therefore not be worse in spite of partial renunciation on its rights pursuant to the EEG than if he were supplying unlimitedly. If the deviation possibility of Art. 8 para 3 EEG is used properly, this would permit an effective economic integration of individual installations into the grid to be achieved without undermining the financial interests of the feeder. In this sense, Art. 8 para 3 EEG also serves as a fundamental objective of the EEG to minimize the economic costs of energy generation.

17.3.2.4 Provision of Grid Security

The rapid development of renewable energy on one side and the existing partial lack of suitable transmission lines on the other side have increasingly led to

[21] A form-based deviation in terms and conditions of the grid operator separates itself out because the provision of Art. 8 para 3 EEG is in obvious contradiction of the aim to achieve a better integration of installation into the grid.

situations in Germany where grid security is threatened by excess supply. In order to prevent damage to the grid and to ensure the reliability of electricity supply, the so-called 'generation management' instrument was earmarked in the EEG of 2004 (Salje 2008). During a temporal overload of the grid by renewable energies (priority principle), those installations that were newly connected to the grid should be disconnected step by step. It was therefore the last-in-first-out or priority principle (Salje 2008). Accordingly, the obligation to install a technical device for remote-controlled feed-in reduction existed only for installations that would be connected to the grid from a specific time, from when—measured from the potential feed-in capacity of the existing installations on the grid—the first theoretical possibility of an overload of the grid with electricity from renewable energy sources was conceivable. Ensuring grid security was thus realized at the expense of individual plants. However, it proves to be problematic that from the time of initial entry, the possibility of a potential grid overload by renewable energies alone would severely hamper further development of renewable energies in terms of the risk for the installation operators, having to accept a far-reaching regulation of its installation.

Basic Principle of Feed-in Management

In order to prevent this while at the same time ensuring the necessary grid security, in Art. 11 EEG from 2009, the legislator has introduced the so-called feed-in management instrument, which is updated by the current version of Art. 11 EEG. Feed-in management signals a departure from the priority principle in the assessment of the requirements necessary for the regulation of supply of the individual installation based on a so-called sensitivity matrix. For this purpose, the explanatory memorandum from the grid operator will clearly describe the interaction between a feed-in change at a grid node and the power flow at a grid operating resource simplified as a linear relationship, the so-called sensitivity factor. The sum of these individual factors produces the sensitivity matrix of the overall grid system which maps the relationship; how the electricity fed into a specific grid node affects the different electricity flows on the grid. On this basis, the grid operator can determine which installations must reduce their feed-in power in order to fix the existing grid congestion. As a result, those installations whose actual feed-in are expected to affect the (n-1)-security strongest would be disconnected. In practice, virtually any installation above the minimum limit of 100 kW may be affected by a reduction in the feed-in, thus all installations must possess appropriate technical equipments for the remote-controlled reduction of the feed-in load. The basic concern of the new regulation on feed-in management is therefore to reduce the regulatory measures to a calculable necessary minimum with respect to the operational security of the grid and thus, largely reduce the operational and economic costs. In addition, potential installation operators would have to ensure a high degree of planning and legal security, pursuant to the old legislation (Federal Ministry for the Environment, Nature Conservation and Nuclear Safety 2009). This will succeed practically only if installations which only connect relatively late to the grid are controlled regularly. They can thus be operated economically because only grid-specific and not chronological criteria are used in an eventual regulatory need.

Practicability

However, feed-in management on the basis of a sensitivity matrix is not without problems because for installation operators who are specifically affected by the regulation it will hardly be predictable when and under which conditions their installations must be adjusted. In addition, it will often not be possible to reliably verify the actual existence of preconditions individually that require the regulation of the installation. Because the existing verification obligation according to Art. 11 para 3 EEG on the existence of a grid-related need for regulation of the installation through feed-in management, which can possibly be abused by the grid operator, must at least be questioned. So the vague sensitivity matrix standard will allow only qualified verifiable statements in individual cases. Added to that, the relativization of the priority principle with reference to the requirements of grid and supply security creates further insecurity based on the identification of a necessity to regulate the installation.

In addition, the sensitivity matrix is by no means static. Rather, it can change depending on the connection situation in the grid. Thus, installations for which there exist specific needs for regulation may vary. But this stands against the legislator's aim to create greater planning security for installation operators, through the departure from the priority principle as basis of feed-in management, because the eventual regulation need will not place the installation operator in the position to make concrete estimates on his installation. However, in practice, this situation is mitigated by the fact that the legislator has provided in Art. 12 EEG a hardship clause to the effect that the installation operator is compensated for the losses incurred by a feed-in reduction.

Hardship Clause

If the installation operator cannot feed all the electricity it generates into the grid due to a necessary measure for feed-in management, he has to be compensated. The EEG provides in Art. 12 para 1 a legal right to reparation (Federal Ministry for the Environment, Nature Conservation and Nuclear Safety 2009). The basis for determining the level of compensation is either through explicit agreements between the grid operator and the feeder or, in the absence of such agreements, through statutory feed-in tariffs under the EEG. The hardship clause therefore helps to minimize the financial consequences of feed-in management on the installation operator in order to preserve at least the claim for the payment of remuneration during the installation control periods. Nevertheless the difficult problem of enforceability of the claim for compensation in the form of fictitious feed-in revenues arises in practice if no proper rule was negotiated with the grid operator in advance. For fictitious feed-in revenues are not verifiable through concrete measurements as is the case with the actual feed-in quantities, giving rise to fears of dispute over the concrete amount of compensation in individual cases. Because of this, the hardship clause is a discrimination of the installation operator in relation to the possibility of a complete feed-in.

Furthermore, the hardship clause in terms of the objective of EEG appears not to minimize the economic costs of energy generation because according to Art. 12 para 2 EEG, the grid operator can reschedule payment on grid charges pursuant to Art. 12 para 1. With the understanding that in order to maintain grid security the regulation of EEG installations is contemplated so as to prevent the breakdown of the grid system. However, if conventional energy generation systems are still connected to the grid, then increased grid charges are equally a consequence of the continued supply of electricity from conventional energy carriers. In this case, the consumer finances not only the renewable energies but also overwhelmingly finances the fossil and nuclear energy carriers due to the increased fees as long as they remain connected to the grid during a (temporary) connection for the purchase of renewable electricity. One can therefore ask whether the increased grid charges are in fact direct results of increased supply from renewable energy sources or whether it is the interest of the grid operator presented under the guise of grid security to optimally operate their installations which generate electricity from conventional sources of energy. But this is too often happily left out in the public discussion.

17.3.3 Payment Claim

In direct relation to the legal purchase obligation, the EEG also standardizes in Art. 16 para 1 the rights of the feeder to demand payment for electricity fed into the grid from a renewable energy installation.

17.3.3.1 Principle

The fed-in electricity from renewable energy sources is therefore to be remunerated in accordance with Art. 18–33 EEG—which set the minimum compensation depending on the specific type of generation. According to Art. 16 in conjunction with Art. 18–33 EEG the (contractual) authority of the grid operator will intervene when they have to remunerate the injected electricity according to the law.

Precondition for the payment claim is that the fed-in electricity has been generated exclusively from renewable energy sources (exclusivity principle) and the installation operator fulfils his obligations under Art. 6 EEG, in particular installations with a feed-in capacity of more than 100 kW equipped with a technical device for remote-controlled feed-in reduction by the grid operator.

The claim for payment also explicitly exists according to Art. 12 para 3 EEG even when the power is initially stored. Thus, the EEG takes into account the changing technical requirements in practice, according to which the storage or storability of the renewably generated electricity will in the future-oriented energy concept be an essential criteria of efficiency.

17.3.3.2 Feed-in Obligation

In association with the payment obligation of the grid operator, the EEG of 2009 equally obliges the installation operator to actually feed all the electricity generated by the installation into the grid, as long as he claims remuneration in accordance with the EEG. This standardization of the feed-in obligation takes into account the changing circumstances in the energy market. In particular, the drastic price increases for raw materials of electric energy recently has increasingly led to an initial competiveness of specific renewable energy generation forms. This applies particularly to the energetic use of wind energy. With respect to specific market situations on the open market, some can achieve higher fees than those under the EEG. Basically, there would be an interest of the producer to offer its electricity in the short term on the open market which on the downside however means a significant loss of predictability for the grid operators (Maslaton and Zschiegner 2009). The standardization of feed-in obligation of installation operators by the legislator is intended to prevent this. The feeder is thus obliged—as long as he claims remuneration in accordance with the EEG—to feed all the electricity generated by his installation into the grid.

17.3.3.3 Legal Compensation Rates

The EEG contains numerous special provisions with regard to the concrete amount of the remuneration for the supplied electricity from renewable energy carriers. This is basically to distinguish between the general provisions of Art. 18–22 EEG applicable to all generation forms and the installation-specific provisions of Art. 23–33 EEG.

General Principles of Remuneration

The remuneration of electricity from installations generating energy from renewable carriers takes places through installed payments according to Art. 18 para 1 EEG. These payments relate to the corresponding threshold values for the specific installation type in accordance with Art. 23 et seq. EEG. If an installation exceeds a threshold value, a differentiated compensation takes place proportionally for each performance level achieved. This provides for a higher fairness in remuneration because it will prevent sudden rise in payment claims when the performance limits for the individual compensation levels are exceeded (Federal Ministry for the Environment, Nature Conservation and Nuclear Safety 2009). Remuneration as a function of performance tranches ensures that smaller installations can also be operated economically without being cheated financially by large installations.

In temporal terms, Art. 21 para 2 EEG standardizes the demand for minimum remuneration for a period of 20 calendar years including the commissioning year. A relevant commissioning according to Art. 21 para 1 EEG will be the time that electricity generated exclusively from renewable energy sources is first fed into the grid system. This should in particular ensure that no earlier start of operation is simulated by the (partial) use of conventional energy sources so as to avoid the consequences of the legal digression of compensation rates—perhaps by using diesel in the generator of a biomass installation for test purposes and so will produce electricity.

In order to accelerate technological progress, the minimum remuneration is subject to an annual digression according to Art. 20 EEG. The entitlement to legal remuneration is thus reduced, depending on the specific year the installation started operating. An economically profitable installation therefore necessarily requires constant improvement of the facilities, particularly in terms of their efficiency. Thus, the reduction contributes significantly to the achievement of the purpose of the Act: the further development of technologies for the generation of electricity from renewable energy sources.

Installation-Dependent Remuneration Regulations

The specific remuneration schemes for each installation type can be found in Art. 23 et seq. EEG. The distinction between the respective generation forms is justified by the different state of development of the underlying technologies on one hand and the various technical uses and applications on the other hand. Thus, the compensation rules do not regulate exclusively only the remuneration rate but at the same time indirectly include control mechanisms to promote technical development as long as the use of certain technologies bonus payments are provided.

17.3.3.4 Future Security

The development of installation for the generation of electricity from renewable energy sources is usually associated with significant investments, which is why the installation operator has a great interest to ensure the financial viability of the installation. Since this will essentially be realized through the receipt of the statutory feed-in tariff, this raises in practice the question of their future security. This pertains in particular to the question whether the legislator reduces the compensation rates for previous remuneration periods and can change the prospective compensation claim for installations already in operation with respect to their actual amount and/or duration, to the detriment of the installation operator.

The general constitutional principles on the admissibility of legal repercussions must be the standard of review for these issues. Hence, it is necessary to distinguish between so-called genuine and false repercussions and a general fidelity clause with regards to the legislation.

The existence of a genuine repercussion excludes in principle a subsequent law amendment to the detriment of the norm addressees.[22] A genuine repercussion exists if legal intervention is made subsequently on already past and completed offenses.[23] The review standard of the EEG compensation claims basically concerns the previous billing periods. In this case, no belated intervention can be made at the expense of the feeder.

Contrarily, the law on false repercussion intervenes on legal bases in already begun but not completed offenses, which must have been charged with the offence by definition.[24] In the existing context, this regularly concerns the prevailing accounting period in each case according to the EEG. In this situation, amendment of the law to the detriment of the norm addressees is in principle permissible even if belated.[25] Provided no legitimate expectation was created by the regulation, which encouraged the addressee into specific investments. In these situations, the fidelity clause outweighs the interest of the public in the modification of the standards, so that a fidelity clause to that extent is also guaranteed (Jarass and Pieroth 2009). One will be able to approve such a trust worthy of protection for installation operators according to the EEG. The embodiment of the feeder-grid operator relationship has to motivate the installation operator to invest extensively in renewable energies via the potentials of the EEG. To that extent it can be assumed that there exists a fundamental fidelity clause in favour of the installation operator according to the EEG. It might, however, be questionable if future accounting periods with regards to established installations are also channelled into this fidelity clause (Maslaton and Zschiegner 2009). In any case, one will at least be able to deduce the proportionality from the constitutional requirement that in cases in which the legislator motivates the norm addressees to invest over longer periods through the granting of certain financial incentives, which represents a control measure in this case, these granted advantages should not be withdrawn.[26]

There is basically no fidelity clause for offences based solely in the future. The norm addressee cannot rely regularly on the continued existence of a system favorable to him. This concerns in particular the question of whether the potential installation operator can rely on the fidelity clause as long as the investment project has not yet been realized. In principle, this is to be rejected. Therefore, if the legal situation changes to the detriment of the feeder before his installation is connected

[22] Federal Constitutional Court Decisions 13, 261 (272); 45, 142 (173).

[23] Federal Constitutional Court Decisions 57, 361 (391); 68, 287 (306); 72, 175 (196).

[24] Federal Constitutional Court Decisions 51, 356 (362); 69, 272 (309); 72, 141 (154).

[25] Federal Constitutional Court Decisions 63, 152 (175); 72, 141 (154).

[26] From the authors' perspective, it can be doubted whether the reduction of feed-in tariff via the first amendment EEG, FLG I. 2010 I, p. 1170 et seq. is enough.

to the grid system, then he is completely bound in principle to the new legal situation.

17.4 Summary and Outlook

In summary, it should be noted that the German legislator has, through the Renewable Energy Resource Act, created important legal framework conditions to enable the desired energy policy change. The rapid increase of installed capacity from renewable energies in Germany is both a consequence and an impressive testament to the functionality of this legal instrument. In addition, the 'German way' which renounces unilateral subsidisation of renewable energies and intervenes in a regulatory manner in the electricity market, enables renewable energies to be able to compete in the market, and is particularly useful in terms of longevity. For through the various control mechanisms such as the assorted bonus payments and the digression of the minimum remuneration, the continuous development of technologies for alternative energy sources will be actively managed and promoted. In so doing, it will prevent the EEG from being a 'comfortable pillow' for installation operators and developers. The enormous practical benefit of this concept is proven by the fact that Germany is currently one of the world's leading nations with regard to the development and manufacture of technologies and installations for the generation of energy from renewable energy sources.

However, if the energy change is to be successfully completed, the path chosen by the EEG should be consistently pursued. In addition, it is necessary that the legislator in the long term holds to its basic decision taken in the EEG and thus creates the actual conditions for more extensive investments in the renewable energy sector in Germany. This admonition to a proactive, consistent and predictable energy policy applies particularly in light of the 'Zigzag course' of the current legislative period, which is barely presented convincingly as a major policy change and which has led to a noticeable change in the development of renewable energies in Germany. The legislator must once more clearly bear in mind that the energy policy change will be supported to a considerable part by small and micro enterprises with tremendous economic investment volumes—which to a certain degree threaten their very existence. These entities are already dependent on a long-term and above all predictable development of legislation in the energy sector due particularly to the extent of their limited economic resources.

Biography André Zschiegner studied law at the University of Leipzig. In the last 3 years he worked as a research assistant at the Department of Civil and Public Law with references to environmental and European Law at the Brandenburg University of Technology Cottbus-Senftenberg.
In particularly, he is concerned with questions of law of renewable energies.

Emmanuel Wanki A. is an academic Staff and at the same time a PhD researcher at the Department of Civil and Public Law with References to the Law of Europe and the Environment, Brandenburg University of Technology Cottbus-Senftenberg, Germany.

References

Ellenberger J (2014) Commentary on the Art. 104. In: Palandt O (ed) Bürgerliches Gesetzbuch – Kommentar (German Civil Code). C.H. Beck, Munich

Federal Ministry for the Environment, Nature Conservation and Nuclear Safety (2009) The explanation to the EEG 2009. Federal Ministry for the Environment, Nature Conservation and Nuclear Safety, Berlin

Jarass HD, Pieroth B (2009) Basic law for the federal republic of Germany: GG. C. H. Beck, Munich

Maslaton M, Zschiegner A (2009) Handbook of the law of photovoltaics. Verlag Für Alternatives Energierecht, Leipzig

Salje P (2008) EEG 2008: renewable energy sources act, 2nd edn. Carl Heymanns, Cologne

Chapter 18
Climate Change Effects on Agriculture and Water Resources Availability in Syria

Bachar Ibrahim

18.1 Introduction

The topic of climate change is nowadays announced as one of the most important subjects and challenges at the global level. Climate change is expected to have many strong effects on the global environment. Many studies show that climate change has caused a decrease in the agricultural production and in water storage in the Middle East region. For example, in the years 2007–2008, drought took place in Syria that led to a decrease in wheat production by 78.9 % in most of its areas and it reached about zero in rain-fed areas, while in the irrigated areas the average yield dropped by 31 % due to the lack of water needed for irrigation (UNDP 2008). There has been a decrease in groundwater because of a shorter period of the recharge season and the drop in water retention as of snow (Döll and Flörke 2005). Such a decrease of agricultural products and water shortage would definitely impede economic growth and create difficulties against achieving sustainable development in the countries of the region in general (Droubi 2009).

The recurrent drought led to a reduction of the available water supplies and to negative impacts on the quality of water, and consequently to aggravating problems related to water resources management in the country. At present, most Syrian cities are suffering from a water supply shortage.

A preliminary assessment has revealed nationwide changes in rainfall patterns and fluctuations in temperature during the past five decades. Over the past years, average annual rainfall has fallen dramatically in the main agricultural areas. As a result, the country has suffered from a lack of rain and the effects of drought in the long term (National Communication 2009).

Climate change will have negative impacts on the land use patterns, accelerate the pace of land degradation, and increase the risks of drought. In the Eastern region

B. Ibrahim (✉)
Department of Rural Engineering, Damascus University, Damascus, Syria
e-mail: bibrahim@scs-net.org

E. Albrecht et al. (eds.), *Implementing Adaptation Strategies by Legal, Economic and Planning Instruments on Climate Change*, Environmental Protection in the European Union 4, DOI 10.1007/978-3-540-77614-7_18, © Springer-Verlag Berlin Heidelberg 2014

of Syria, heat waves and dust storms have become a nightmare to the residents, as they are increasing. The rise of sea level is expected to overflow the lowlands on the Syrian coast (Meslmani and Faour 2009). Moreover, climate change will have impacts on health services and on other economic sectors and natural ecosystems. Due to the above mentioned facts and based on prophylactic principles, Syria has been concerned about issues related to climate change and how to deal with its causes and consequences within the framework of international activities. Syria had participated in relevant negotiations under the United Nations Framework Convention on Climate Change and the Kyoto Protocol (Droubi 2009).

18.2 Overview

The Syrian Arab Republic lies in Western Asia, on the Eastern shores of the Mediterranean Sea. Syria is bordered in the North by Turkey, in the East by Iraq, in the South by Jordan, and in the West by Lebanon and the Mediterranean Sea. Geographically, it is located in the middle of arid and semi-arid areas. Its climate differs among the coastal region, the desert in the middle and the forestry areas in the North and Northwestern regions of the country. The surface area of the country is about 185.518 km^2, out of which 32.2 % is arable land (six million ha), and 45 % steppe and pasture lands. The country is divided into 14 governorates. Figure 18.1 shows the distribution of those governorates in the county.

18.2.1 Soil

Soil types and soil structure have a direct impact on agriculture. For example, the newly-formed basaltic rocks in Hawran plain south of the country, where the area is overlain by this type of nude superficial geology, reduce the area of agricultural land. The Gypsum rocks overlie soil in the Jazira region and in the region of the Al-Bishri Mountains. The sand overlies its surface that is subject to wind erosion and the formation of drifting sands (Droubi 2009). Figure 18.2 shows the different types of soil in Syria (Tavernier et al. 1980).

18.2.2 Climate

The climate in Syria is Mediterranean. There are four different seasons i.e. winter which is cold and wet, summer which is warm and dry, and spring as well as autumn. The precipitation range is between 100 and 1,400 mm/year. Syria is divided into five different climatic zones: the very arid, arid, humid, semi-arid and semi-humid as shown in Fig. 18.3.

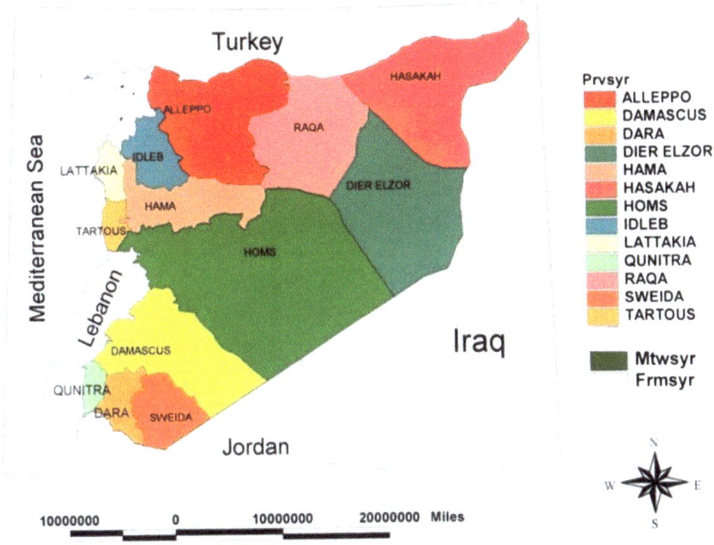

Fig. 18.1 The governorates of Syria. Source FAO (2004)

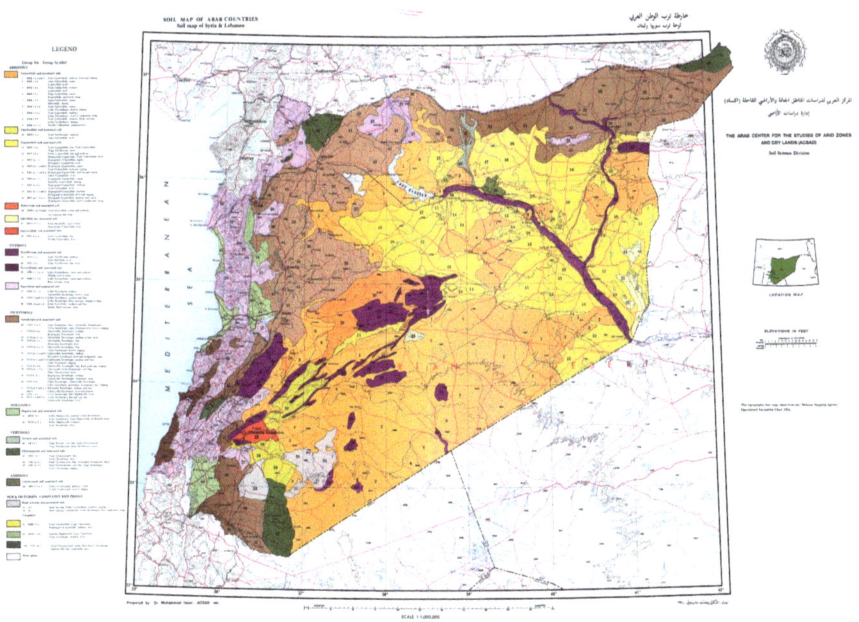

Fig. 18.2 Soil taxonomy and the soil map of Syria and Lebanon. Source the Arab Center for the Studies of Arid Zones and Dry Lands (ACSAD)

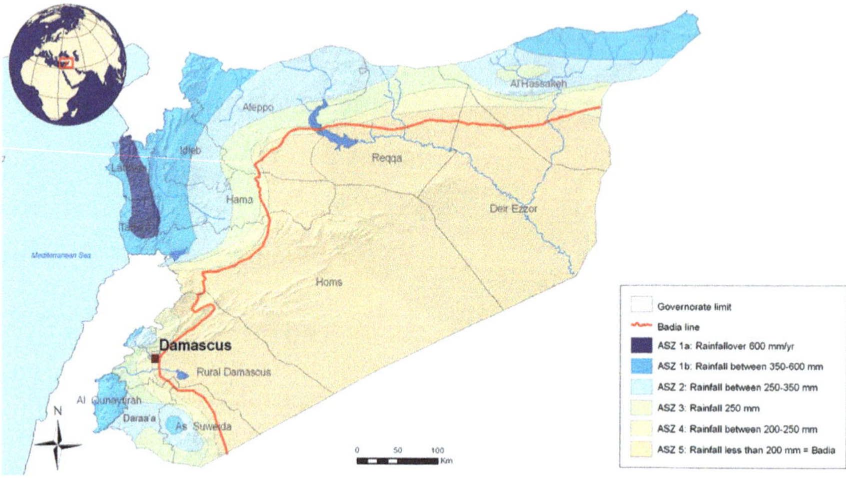

Fig. 18.3 The different climatic zones in Syria. http://www.ifad.org/pub/pn/syria.pdf. Source IFAD (2010)

Table 18.1 The different cultivated areas and plants in the several stabilization zones in Syria (Drubi 2009)

Stabilization zones	Annual rainfall	Area (ha)	Cultivated plant
1	Over 600	2,701,000	Field crops
1.1	Over 350		
1.2	350–600		Wheat, legumes and summer crops
2	250–350	2,475,000	Barely, wheat, legumes and summer crops
3	250	1,303,000	Barely legumes
4	200–250	1,830,000	Barely
5	Less than 200	10,209,000	Prohibited cropping

Syria is divided in terms of its agriculture into different stabilization zones related to different crops cultivation areas, as illustrated in Table 18.1.

18.3 Climate Change Global Impacts in the Mediterranean Region

The Middle East region suffers from a big problem with water-stresses and water scarcity which may be affected by climate change. The effects of climate change on the Eastern Mediterranean region was analyzed by Alpert et al. (2008) and found the following:

1. The average temperature over the whole Mediterranean area has increased by 1.5–4 °C in the last 100 years,

2. The precipitation has had a negative trend in the last 50 years,
   Weiß et al. (2007) expected that the drought events in the Middle East may increase ten times in the next 100 years.

The climate change impact on precipitation and temperature was evaluated within the framework of the national communication 2009 which indicated an increase in summer temperature in all climate stations in the country and an increase in the yearly precipitation amount in coastal and Western regions. There have been changes in both winter and autumn season, and precipitation in winter in the northern and north eastern zones of Syria showed some decrease in the last five decades.

18.3.1 Climate Change Impacts on Water Resources in Syria

The total available water resources of the country are estimated at around 17,454 mm^3/year including the treated waste water and the drainage water from agriculture, 10,959 mm^3 of which is surface water and 6,044 mm^3 of which is groundwater. The total demand is about 17,828 mm^3/year water. Table 18.2 illustrates the evaluation of the average annual water balance of the seven different Syrian water basins as calculated by the Ministry of Irrigation for the year 2003.

Table 18.2 shows that most of the water basins in Syria are suffering from water shortage. According to Döll and Flörke (2005), the change of groundwater recharge between 1961 and 1990 and the 2050s (2041–2070) could be more than minus 30 %. The overall water deficit in the country is growing. In the years 2001–2002, the average value deficit was 16 % more than in the years 1992–2000 (Rabboh 2007).

The climate change projection for the upper Euphrates and Tigris watershed area shows that major reduction changes in snow water equivalent may occur in the stream flow for these two rivers. The reduction may reach up to 100 mm in snow water equivalent (Onol and Semazzi 2006). The model-derived climate sensitivity of the Euphrates River discharge shows that a 25 % increase or decrease in precipitation rises or lowers the discharge profile of the river while keeping its hydrograph shape unchanged (Smith et al. 2000). This prediction means that the annual discharge rises to 40,655 mm^3 or drops to 15,751 mm^3 compared to the reference value of 27,048 mm^3. This is a 50 % rise and a 42 % drop respectively, nearly twice the imposed percentage change in precipitation. Other studies predict around a 10–25 % reduction in river runoff in the upper Euphrates and Tigris basin in 2070 compared to the average flow of the year 2000 (Lehner et al. 2001).

Table 18.2 Water balance of an average year of water basins in Syria

Water balance components	Unit	Khabour and Tigris	Euphrates and Aleppo	Orontes	Coastal basin	Barada and Aawaj	Yarmouk	Badia basin	Total
Basin area in Syria	km^2	21,129	51,238	21,624	5,049	8,630	6,724	70,786	185,180
Basin area/area of Syria	%	11.7	28.3	10.1	2.8	4.8	3.2	39.1	100
Surface water	Million m^3	788	7,105	1110	1,557	12	180	163	10,915
Ground water	Million m^3	1,600	771	1,607	778	838	267	183	6,044
Surface and ground water	Million m^3	2,388	7,876	2,717	2,335	850	447	346	16,959
Degree of organization	Million m^3	95	95	85	65	90	85	60	575
Available water resources	Million m^3	2,269	7,482	2,310	1,518	765	380	208	14,932
Recycling of treated drainage water	Million m^3	95	306	325	0	260	72	35	1,120
Refunded agricultural drainage	Million m^3	430	575	210	57	89	32	0	1,402
Total water resources	Million m^3	2,794	8,363	2,872	1,575	1,123	484	243	17,454
Demand on irrigation water	Million m^3	4,300	5,755	2,093	566	983	400	68	14,165
Demand on drinking water	Million m^3	38	322	240	124	270	89	44	1,127
Demand on water for industry	Million m^3	89	86	229	85	76	7	2	574
Evaporation	Million m^3	132	1,614	148	16	6	31	15	1,962
Total demand	Million m^3	4,559	7,777	2,710	791	1,335	527	129	17,828
water balance	Million m^3	1,765	586	162	784	−212	−43	114	−374

http://environ.chemeng.ntua.gr/ineco/Default.aspx?t=318

18.3.2 Climate Change Impacts on the Agricultural Sector

Agriculture is one of the most important economic sectors in Syria and provides nearly 25 % of gross domestic product (GDP). Agricultural activity in Syria is considered as the most important production activity (Masayuki et al. 2012). Agricultural exports have a big share in foreign trading and contribute to bringing in foreign currencies. Furthermore, the agricultural sector affords a lot of raw materials for other economical and industrial sectors. It also contributes to employing manpower and serves as a source of livelihood for a significant portion of the population. The agricultural sector occupies the second rank after oil in export revenue in the balance of Syrian export (Statistical Abstract 2004).

The two most important crops in the country are wheat and cotton. Wheat is considered the most important and strategic crop in Syrian agriculture (NAPC 2002). It occupies 70 % of the irrigated land devoted to strategic crops (National 2009). The climate change impact on the agriculture, as before mentioned, causes reduction of the agricultural productivity. More detailed facts on this can be found in the results of the study conducted within the framework of the national communication of Syria submitted to the UNFCCC in 2009. This was done using the CROPWAT model which is an irrigation management model developed by the FAO Land and Water Management Division (Smith 1992). It was used in evaluating crop water requirements and irrigation needs. It was also utilized to assess the effect of climate change on wheat (irrigated and rainfed) yield and on cotton yield in Hassakeh governorate.

The results showed that wheat requirements from water are 563 mm annually. However, actual crop water use is estimated by 402 mm. Consequently, the difference between actual and potential evapo-transpiration is 161 mm. This indicates that wheat grows under water stress under the current crop management system and the diminution in yields due to this stress is about 30 %. Cotton water requirements are 1,287 mm annually and the actual crop water use is estimated by 1,169 mm annually and the crop production under this condition of water shortage will be reduced by 7.5 %.

18.4 Adaptation Measurers to Climate Change in Agricultural and Water Sector

The conducted adaptation measurement to climate change effects in the agricultural sector were: (1) increase the irrigation water use efficiently through conducting a national program for promoting modern irrigation approaches and convince farmers to change from inefficient traditional irrigation systems to advanced efficient modern systems. For this goal, a fund programmed with 52 billion pounds[1] was

[1] 52.000.000 SYP = 338.000 € (2009: 1 SYP = 0,0065 €).

established for financing farmers. (2) Developing the capacity of the department of extension services by conducting advanced training program in the field of agriculture and water resources management. (3) New water policies have been adopted by the ministry of irrigation based mainly on water demand management rather than on water supply management, reducing water loss and reallocating water to crops with higher economic value.

18.5 Conclusion and Suggestions

In order to address the consequences of climate change on the agricultural sector, it is not adequate to handle the problem of climate change only by trying to solve the problem of water shortage and how to save water efficiently. Actually the following measures should be taken: (1) Surface water resources should be used in an optimal way, (2) improvement of ground water management, (3) imposing control on the use of ground water resources, (4) protection of environment from pollution, (5) treatment of sewage water, (6) determining and growing plant species that can bear drought and, (7) increase rain effectiveness through conservation farming and water harvesting.

Biography Professor at University of Damascus, Faculty of Agriculture Department of Rural Engineering. Department Head-Rural Engineering, responsible for teaching and research. Licensed Syrian EIA expert and listed EIA trainer. Email: bibrahim@scs-net.org, Damascus-Syria.

References

Alpert P, Krichak KO, Shafir H, Haim D, Osetinsky I (2008) Climatic trends to extremes employing regional modeling and statistical interpretation over the E. Mediterranean. Global and planetary change, vol 63. Elsevier, Amsterdam, pp 163–170

Döll P, Flörke M (2005) Global-scale estimating of diffuse groundwater recharge. Frankfurt hydrology paper 03. Institute of Physical Geography, Frankfurt University

Droubi A (2009) Climate change, water and the policy-making process in the Levant and North Africa, closed workshop with leading water experts from the Issam Fares Institute for Public Policy and International Affairs American University of Beirut

FAO (2004) Map 111: Administrative map of Syria: http://www.google.de/imgres?hl=de&sa=X&tbo=d&biw=1472&bih=664&tbm=isch&tbnid=fHBOFUIpkEbhCM:&imgrefurl=http://www.apipnm.org/swlwpnr/reports/y_nr/z_sy/symp111.htm&docid=33D91RaOpjeTHM&imgurl=http://www.apipnm.org/swlwpnr/reports/y_nr/z_sy/sy_map/symp111.jpg&w=724&h=565&ei=lr4YUbLYGZDLswa20oGIDw&zoom=1&iact=hc&vpx=863&vpy=127&dur=1021&hovh=198&hovw=254&tx=130&ty=89&sig=110028906626043686380&page=1&tbnh=144&tbnw=190&start=0&ndsp=33&ved=1t:429,r:5,s:0,i:94. Accessed 11 Nov 2013

IFAD—International Fund for Agricultural Development (2010) Syrian Arab republic: thematic study on land reclamation through de-rocking

Lehner B, Henrichs T, Döll P, Alcamo J (2001) Euro Wasser—model-based assessment of European water resources and hydrology in the face of global change, vol 5, World Water Series. Center for Environmental Systems Research, University of Kassel, Kassel

Masayuki T, Hideyuki K, Yuka A, Ibrahim B, Arslan A (2012) The 2nd final draft_120319_Joint Terminal Evaluation Report of the Project Promoting Modern Irrigation in Syria

Meslmani Y, Faour G (2009) Syrian sea level rise vulnerability assessment 2000-2100 (GIS). (INC-SY_V&A_Syrian sea level rise). United Nation Development Program (UNDP)/GCEA, Damascus, Syria

NAPC (National Agricultural Policy Center) (2002) Syrian agricultural trade. Ministry of Agriculture and Agrarian Reform, Syria

National Communication (2009) Enabling activities for preparation of Syria's Initial National Communication to UNCCCF

Onol B, Semazzi F (2006) Regional impacts on climate change on water resources over Eastern Mediterranean: Euphrates-Tigris basin. In: Proceedings of 18th conference on climate variability and change, 86th AMS meeting USA

Rabboh A (2007) Water demand management in Syria, 3rd Regional Workshop on: water and sustainable development in the Mediterranean water demand management, progress and policies. Blue Plan UNEP/MAP, Zaragoza, Spain

Smith M (1992) (CROPWAT), A computer program for irrigation planning and management. Irrigation and Drainage Paper 46, FAO, Rome, Italy

Smith RB, Foster J, Kouchoukos N, Gluhosky PA, Young R, Pauw ED (2000) Spatial analysis of climate, landscape, and hydrology in the middle east: modeling and remote sensing. Center for earth observation report No.2. Yale University, New Haven, USA

Statistical Abstract (2004) The annual agricultural statistical abstract. Ministry of Agriculture and agrarian Reform, Syrian Arab Republic

Tavernier R, Osman A, Ilaiwi M (1980) Usda soil taxonomy and the soil map of Syria and Lebanon. Soil science division, ACSAD/SS/ 14/80. (map1: 1,000,000) ACSAD: The Arab Center for the Studies of Arid Zones and Dry Lands, Damascus

UNDP (2008) Syria drought appeal, 36

Weiß M, Flörke M, Menzel L, Alcamo J (2007) Model based scenarios of Mediterranean droughts. Adv Geosci 12:145–151

Chapter 19
Crowdsourcing and Climate Change: Applications of Collaborative Information Systems for Monitoring and Response

Hendrik Send, Anna Riedel, and Anna Hansch

19.1 Introduction

In this paper we explore how use can be made of collaborative information systems for monitoring and coordination in the context of climate change and its effects. We discuss how citizens participate in such initiatives and how participation could be increased.

Researchers in the field of climate change adaption and humanitarian action call for support in coordination and collaboration through systems that enable accurate and real-time information. In the last decade a new type of information system has emerged that may be defined as social media. Instances of social media systems have been shown to successfully enable the coordination of large amounts of information in personal, enterprise, non-governmental and governmental settings. We then introduce three case examples of information systems that are based on the social media paradigm to collect data in the environment. Each of the cases deals with different stages of climate change from monitoring to response. Also each of the cases depicts a solution to challenges of social media systems. We discuss these as tentative applications of social media to the field of climate change adaption. Hereafter we look into reasons for the difficulties to actuate user contributions and derive measures to increase participation in the field of climate change adaption from research in open innovation.

H. Send (✉) • A. Hansch
Humboldt Institut für Internet und Gesellschaft, Bebelplatz 1, 10099 Berlin, Germany
e-mail: hendrik.send@hiig.de

A. Riedel
Institute of Electronic Business, Hardenbergstraße 19, 10623 Berlin

E. Albrecht et al. (eds.), *Implementing Adaptation Strategies by Legal, Economic and Planning Instruments on Climate Change*, Environmental Protection in the European Union 4, DOI 10.1007/978-3-540-77614-7_19, © Springer-Verlag Berlin Heidelberg 2014

19.2 Challenges of Climate Change

Emerging results point out that acute (i.e. severe weather events) as well as chronic climate effects (i.e. long-term changes in the environment) can have extreme effects that strike in particular those countries that are already suffering of poverty and marginalization (Solomon et al. 2007).

Governments over the world are therefore confronted with the challenges of climate change. Especially in developing countries the infrastructure does not always support an efficient information flow, which is necessary to provide the institutions in charge with essential data in order to survey the situation, collaborate with other institutions and, at the best, to act anticipatory.

Other than that, governments face the following five challenges, identified by Perlmutter and Rothstein (Perlmutter and Rothstein 2010: 9):

1. Powerful and rich interest groups have been and still are an obstacle to major reform of the existing energy economy.
2. There are strong divergences of interests between the developed countries and those of the developing world, who distrust any notion of moving away from a development strategy based on rapid industrialization and intensive use of conventional sources of energy.
3. The potential costs are vast and unappealing in a period of economic turmoil and fear.
4. Public opinion is not (yet?) a powerful voice for action and sacrifice.
5. No political system is good at taking long-run needs seriously, that is, paying heavily now to ensure future benefits or to avoid dangers that may not ever develop. The easier option is to 'muddle through', to hope that something will turn up, or someone else will bear the brunt of whatever transpires.

On one hand, effective national and international policy-making is more and more dependent on decentralized information, while on the other hand citizens begin to form an oppositional public sphere, which speaks up for topics such as sustainability and environmental protection.

The perception of ICTs' role in climate change is twofold in the contemporary discussion of climate change: They contribute to the increase in energy and resource consumption through direct and indirect effects, while at the same time being seen as 'enabling tools' because of their ability to augment the effectiveness and efficiency of development programs (Kalas and Finlay 2009) if well integrated.

At the same time researchers in the field of humanitarian action call for systems that enable accurate and real-time information to be gathered and managed efficiently (Greenough et al. 2009: 206). These tools may be used in three different phases of climate adaptation: foresight, monitoring and response.

19.3 Crowdsourcing: Social Media and Participation

The term social media describes a set of methods for web-based collaboration that features highly accessible and scalable mechanisms. The key difference to classical publishing is that social media platforms usually allow all users that access content on the platform—or a large group of users—to contribute. This has been a paradigmatic shift for information management on the internet because before, websites followed the distinction of classical mass media between a small group of professional editors or writers and a large group or readers or recipients. The concept of social media has propagated to personal, enterprise and governmental and non-governmental applications.

Recently, this phenomenon has attracted great interest. First, because group solutions to problems in different areas of application have been shown to often be superior to those made by individuals (Surowiecki 2004). Second, the advent of the internet has made the forming and coordination of groups and the communication within groups 'ridiculously easy' (Shirky 2008). While communication on the internet will not increase to quality of solutions it does greatly increase the speed of information exchange and problem-solving in distributed groups. Given this potential, it is of great interest how distributed communication can be leveraged in different areas of application (Hoffmann 2009).

Currently problems of very different characteristics are being approached by this method. As for Kazman and Chen, crowdsourcing is a way to tap into the 'productivity, wisdom and creativity' of a distributed workforce (Kazman and Chen 2009).

The research of knowledge sharing and contribution has been intensified after two subsequent studies proving the social media encyclopedia Wikipedia to be more exact and detailed than the comprehensive Encyclopedia Britannica and the German Brockhaus Enzyklopädie (Giles 2005). Because social media seems to spur contribution by users and facilitate the collection of knowledge and creation of innovations, it has been quickly adopted by enterprises throughout the world.

As the complexity of products, technologies and phenomena increases, it becomes less and less probable for the individual to solve problems on its own. On a larger scale enterprises have experienced the same. The strategic change of Procter and Gamble's innovation strategy from a closed shop model to the 'connect and develop' strategy in the year 2000 has been the starting point of an industrial development to 'tap the global brain' (Nambisan and Sawhney 2008).

19.3.1 Challenges of Crowdsourcing

While crowdsourcing offers new intriguing ways to approach distributed communication and problem-solving, it is prone to several challenges.

As more and more enterprises discover the advantages of open innovation initiatives the users and their ideas are beginning to also become a resource that has to be gained and well attended to. Researchers in the field of innovation management and online communities are looking into the incentives for user activities in online collaboration.

The validation and evaluation of content that stems from crowdsourcing processes remains one of the challenges in the field. Several approaches have proven workable in different areas. For instance, the crowdsourcing encyclopedia Wikipedia features a fine-grained system of different levels of rights. Users partly achieve more rights in a meritocratious system: the more contributions they have made the more rights they earn. Also for higher status-levels elections are held to probe the 'social compatibility' of contestants within the user crowd. In the case of Ushahidi, a crowdsourcing platform to document disasters and direct help, users can freely contribute information about disasters. Members of NGOs, who are considered reliable information sources, are then asked to visit the spot and verify the report. Also commercial firms offer to verify user-generated content by aggregating and comparing answers from different users so that a higher level of reliability can be achieved.

The amateur plays a great role in crowdsourcing and is often seen as the main contributor in crowdsourcing processes (Hoffmann 2009). Nevertheless, there are also platforms with mainly professional users like the innovation crowdsourcing platform InnoCentive. In a study in 2007 over 65 % of the InnoCentive community members held a Ph.D. (Brabham 2008). Also practitioners are looking into the possibilities of crowdsourcing within organizational settings as in enterprises. Here the crowd consists of knowledge workers who are given new tools like blogs and wikis to improve information transfer, innovation and communication (McAfee 2006).

Incentives and motivation play a great role in crowdsourcing projects because they depend on attracting and motivating a crowd of voluntary problem solvers. While remuneration is considered to be the most important currency in a classical working condition, non-monetary aspects like the love, glory and fun are very important (Malone et al. 2010). Researchers take a lot of interest in understanding the processes and motivations of open-source software programmers who have brought up very competitive programs and distributed them freely. When dealing with rather mechanical tasks that cannot offer intrinsic motivation a monetary remuneration is necessary. In this case Hoffmann (2009) argues that it is enough to offer a few cents for very simple problems while more complex problems like the classification of data will also require better funding.

Crowdsourcing depends on many competencies and resources that are often not found in organizations and enterprises (Diener and Piller 2009) and can only be achieved by an intense organizational learning phase. An important issue is how to define the tasks that are being given to the crowd. If the task is to complex or important information is missing, the solutions will not be satisfactory. However, if only a small part of a problem is given to the public, they can only work on this small part and therefore not contribute to a comprehensive solution (Keupp and Gassmann 2009). Currently many intermediaries facilitate the process of crowdsourcing and help organizations to take the necessary prerequisites.

An ongoing controversy is about whether the remuneration is fair or what amount would be fair for which task. The users of *Amazon's Mechanical Turk* work for very low but fixed remuneration. But many crowdsourcing platforms are based on a competitive winner-takes-all model. In the end, only the contributors of the winning solutions will receive price money while sometimes over 1,000 ideas have been posted. As a result, a discussion about 'speculative work' has arisen (Howe 2009).

19.3.2 Crowdsourcing in the Context of Climate Change

We present three example cases that depict how crowdsourcing can be used to collaborate effectively in the often distributed structures of volunteers, stakeholders and affected persons. Each of the cases shows how the active groups leveraged the possibilities of crowdsourcing while tackling its challenges at the same time. Notably in the San Diego Wildfire case not one single organization supported the interaction—it rather emerged on different platforms—while in the BirdWatch 2010 and Ushahidi project one organization facilitates all action.

19.4 Example Case: Collective Bird Monitoring as Data Set for the Biodiversity Indicator

Climate change threatens biodiversity and the richness in species. The year 2010 was declared the International Year of Biodiversity by the United Nations, in order to increase awareness of the rapidly growing extinction of species (Rands et al. 2010). According to the International Union for Conservation of Nature (IUCN) the loss of biodiversity is currently 1,000–10,000 times higher than it would be without human influence. Therefore, almost one third of the species is threatened by extinction (International Union for Conservation of Nature 2010). However, the variety of living organisms and ecological systems is the basis for human life and the future of the next generations.

19.4.1 Biodiversity Indicator in Germany

In Germany, the sustainable development indicator for biodiversity provides information and foresight about the condition of the environment. The calculation of this indicator is based on data about the development of species of birds. As many species of birds are bound to intact biotopes the indicator also represents the development of other species. Therefore, the national collective action bird monitoring is the basis for creating public awareness for conservation. Annually, about 5,000 volunteers work more than 200,000 h watching birds (Dachverband Deutscher Avifaunisten 2010). Their data is collected in reams of regional bird and nature conservation associations. The NGO Naturschutzbund Deutschland e. V. (NABU) is one of the biggest of those associations. Besides the political work for conservation, NABU wants to inspire people to commit themselves to environment protection through collaborative activities. Nationwide there are different projects combined with local and national cross-channel marketing campaigns to catch public attention.

19.4.2 BirdWatch 2010 Campaign

Gathering data of bird monitoring is a collective amateur action. Bird watching itself is a local activity and a single person or a smaller group can only observe birds within a limited space. But by putting together all the information, a valid report about the population of a species can be produced. In order to merge the contributed outcome of the bird monitoring groups, new information technology tools are used by NABU. Like in web 2.0 information is created by the users and shared via digital communication tools.

Every year on the second and third of October the *(European) BirdWatch* takes place. According to NABU it is the biggest bird-monitoring event worldwide. In 2010 volunteers in 28 European countries gathered for two days to monitor migratory birds. Without the collective power of all observers in the peak of bird migration there wouldn't be the chance to get any valid monitoring results. The outcome is collected and shared in an online database. In Germany, birds were monitored in 120 places and almost 200 species were seen (NABU 2010).

A similar action was launched in Mai 2011 in Germany. From May 13–15 the new campaign *Stunde der Gartenvögel* (hour of the garden birds) was launched. Volunteers were asked to monitor the garden birds in their neighborhood to get local data about the population of native birds to improve their habitat conservation (Naturschutz 2011). The monitoring results could be registered in an online database via mobile application. *Stunde der Gartenvögel* was accompanied by social media coverage. NABU shared through its Twitter channel the monitoring experience and important findings about the local populations. In this participatory action more than 27,000 amateurs registered almost one million garden birds (NABU 2011).

To improve the validity of the foresight data, the merging of all national information via ICTs' is initiated by several cooperating ornithological professional associations. Their aim is to set up the Internet portal ornitho.de to build up regional capacity for editing the contributed monitoring data. The network was established in 2011. They intent to make the results available for all interested volunteers who improve the quality of the data by correcting wrongly identified birds. Thus a nationwide database for bird monitoring is created and so the collective powers of bird experts and volunteers can be used to create a comprehensive overview of the bird population in Germany (Wahl 2010).

By using the collective power of crowdsourcing for bird monitoring a very important contribution for nature conservation can be made. The data about the local bird population is elementary for the regional condition of the species and thus the environmental protection for biotopes can be strengthened. Moreover, the local data collected by volunteers is crucial for the compilation of the sustainable development indicator for biodiversity. Consequently, governmental conservation and the struggle against climate change are based on regional amateur contribution. Through new digital media tools like mobile applications the collaboration of the volunteers and the collection of data by the associations is getting much easier. Social media gives the birdwatchers the chance to network and share their experiences. By expanding the use of social interaction tools, cooperation with associations can be improved and the community of the birdwatchers can be reinforced.

19.5 Example Case: Volunteer Movement in the San Diego Wildfires

San Diego is the second-largest city in California and is located in the southwest of the federal state. In 2007, a terrible series of wildfires occurred in San Diego County, starting on October 21. Due to extremely hot weather, drought and the strong Santa Ana winds the fires spread out on about 13 % of the county's total land area. Many people died or were injured, 1,600 homes were destroyed and hundreds of thousands houses were evacuated (Majchrzak and Birnbaum-More 2011).

Besides the disaster response there was a large volunteer movement that played an important role in monitoring and organizing the fight against the firestorm.

19.5.1 Information Supply by San Diego State University and KPBS Public Broadcasting

Characteristic for the voluntary monitoring service was the creation of knowledge with the help of quickly adapted web 2.0-technologies. During the crisis, the volunteers became involved in two different networks that supported each other in the course of the disaster.

19.5.2 San Diego State University

The first was a group of citizens participating in a project at San Diego State University. In a cooperation of the Geography Department and the Visualization Center satellite images and wind overlays were uploaded on Geographical Information System (GIS) maps in order to constantly inform the public about the ongoing fires and to give an idea about how the fire could develop. As this overview was very important for the people, the map was provided as a print version (also for each fire) to take away. After the campus was closed due to the fires, the group of volunteers coordinated and documented their activities in a wiki (Majchrzak and Birnbaum-More 2011).

19.5.3 Local Public Broadcasting Station KPBS

The second important volunteer group gathered around the local public broadcasting station KPBS. Besides KPBS, Google and later on the San Diego State University's Geography Department and reams of citizens supported the collective online monitoring of the wildfires. The small radio station put up quickly adapted common online tools like a Google MyMap linked from their website to show the extent of the fire, evacuation routes and shelters. Herein, coordination of activities was visualized and accessing as well as comprehending information was simplified. During the firestorms Google supported KPBS continually to keep up the fire map. Since the number of fires grew, KPBS used the twitter account @kpbsnews to extend the information flow (Poulson 2007). As a reaction to both information services people started to email their feedback. This collective power-sharing through crowdsourcing provided significant monitoring information about the wildfires. During the time of the fires 1.5 million people viewed KPBS MyMap and about 1,000 made use of the twitter feed (Majchrzak and Birnbaum-More 2011).

19.5.4 Accessibility of Information

The information that was made accessible by San Diego State University and KPBS was very important for the people living in San Diego County. The situation during the firestorms was changing rapidly and traditional media could not satisfy the need for highly topical local news. The volunteer workers recognized the potential of web 2.0 technologies for collaboration and coordination and adapted them for monitoring the situation. They started organizing themselves quickly and gathered available information about the development of the fires as well as of the official relief efforts. Providing easy access to the collective online knowledge creation tools, a contact point for the people was made available. People in the region were not mere recipients; they also corrected or reported incidents. Thus, based on web 2.0 tools, users both created and accessed content. The big challenge was to keep the technical infrastructure working despite the increasing traffic. Demands were also managed by power sharing with partners. Using the wisdom of the crowds, volunteers in San Diego could improve the local disaster response capacity and therefore supported the fire brigade and the rescue teams. The collective vision to monitor and fight the firestorms could be realized when the last fire was finally put out after almost three weeks.

19.6 Example Case: Coordination of Disaster Response Via Digital Media After the 2010 Haiti Earthquake

On January 12th 2010, Haiti was hit by a disastrous earthquake of magnitude 7.0 on the Richter scale. The epicenter was about 25 km away from the capital Port-au-Prince (U.S. Geological Survey 2010). It is estimated that up to 300,000 people were killed and hundreds of thousands left homeless. According to the Human Development Index, Haiti belongs to the least developed countries in the world and is the poorest country in the Americas (UNDP 2010). As the infrastructure was poor to begin with, the damage caused by the earthquake was devastating. The local infrastructure that would have been necessary to respond to the catastrophe almost completely broke down and so the Haitian people were not able to take initiative for disaster and crisis management. Thus, immediately after the earthquake the action of international relief organizations was initiated. Although the incident cannot be attributed to climate change the scenario serves as a useful example of coordination in an environment crisis.

19.6.1 From SMS Broadcasting to Crisis-Mapping Platforms

When the international community arrived in the massively destroyed country, efforts to coordinate the supporting measures had to be made. In order to localize and verify emergency cases, use was made of innovative information technology tools and applications. Within a few days, an ecosystem of traditional media and new digital media tools for collective online knowledge creation were established.

First of all, the reconstruction and integration of local radio stations was an important part of setting up and coordinating humanitarian response. In order to bridge the time until local radio stations could take up their work again, an information program was developed by the emergency responders. Returning to work, the radio stations played an important role for supporting the reconstruction works in the regional communities (Nelson and Sigal 2011).

19.6.2 Red Cross Short Message Service

The International Federation of the Red Cross used Short Message Service to text information about health issues. Furthermore, Mission 4636 was launched where citizens could use the short code 4636 to call search and rescue operations. But after a while the roughly 800,000 requests also covered other concerns like food, shelter and missing people. As the messages were written mainly in Haitian creole, local volunteers translated, categorized and passed on the inquiries to the relief organisations in a collective effort. Furthermore, the information was also geo-located and formed the database for mapping and verifying incidents on crowdsourcing platforms like Ushahidi.

Volunteer workers also initiated the launch of the crisis-mapping platform haiti.ushahidi.com. The published content was not only sent by Mission 4636, but also gathered from different social media sources like Facebook, Twitter or weblogs. The gained information about emergencies, security or needs like water or shelter was sorted out and arranged on online maps through participatory action. So the real-time data could be shared with the emergency responders and became the standard map used for coordinating their rescue work. The website was accessible both for aid organizations and the public, and it was also used by the US Marines and the US Cost Guard (Nelson and Sigal 2011). Apart from Ushahidi, there were other attempts using crowdsourced maps like the Humanitarian Open Street Map that pursued the target to provide the disaster responders help to find the way to the operational areas (ibid).

After the devastating earthquake in Haiti, international aid organizations directly started their operations. Trying to begin with the disaster response, a variety of organizations and rescue teams that arrived needed to get along in a destroyed infrastructure, so it was very difficult to set up their services. In this extreme situation new forms of collaboration arose, building on collective action and problem-solving through innovative information technology tools. Being based on SMS broadcasting and an immediate involvement of local media, initiatives making use of new digital media developed. Having a strong team of volunteers, data could be gathered, proved and made available on crowdsourced maps. Thereby, essential tools for the coordination of the search and rescue-teams were created. This was made possible through the enormous collaborative effort of the volunteers. Struggling with the poor infrastructure, other partners like telecommunication companies or the Tufts University Boston volunteered to provide technical support. Thus, 'Haiti became the first real-world crisis laboratory for several media platforms that had only recently emerged' (ibid). Facilitated by local media, Haitian citizens participated in the collective disaster relief and did build up a new regional capacity to provide first aid and prepare the reconstruction works of their homes.

19.7 Discussion

In this article we show that new forms of collaboration can make a valuable contribution to dealing with climate change. We only presented a highly eclectic set of three example cases to illustrate the possibilities of crowdsourcing with regards to climate change. Neither all challenges nor all benefits of the application have been discussed in this article; the aim was rather to present an exploratory introduction to the example cases together with the depiction of social media.

Ospina and Heeks (2010: 17) identify the areas of intersection between ICTs and climate change monitoring and development: data capture, processing, presentation and dissemination. We applied the presented example cases to these areas and provided an outlook.

Regarding data capture each of the presented case examples stands for a different aspect in which crowdsourcing might complement governmental or centralized methods. The BirdWatch database allows a heterogeneous group of regionally dispersed experts to contribute their information. The fire map during the San Diego wildfires provided by KPBS relied on very few sources of real-time data at the actual place of the incidents. And the Ushahidi platform for Haiti overcame the obstacles of a damaged infrastructure by making SMS and telephone calls a viable medium for citizens across the country to report incidents and emergencies along with the monitoring of other social media sources. These possibilities to connect distributed data sources can be applied to monitor many climate change-related phenomena on a long-term basis or in real-time. Examples of the phenomena might be the availability of water, water quality, or biodiversity.

Regarding processing and presentation of data, the BirdWatch campaign enables the monitoring of bird population represented by sighting numbers on the level of species and regions as well as in the form of maps. Furthermore, the value of the San Diego wildfire platform stems from the presentation of fire propagation data on maps that could be easily interpreted and taken into account for individual action. Also Ushahidi makes the data actionable through the aggregation of several data sources and the verification of raw reports by NGO employees who were directed to possible incidents.

Finally the dissemination of the aggregated data to the concerned audience is the third feature that makes crowdsourcing valuable. While the BirdWatch data is being used by experts and even government officials, the KPBS wildfire maps were helpful because they were directly accessible by any individual with a network connection who was directly or indirectly affected by the fires. As depicted in the Ushahidi case, even international rescue teams could act upon the information given on the platform. Therefore, crowdsourcing involves the possibility to reach directly affected individuals as well as experts or official decision-makers.

Sticking to the hypothetical example of water management issues, a crowdsourcing platform could make it possible for those affected to plan the usage of the resources, for experts or helpers to direct guidance and help to areas of need as well as for governmental representatives to base decisions upon acute and up-to-date data.

In general, more extreme weather events will lead to an increased need to monitor information and distribute it to those affected. The areas of application are probably as extensive as the effects of global warming. Be it the change of sea levels or the melting of ice covers, changes in rainfall patterns or water resources. Regarding life animals possible applications are biodiversity issues as well as diseases and animal migration patterns that can be investigated closer through means of crowdsourcing. Future research could provide a valuable contribution through a collection and interrelation of a comprehensive set of scenarios for crowdsourcing in the context of climate change, thereby establishing the success factors.

As the practices associated with crowdsourcing become more common and more users have learned how to contribute to distributed online communication forms, crowdsourcing might even become a regional competence that can help to monitor climate-related phenomena or deal with the consequences of climate change.

A challenging problem will remain that the emerging regions in the world still suffer from a 'digital divide' and do not offer the infrastructure, affordability and skills necessary to profit from ICT as much as advanced economies do (Dutta and Bilbao-Osorio 2012).

Biography Dr. Hendrik Send is project leader in the research area Internet-enabled Innovation at the Humboldt Institute for Internet and Society. Also, he is Professor at the Anhalt University of Applied Sciences and organizer of the master program Digital Communication.

Since fall 2012 Dr. Anna Riedel is Head of Research and Consultancy at iDeers Consulting GmbH, which was jointly founded by the Institute of Electronic Business (IEB) and the Hirschen Group. In the same year, Anna Riedel's PhD thesis was published. Previously, she worked as a research assistant and project manager at the IEB. Anna Riedel has acquired a degree in Business Administration (Dipl.-Kauffrau FH) in 2006 and a Master (M.A.) in European Culture and Economics. Until her appointment at the IEB she worked as an online Marketing Manager with an Internet start-up in Berlin.

Anna Hansch works as a project coordinator of the research project Innovation and Entrepreneurship at the Alexander von Humboldt Institute for Internet and Society (HIIG). At the HIIG and the Berlin University of the Arts she is writing her doctoral thesis. Anna Hansch studied Literary, Cultural and Media Studies (B.A.) at the University of Siegen and Applied Literary Studies (M.A.) at Freie Universität Berlin.

References

Brabham DC (2008) Moving the crowd at iStockphoto: the composition of the crowd and motivations for participation in a crowdsourcing application. First Monday 13(6). Retrieved 5 Dec 2011. doi:10.5210/fm.v13i6.2159

Dachverband Deutscher Avifaunisten e.V. (2010) Birds in Germany—2010. Retrieved December 12, 2011, from http://www.dda-web.de/downloads/texts/publications/statusreport2010_ebook.pdf

Diener K, Piller F (2009) The Market for Open Innovation Increasing the efficiency and effectiveness of the innovation process. Group, (c), 144. Retrieved from http://mass-customization.blogs.com/files/extract_the-market-of-open-innovation_2010-report.pdf

Dutta S, Bilbao-Osorio B (2012) The global information technology report 2012: living in a hyperconnected world. World Economic Forum. Retrieved 15 June 2012. http://www3.weforum.org/docs/Global_IT_Report_2012.pdf

Giles J (2005) Internet encyclopedias go head to head. Nature 438(7070):900–901

Greenough PG, Chan JL, Meier P, Bateman L, Dutta S (2009) Applied technologies in humanitarian assistance: report of the 2009 applied technology working group. Prehosp Disaster Med 24(2):206–209

Hoffmann L (2009) Crowd control. Commun ACM 52(3):16–17

Howe J (2009) Crowdsourcing: why the power of the crowd is driving the future of business. Three Rivers Press, New York, NY

International Union for Conservation of Nature (2010) About the biodiversity crisis. Retrieved January 16, 2012, from http://www.iucn.org/what/tpas/biodiversity/

Kalas PP, Finlay A (2009) Planting the knowledge seed: adapting to climate change using ICTs. Building communication opportunities (BCO) alliance. Online at http://www.bcoalliance.org/Climate-Change. Accessed 14 Dec 2011

Kazman R, Chen H (2009) The metropolis model a new logic for development of crowdsourced systems. Commun ACM 52(7):76–84

Keupp MM, Gassmann O (2009) Determinants and archetype users of open innovation. RandD Manag 39(4):331–341

Majchrzak A, Birnbaum-More P (2011) Emergency! Web 2.0 to the rescue! Commun ACM 54 (4):125

Malone TW, Laubacher R, Dellarocas C (2010) The collective intelligence genome. MIT Sloan Manag Rev 51(3):21–31

McAfee AP (2006) Enterprise 2.0: the dawn of emergent collaboration. Eng Manag Rev IEEE 34 (3):38–38

Mission 4636 (n.d.) Retrieved January 17, 2012, from http://www.mission4636.org/

NABU – Naturschutzbund Deutschland e.V. (2010) Birdwatch 2010—the great variety. Retrieved December 17, 2011, from http://www.nabu.de/aktionenundprojekte/birdwatch/ergebnisse/12865.html

NABU – Naturschutzbund Deutschland e.V. (2011) Nationwide, fewer blackbirds and chaffinches. Retrieved January 17, 2012, from http://www.nabu.de/aktionenundprojekte/stundedergarten voegel/ergebnisse/13801.html

Nambisan S, Sawhney MS (2008) The global brain: your roadmap for innovating faster and smarter in a networked world. Wharton School Publication, Upper Saddle River, NJ

Naturschutz Heute (2011) Hear and see. Retrieved December 12, 2011, from http://www.nabu.de/imperia/md/content/nabude/nh/nh211.pdf

Nelson A, Sigal I (2011) Media, information systems and communities: lessons from Haiti. Retrieved December 12, 2011, from: http://www.knightfoundation.org/media/uploads/publica tion_pdfs/KF_Haiti_Report_English.pdf

KPBS Online (2007) 2007 San Diego County fires. Retrieved January 16, 2012, from http://bit.ly/HORoGe

Ospina AV, Heeks R (2010) Unveiling the links between ICTs and climate change in developing countries: a scoping study. Retrieved June 12, 2012, from: http://www.infoandina.org/sites/default/files/recursos/a%20scoping%20study.pdf

Perlmutter DD, Rothstein RL (2010) The challenge of climate change: which way now? Wiley, Chichester. doi:10.1002/9781444328523

Poulson K (2007) Firsthand reports from California wildfires pour through twitter. Message posted to http://www.wired.com/threatlevel/2007/10/firsthand-repor/

Rands MRW, Adams WM, Bennun L, Butchart SHM, Clements A, Coomes D (2010) Biodiversity conservation: challenges beyond 2010. Science 329(5997):1298–1303

Shirky C (2008) Here comes everybody: the power of organizing without organizations. Penguin Press, London, p 86, 336

Solomon S, Qin D, Manning M, Chen Z, Marquis M, Averyt KB (2007) IPCC, 2007: summary for policymakers. Climate change 2007: the physical science basis. Contribution of Working Group I to the Fourth Assessment Report of the Intergovernmental Panel on Climate Change. Cambridge University Press, Cambridge

Surowiecki J (2004) The wisdom of crowds. Random House Inc, New York, NY, p 58, 640

U.S. Geological Survey (2010) Earthquake details. Retrieved January 18, 2012, from http://earthquake.usgs.gov/earthquakes/eqinthenews/2010/us2010rja6/

United Nations Development Programme/UNDP (2010) Human development report 2010. The real wealth of nations: pathways to human development. Retrieved January 18, 2012, from http://hdr.undp.org/en/media/HDR_2010_EN_Complete_reprint.pdf

Wahl J (2010) Ornitho.de – making great strides! Retrieved December 16, 2011, from http://www.dda-web.de/downloads/texts/publications/falke/falke_2010_12_ornitho.pdf

Chapter 20
Beside Adaptation: Concepts for the Future

Simon Spyra and Eike Albrecht

20.1 Introduction

Maybe it was God who in the beginning created heaven and earth (Gen 1,1), who created mankind (Gen 2,7), who committed nature to us (Gen 2,15). Whether it is a question of creed or not: Our atmosphere is unique. Its special consistence makes it possible for the temperature on the earth's surface to constantly stay at a certain level, a level at which our environment can unfold and the Garden of Eden can be preserved and protected. And if God has committed nature to mankind in equal shares, all humans bear responsibility for protecting and saving the Garden of Eden. And if not, the same is true by the principle of equity.

However, it is not given that everybody is able to make a contribution to the same extent and that someone who is capable of contributing (more) will necessarily do so. It is likewise uncertain whether we are capable of fulfilling the biblical mission at all, and even whether we are able to keep the extent of climate change at a tolerable level. What has been caused so far is irreversible.[1]

[1] Cf. The Intergovernmental Panel on Climate Change (IPCC) presented an overall special report on "Managing the Risks of Extreme Events and Disasters to Advance Climate Change Adaptation (SREX)". The 600 pages report in English is available online and can be downloaded at http://www.ipcc-wg2.gov/SREX. A printed version is available since May 2012 from the IPCC office at http://www.ipcc.ch. For further interest: the key statements of the SREX, which were comprised by the Ministry of Education and Science (BMBF), the Ministry of Environment, Nature Conservation and Nuclear Safety (BMU) in cooperation with the German IPCC coordinating group, are available at http://www.de-ipcc.de/ (as of 01.10.2012).

S. Spyra (✉) • E. Albrecht
Department of Civil and Public Law, Brandenburg University of Technology Cottbus-Senftenberg, Erich-Weinert-Str. 1, 03046 Cottbus, Brandenburg, Germany
e-mail: spyrasim@tu-cottbus.de; albrecht@tu-cottbus.de

E. Albrecht et al. (eds.), *Implementing Adaptation Strategies by Legal, Economic and Planning Instruments on Climate Change*, Environmental Protection in the European Union 4, DOI 10.1007/978-3-540-77614-7_20,

Adaptation seems to be the only possible way. It is a distinctive human quality to adapt to the climate, thus adapting to a changing climate can be considered typically human behaviour. The causes for climate change are multifactorial. The approaches to counter climate change are multifactorial: an adaptation through mitigation with the help of economic principles, the law of supply and demand. Several countries are convinced of the efficiency of this approach, other countries might follow so that a worldwide network of the existing systems can be considered. First attempts in this direction have already been made. In any case, adaptation alone is not enough. Mitigation is necessary, and—if combined with adaptation—could be part of the solution. But adaptation and mitigation are expensive and the finances needed for adaptation and mitigation will increase. Thus, economical instruments may help to direct the investments to those places and those activities with the largest effect at lowest costs.

This article presents a short overview of the initiatives and mechanisms that practice climate protection on the basis of economic principles. One of the major instruments in the climate change regulatory framework is emission trading; therefore main emphasis of the contribution is laid to this instrument. Further problem-solving approaches will also be depicted.

20.2 The European Emission Trading System (EU ETS)

At the moment, the EU ETS is the biggest market for emission rights worldwide. It comprises more than 90 % of the trading volume worldwide (Kossoy and Guigon 2012, 17).

Apart from the EU ETS, however, there are some other initiatives which are implementing the principle of the EU ETS. As a result, several national and regional CO_2-reducing initiatives and market mechanisms have emerged in industrial and developing countries since the introduction of the EU ETS. In Australia a law has recently been passed which establishes a country-wide trading system according to the cap-and-trade approach by the year 2015 and which is expected to cover roughly 60 % of the annual Australian greenhouse gas emissions (Australian Government 2012a).

According to recent statements, the European Union and Australia will combine their CO_2 emission trading systems (European Commission 2012). With the Emissions Trading Directive coming into force, the legal ways to do this are in principle already in place on the European side. Thus, Art. 25 allows the combination with other systems for the trading of greenhouse gas emissions (Art. 25 Directive 2003/87/EC).

Québec has also prepared its own cap-and-trade plan on the basis of an ordinance,[2] and as part of the Western Climate Initiative (WCI) a link to the Californian system is provided (WCI 2011).

[2] Details on this ordinance are available on the site of the Ministry of Environment at http://www.mddep.gouv.qc.ca/changements/carbone/Systeme-plafonnement-droits-GES-en.htm (as of: 12.11.2013).

In Mexico and South Korea, too, climate protection laws have been passed by parliament (Kossoy and Guigon 2012, 73). Particular attention should also be paid to China, whose number of regional cap-and-trade schemes could form the basis for a nationwide regulation in the coming years.

In the following, individual country initiatives are examined in more detail.

20.3 The Countries in Detail

In the following section, the activities of selected and most important economics of the world in respect to climate change are represented.

20.3.1 Australia

In November 2011, the Australian Parliament passed the "Clean Energy Future Package"[3] containing regulations that will establish emission trading in Australia. On the basis of the Carbon Price Mechanism (CPM), the Australian enterprises must pay a CO_2 tax of 25 Australian Dollars per tonne CO_2 until June 2015.[4] After that the price shall be determined in line with the market (Australian Government 2012b). The CPM is expected to cover approximately 500 business units that are responsible for 60 % of Australia's greenhouse gas emissions from the sectors of power generation, industrial facilities, fugitive emissions and landfills (Australian Government 2012b).

The CPM is designed for linking to other emission trading systems that operate internationally (Kossoy and Guigon 2012, 77). Accordingly, the intention has recently been declared to grant Australian companies after 2015 the right to provide CO_2 allowances, which they acquired on the European market, in order to meet their legal emission tax obligation (European Commission 2012, 1).

[3] Detailed information on the "Clean Energy Future Package "is available at http://www.cleanenergyfuture.gov.au/ (as of 12.11.2013).

[4] Meanwhile, the new Australian Government, already promised within the election campaign, will withdraw from the CO_2-Tax (Source: Liberal Party of Australia, available at http://www.liberal.org.au/latest-news/2013/09/10/tony-abbott-interview-david-koch-and-samantha-armytage-sunrise (as of 12.11.2013)).

20.3.2 New Zealand

New Zealand introduced its emission trading system with effect from 1 July 2010. Annually, about 30 million tones of CO_2 are offset through this system.[5] In contrast to other emission trading schemes, in particular the EU ETS, which initially only included CO_2-emissions from industry, the New Zealand system has from the beginning also included the sectors of waste management, fisheries, forestry as well as agriculture (CO_2-Handel.de 2012).

20.3.3 North America

In 2007 the "Western Climate Initiative (WCI)" was launched that brought together the Canadian provinces of British Columbia, Manitoba, Ontario, Québec and the U.S. State of California to design through their joint efforts harmonized cap-and-trade rules with the intention to enact the ideas found into respective national laws, to implement these and integrate them into the respective legal systems (Kossoy and Guigon 2012, 83ff.).

20.3.3.1 California

As of 1 January 2013, regulations are in force to establish a Californian emission trading scheme. In the first one-year commitment period till 2014 large stationary emitters of the industrial and power sector are comprised that emit at least 25,000 tCO_2eq per year (Kossoy and Guigon 2012, 84). In the second commitment period from 2015 onwards, the sectors transport, buildings and industrial fuels will also be included (ibid.).

20.3.3.2 Québec

In November 2009 Québec adopted the goal of reducing its greenhouse gas emissions by 2020 to 20 % below the base level of 1990. As a key tool for achieving this, a cap-and-trade program, which was developed in the context of the WCI and adopted in December 2011, shall also be used.[6] Since 2013 industrial and power plants with emissions of more than 25,000 tCO_2eq per year have been recorded.

---

[5] Cf. and for further information the New Zealand government, available at http://www.climatechange.govt.nz/emissions-trading-scheme/ (as of 12.11.2013).

[6] This cap-and-trade program is part of "Québec's Climate Change Plan". (Source: Ministry of Sustainable Development, Environment and Parks of Québec 2008).

Fuel dealers in the transport and construction sectors will be included in the trading system from 2015 (Government of Québec 2012).

20.3.3.3 Alberta

As early as 2007 Alberta adopted a compulsory trading system for greenhouse gas emissions depending on emission intensity on the basis of the first Canadian greenhouse gas reduction legislation. According to the legislation of the Climate Change Emissions Management Act (CCEMA), approximately 100 plant operators with annual emissions of more than 100,000 tCO_2eq have to reduce their emissions by 12 % every year, compared to the baseline period of 2003–2005 (Province of Alberta 2009).

20.3.3.4 British Columbia

The "Greenhouse Gas Reduction Targets Act" (GGRTA) that came into force in 2007 commits the government of British Columbia to reduce their greenhouse gas emissions by 33 % in the period 2007–2020 and by at least 80 % by 2050 (Government of British Columbia 2007). Especially public sector organisations[7] are also obliged by this law to adhere to standardized climate neutrality for the year 2010 and each subsequent calendar year thereafter (ibid.).

20.3.4 Republic of Korea (South Korea)

Early in 2010 the Republic of Korea adopted the legal framework for emission reductions and green growth (Republic of Korea 2010). It is intended to establish a national emission trading system as key instrument of the national policy on climate protection policy (United Nations Economic and Social Commission for Asia and the Pacific (ESCAP) 2012). In the beginning of 2015 the South Korean emission trading system will start. The aggregate trading system will comprise companies that emit more than 125,000 tCO_2eq per year and plants which emit more than 25,000 tCO_2eq per year (Kossoy and Guigon 2012, 91).

[7] Detailed information on "Clean Energy Future Package" is available at http://www.cleanenergyfuture.gov.au/ (as of 12.11.2013).

20.3.5 Mexico

In support of the Mexican goal to reduce greenhouse gas emissions by 30 % by the year 2020, a climate protection law was passed in April 2012. As framework legislation it opens up scope for the government to encourage the development of mitigation and adaptation measures. This framework legislation gives the government the authority to implement programs, policies and measures to avoid greenhouse gases. This also includes an emission trading system (Kossoy and Guigon 2012, 92).

20.3.6 China

As early as 2009 the first domestic carbon-reducing market mechanisms developed that were triggered by voluntary initiatives. In order to record emissions of agriculture and forestry, the "Panda Standard" emerged in 2009 and the "China Green Carbon Foundation" (CGCF) in 2010.[8] In the municipality of Tianjin a voluntary market for heat suppliers of residential buildings was launched in 2010, based on the intensity of emissions (Kejun 2012, 97).

In March 2012 the government in Beijing was the first of China's provinces and municipalities to put up for discussion a draft of a proposed mandatory emission trading system, the operational phase of which is to begin in April 2014 (Kejun 2012, 98). The sectors to be covered by the ETS, which will be functioning then, have not been explicitly listed yet, but it is clear that companies are comprised that emitted more than 10,000 tCO_2eq per year in the years 2009–2011 (ibid.).

20.3.7 Japan

The Japanese carbon market can be divided into different sectors. In addition to the first sector, at international level, there is a second that started a cap-and-trade scheme as a local trading system. This covers the main facilities and buildings within Greater Tokyo having its first commitment period from 2010 to 2014 and a second from 2015 to 2019.

Third, at the national level, the Ministry of Environment launched the Japan Voluntary Emissions Trading Scheme (JVETS) in 2005. As part of the JVETS, the voluntarily participating organizations have to adopt legally binding CO_2 reduction

[8] The Panda Standard is a certification scheme for domestic and forestry offset projects initiated by the China Beijing Environment Exchange, BlueNext, Winrock International, and the Asian Development Bank. The China Green Carbon Foundation was launched in 2010 by China's State Forestry Administration. *Source*: Wang (2011).

targets. Fourth, at the national level two voluntary crediting systems exist in parallel. The first is a Domestic Credit System which was introduced in October 2008. In this program large enterprises offer technology, financing and other support for small and medium businesses, but also for civil society (businesses and households), transport and other sectors and verify a reduction in greenhouse gas emissions achieved through credits. In addition, the Ministry of Environment established the "Japan Verified Emission Reduction (J-VER) Scheme". It is a system for verifying credits generated by reducing/removing greenhouse gases through national compensation projects. Japan's involvement at international level has been declining, despite its primary origin. Thus Japan does not participate in the second commitment period of the Kyoto Protocol (Mizuno 2012, 102).

20.3.8 Switzerland

In the revised CO_2-Act,[9] Switzerland's climate change objectives and measures until 2020 are fixed. In Articles 15–21, the new version defines a Swiss emission trading system which is "highly compatible with the EU ETS, and which is to be linked with the EU ETS through bilateral agreements". The CH ETS has only a few significant deviations from EU ETS,[10] so that the CH ETS is designed according to the cap-and-trade principle and has a pre-determined absolute cap on emissions.[11] Comparable with the EU ETS, the emissions cap is reduced by 1.74 % each year.

20.4 Other Initiatives

In addition, for example, Belarus, Chile, Costa Rica, Colombia, Indonesia, Jordan, Kazakhstan, Morocco, South Africa, Thailand, Turkey, Ukraine or Vietnam have developed for their economic development mechanisms in the area of greenhouse gas reduction, renewable energy and energy efficiency (Kossoy and Guigon 2012, 104).

All these initiatives could be linked to each other, thus further contributing to reducing greenhouse gas emissions. A link will be even more important in view of the fact that the prospects established by international treaties are not good.

[9] The bill is available on the sites of the Federal Administration for the Environment (BAFU) at http://www.bafu.admin.ch/klima/00493/06577/12209/index.html?lang=de (as of 12.11.2013).

[10] Thus fossil-thermal power plants, that need to offset their emissions in accordance with Article 22 of the Law on CO_2, are not subject to the CH-EHS. In contrast to most EU countries, however, waste incineration plants are included in the CH ETS and medium-sized enterprises are given the opportunity to participate voluntarily. Pending a link with the EU ETS the auctioning is also regulated differently.

[11] Cf. Art. 18 CO_2-Law new version.

The climate summit in Doha has ended with a rather weak compromise and agreed an extension of the Kyoto Protocol into a second commitment period. But in this second Kyoto phase (KP II), which started on 01.01.2013 and runs until the end of 2020, only nine other countries, apart from the 28 EU member states, commit themselves to fixed reduction targets.

Furthermore, the EU only represents probably 11 % of the global emissions of today; all of the Kyoto countries for the second commitment period emit only 15 or maybe 16 % of global greenhouse gas (Hedegaard 2011). On the other hand, the former Kyoto countries Japan, New Zealand and Russia no longer have set climate targets for the second commitment period. Canada has left the group with notification from 15 December 2011.[12] The USA, which have signed but have never ratified the Kyoto Protocol, still do not belong to the group.

Comparatively the behaviour at international level is in parts highly contradictory to the initiatives and efforts at national level. Just to mention the situation in the two largest emitting countries: Between 2005 and 2011, the USA have reduced, even though not obliged by international agreements, their greenhouse gases emission by 6.9 % (EPA 2013) and rank worldwide number two regarding the development of renewable energy projects. China, even though meanwhile the biggest emitter of greenhouse gases in the world with a dramatic increase in even the per capita emission, is leading this list and is the biggest developer of renewable energy projects. Also recent policy changes give the impression that China's new government under Premier Minister Li Keqiang is obviously moving forward to a more efficient and environment and climate friendly policy, as a recently published directive "Opinions of the State Council on Accelerating the Development of Environmental Industry" indicates (Traufetter and Zand 2013).

This discrepancy of weak international commitment and rather progressive national action continues to intensify, because in the context of the second period of the Kyoto Protocol states are only allowed to trade emission credits on a very limited scale. What will happen to such pollution allowances after 2020 has not been settled so far.

A timetable for negotiating a comprehensive climate contract by 2015 was agreed upon. The contract is to be completed by that year so that it can enter into force in 2020. The USA as a party to the UNFCCC is participating in the negotiations for the successor protocol to the Kyoto Protocol as well.

[12] Notification C.N.796.2011.TREATIES-1, accessible at http://unfccc.int/files/kyoto_protocol/background/application/pdf/canada.pdf.pdf (as of 14.09.2013).

20.5 Findings and New Concepts

To keep the agreed policy targets of a warming of two degrees above pre-industrial times,[13] the current system, even if well improved, is not sufficient. The possible consequences of a stronger temperature increase are subject to several prognoses and rather well-described (WBGU 2009, 9ff.). On the other hand, any success of reduction in those developed states having fulfilled their obligations in respect to reduction to climate change (like the EU), are more than compensated simply by the increase of emissions of fast developing countries. China is meanwhile the world's biggest emitter; India is in total emitting more greenhouse gases than Germany, Great Britain and France together. This may be explained by the large population in these states. But in several cases even the per capita emissions of non-annex I states exceed the annual per capita emission of highly industrialized states. Examples are Qatar (49.05 t/per capita), the Emirates (24.98 t/per capita), Saudi Arabia (16.57 t/per capita), South Korea (10.40 t/per capita) and meanwhile even China (7.20 t per capita) bringing its per capita emissions within the range of 6–19 tones per capita emissions of the major industrialized countries. These states are still non-annex-I states to the UNFCCC which means in the sense of the agreement they are regarded as developing states. But these countries are in several aspects more comparable with the annex-I-states than with a developing state in the general meaning. South Korea's GDP per capita (based on purchasing power parity 2012) is according to the World Bank statistics[14] 30,801 US$, almost as high as the GDP of the EU (25,500 €[15] = 33,894 US$). The GDP in the Gulf States is even higher (Qatar 2011: 86,507 US$; Kuwait 2011: 49,001 US$; United Arab Emirates 2011: 42,384 US$).[16] China's financial reserves in December 2011 (without gold) were reported being 3.18 trillion US$[17] and are estimated meanwhile more than 3.3 trillion US$ (Urbánek 2013), thus financial reasons for a development change could not be an argument.

---

[13] Copenhagen Accord, Decision 2/CP.15, 18 December 2009, FCCC/CP/2009/11/Add.1, available at http://unfccc.int/resource/docs/2009/cop15/eng/11a01.pdf, (as of 14.09.2013); Cancun Agreements, Decision 1/CP. 16, 10 December 2010, FCCC/CP/2010/7/Add.1, available at http://unfccc.int/resource/docs/2010/cop16/eng/07a01.pdf, (as of 14.09.2013).

[14] Accessible at http://data.worldbank.org/indicator/NY.GDP.PCAP.PP.CD?order=wbapi_data_value_2012+wbapi_data_value+wbapi_data_value-last&sort=desc, (as of 14.09.2013).

[15] Data from Eurostat for 2012, accessible at http://epp.eurostat.ec.europa.eu/tgm/refreshTableAction.do?tab=table&plugin=1&pcode=tec00001&language=en, (as of 14.09.2013).

[16] World Bank data for 2011, accessible at http://data.worldbank.org/indicator/NY.GDP.PCAP.PP.CD, (as of 14.09.2013).

[17] As reported by Chinability, accessible at http://www.chinability.com/Reserves.htm, (as of 14.09.2013).

Canada has not accepted an originally agreed international responsibility and withdrew the Kyoto Protocol just in time before the first commitment period ended.[18] This action became effective for Canada on 15 December 2012 in accordance with Art. 27 para 2 of the Kyoto Protocol. The USA as the largest economy in the world and still one of the richest countries worldwide is reluctant to international binding commitments, as seen at the Cancun Summit from 29 November to 10 December 2010 (Weiss 2010).

All together, this is in the urging matter of climate change policy not acceptable anymore in ethical, political, societal and moral respect. And, the argument, the industrialized states have a historical responsibility for the emissions of greenhouse gases in former times, is not helpful. A historical responsibility for the times before 1990 is difficult to argue. 1990 can be regarded as the earliest year, where the reasons and effects of climate change were obvious and international negotiations for a climate change convention started (WBGU 2011, 39), thus political action of states against the emission of greenhouse gases could be expected (WBGU 2009, 25), but is politically probably not agreeable (WBGU 2009, 27). Regardless to which extent historical responsibilities are accepted by the industrialized states, this does not justify the unlimited emission of greenhouse gases by developing (and developed) states. There is no international customary law allowing states to make the same mistakes as others.

Thus, new ideas are needed and have to be based on the requirement that data reported are accurate and not obviously falsified.[19]

Why not—taking in account the idea of an equal share of the natural resources—setting a certain budget of greenhouse gases emission rights for states or individuals, as proposed by several initiatives. For example, the German Advisory Council Global Change (WBGU) has suggested in a special report "Solving the climate dilemma: The budget approach" (WBGU 2009), a budget approach for states with consideration of different alternatives of calculating the budget of states and discussion on the feasibility and effects of such a new idea.

Similarly to the budget approach, the German chancellor Angela Merkel has launched an initiate to a per capita emission of 4 tonnes. The Indian Economist and Deputy Chairman of the Planning Commission of India opts for a per capita greenhouse gases emission cap of 2 tonnes annually (Rediff online 2009), also experts opt for a maximum per capita emission of 2 tonnes (Focus-online 2008).

A budget approach-based system has several advantages. In Sect. 5.6 of the special report of the German Advisory Council Global Change ten arguments for the budget approach are listed (WBGU 2009, 38ff.):

[18] Notification C.N.796.2011.TREATIES-1, accessible at http://unfccc.int/files/kyoto_protocol/background/application/pdf/canada.pdf.pdf, (as of 14.09.2013).

[19] In 2010 alone, there exists a gap between the sum of GHG emissions of the provinces in China and the official report for China as a whole of 1.4 billon tones of CO_2 (=5 % of the annual global emissions of CO_2) (Source: Dabo Guan et al. 2012, 672 ff).

1. Global responsibility, equity and precaution
2. Radical simplification of climate negotiations
3. Foundation for an historical climate compromise
4. Transparent emission budgets
5. Leeway at national level is linked with accountability at international level
6. Incentives for long-term action
7. Scarcity increases efficiency
8. Climate protection strengthens competitiveness
9. New prospects for international cooperation
10. Definite framework for a low-carbon world economy

These arguments are self-explaining and could be a justification and incentive to try a radical new start in global climate policy.

Biography Simon Spyra is scholarship holder of the State of Brandenburg and Ph.D. candidate at the BTU Cottbus-Senftenberg. The contribution was supported by the Ministry of Science, Research and Culture (MWFK) of Brandenburg at the International Graduate School of BTU Cottbus.
Eike Albrecht is Head of the Department of Civil and Public Law with References to the Law of Europe and the Environment at BTU Cottbus-Senftenberg.

References

Australian Government (2012a) Clean energy future fact sheet, carbon pricing mechanism: who is liable? Available at http://www.cleanenergyfuture.gov.au/500-companies/ (as of 12.11.2013)

Australian Government (2012b) Administration for climate conversation and energy efficiency, 2012. Available at http://www.climatechange.gov.au/government/clean-energy-future/legislation.aspx (as of 12.11.2013)

CO_2-Handel.de (2012) New Zealand will ask agriculture sector to pay up to cover the costs of emissions not until 2015, 03.07.2012. Available at http://www.co2-handel.de/article184_18589.html (as of 12.11.2013)

EPA (2013) United States Environment Protection Agency. National Greenhouse Gas Emissions Data. Available at http://www.epa.gov/climatechange/ghgemissions/usinventoryreport.html (as of 12.11.2013)

European Commission (2012) Press release of 28.08.2012. Available at http://europa.eu/rapid/press-release_IP-12-916_en.htm (as of 12.11.2013)

Focus-online (2008) Merkel: Bemessung des erlaubten CO_2-Ausstoßes auf Pro-Kopf-Basis (*Merkel: Calculation of Allowed CO_2-Emission on Per Capita Basis*), Focus-online from 29 May 2008, available at: http://www.focus.de/politik/diverses/merkel-bemessung-des-erlaubten-co2-ausstosses-auf-pro-kopf-basis_aid_305001.html# (as of 12.11.2013)

Government of British Columbia (2007) Bill 44: Greenhouse Gas Reduction Targets Act (GGRTA) that came into force on 1 January 2008. The bill is available at http://www.leg.bc.ca/38th3rd/3rd_read/gov44-3.htm (as of 12.11.2013)

Government of Québec (2012) Annual caps on greenhouse gas emission units relating to the cap-and-trade system for greenhouse gas emission allowances for the 2013–2020 period, Draft Regulation, Environment Quality Act

Guan D, Liu Z, Geng Y, Lindner S, Hubacek K (2012) The gigatonne gap in China's carbon dioxide inventories. Nature Climate Change 2(9):672–675

Hedegaard C (2011) India: A major player and constructive force in Durban. Article from 16 December 2011, Available at: http://ec.europa.eu/commission_2010-2014/hedegaard/headlines/articles/2011-12-16_01_de.htm. Accessed 13 Sept 2013

Kejun J, Wang W (2012) In: Kossoy A and Guigon P (eds), State and Trends of the Carbon Market 2012, World Bank Report, 2012, Westland Printers, Laurel, MD

Kossoy A, Guigon P (2012) State and Trends of the Carbon Market 2012. World Bank Report, 2012, Westland Printers, Laurel, MD

Ministry of Sustainable Development, Environment and Parks of Québec (2008) 2006–2012 action plan, Québec and climate change, a challenge for the future

Mizuno Y (2012). In: Kossoy A, Guigon P (eds) State and Trends of the Carbon Market 2012, World Bank Report, 2012. Westland Printers, Laurel, MD

Province of Alberta (2009) Climate Change Emissions Management Act, 2007. Available at http://www.qp.alberta.ca/1266.cfm?page=C16P7.cfm&leg_type=Acts&isbncln=9780779740956 (as of 12.11.2013)

Rediff online (2009) Singh: 'There has been a decline in poverty in India', rediff-online from 30 April 2009. Available at: http://business.rediff.com/interview/2009/apr/30/interview-with-montek-singh-ahluwalia.htm (as of 05.09.2013)

Republic of Korea (2010) Article 1 of the Amendments of the Ministry of Government Legislation, Framework Act on Low Carbon, Green Growth, 2010. Available at http://www.uncsd2012.org/content/documents/South%20Korea%20Framework%20Act%20on%20Low%20Carbon%20Green%20Growth%202010.pdf (as of 12.11.2013)

Traufetter G, Zand B (2013) China – Gesundes Wachstum (*China – Healthy Growth*), Der Spiegel 34/2013, p. 86; the "opinions of the state council on accelerating the development of environmental industry" are available at http://english.mep.gov.cn/News_service/infocus/201308/t20130814_257467.htm (as of 12.11.2013)

United Nations Economic and Social Commission for Asia and the Pacific (ESCAP) (2012) Fact sheet low carbon green growth roadmap for Asia and the Pacific, p. 8. Available at http://www.unescap.org/esd/environment/lcgg/documents/roadmap/case_study_fact_sheets/Fact%20Sheets/FS-Cap-and-trade-scheme.pdf (as of 12.11.2013)

Urbánek (2013) China's foreign exchange reserves at the end of 2012 grew to 3.3 trillion, from +700% L.04. KurzyCZ. Retrieved 26 March 2013

Wang W (2011) Linking climate finance to the agriculture and forestry sectors in China, 2011, climate economics in progress. Economica, New York, NY

WBGU (2009) Sondergutachten des Wissenschaftlichen Beirats der Bundesregierung Globale Umweltveränderungen (WBGU): Kassensturz für den Weltklimavertrag – Der Budgetansatz (*Special Report of the German Advisory Council Global Change (WBGU): Solving the Climate Dilemma: The Budget Approach,* available at: http://www.bmbf.de/pubRD/wbgu_sn2009.pdf (as of 05.09.2013)

WBGU (2011) Gutachten des Wissenschaftlichen Beirats der Bundesregierung Globale Umweltveränderungen (WBGU): Welt im Wandel – Gesellschaftsvertrag für eine große Transformation (*Report of the German Advisory Council Global Change (WBGU): Changing World – Social Contract for a Great Transformation*)

Weiss S (2010) Zähes Ringen im Endkampf um neues Klimaabkommen (*Hard Negotiations in the Final Battle for a New Cliamte Change Agreement*), Der Standard, 10.12.2010, available at http://derstandard.at/1291454847137/Zaehes-Ringen-im-Endkampf-um-neues-Klimaabkommen (as of 14.09.2013)

WCI – Western Climate Initiative (2011) Press release of 16.11.2011, available at http://www.westernclimateinitiative.org/news-and-updates/139-quebec-adopts-cap-and-trade-regulation (as of 12.11.2013)

FSC
www.fsc.org
MIX
Papier aus verantwortungsvollen Quellen
Paper from responsible sources
FSC® C105338

If you have any concerns about our products,
you can contact us on
ProductSafety@springernature.com

In case Publisher is established outside the EU,
the EU authorized representative is:
Springer Nature Customer Service Center GmbH
Europaplatz 3, 69115 Heidelberg, Germany

Printed by Libri Plureos GmbH
in Hamburg, Germany